DICTIONARY OF ORIENTAL LITERATURES
VOLUME III

Literatures included in the present volume:

Ancient Near East: Akkadian, Aramaic, Assyrian, Babylonian, Carthaginian, Coptic, Egyptian, Hebrew, Hittite, Mandaic, Persian (Ancient and Middle), Phoenician, Sumerian, Syriac, Ugaritic.

Arab Countries: Classical Arabic, Algerian, Egyptian, Iraqi, Jordanian, Lebanese, Moroccan, Palestinian, Sudanese, Syrian, Tunisian.

Turkey: Turkish, Turkic, Kurdish, Armenian.

Iran: Ancient and Middle Persian, New Persian, Jewish-Persian, Kurdish, Armenian.

Afghanistan: Pashto, Dari.

Soviet East: Abkhazian, Armenian, Avar, Azerbaijan, Chukot, Circassian, Darg, Georgian, Kazakh, Kirghiz, Kurdish, Lezgian, Ossetian, Tatar, Tajik, Turkic, Turkmen, Uzbek, Yakut.

Literatures included in Volume I:

Chinese, Japanese, Korean, Mongolian, Tibetan

Literatures included in Volume II:

India, Pakistan, and Bangladesh: Ancient Indian, Assamese, Baluchi, Bengali, Gujarati, Hindi, Indian literature written in English, Indo-Persian, Kannada, Kashmiri, Maithili, Malayalam, Marathi, Oriya, Panjabi, Pashto, Rajasthani, Sindhi, Tamil, Telugu, Urdu.

Nepali.

Sinhalese.

South-East Asia: Burmese, Cambodian, Javanese, Malay and Indonesian, Philippines, Thai, Vietnamese.

DICTIONARY OF ORIENTAL LITERATURES

VOLUME III
WEST ASIA AND NORTH AFRICA

General Editor
JAROSLAV PRŮŠEK

Volume Editor
JIŘÍ BEČKA

BASIC BOOKS, INC., Publishers
New York

First published in 1974

ISBN 465–01649–9

LC73–82742

Printed in Great Britain
in Times New Roman type
by William Clowes & Sons, Limited
London, Beccles and Colchester

ADVISORY EDITORIAL BOARD

ACKNOWLEDGEMENTS

Our particular thanks are due to the following members of the Advisory Editorial Board for their exceptional interest and willing help: Professor Richard Frye of Harvard University, Professor O. R. Gurney of Oxford University, and Professor Thomas M. Johnstone and D. M. Lang of the School of Oriental and African Studies, University of London. Among our Czech colleagues we are particularly grateful to Dr Dušan Zbavitel, who in addition to editing Volume II gave unstinting help and advice during work on the other two volumes. Most of the Czech contributions were translated into English by Mrs Iris Urwin, to whom our warmest thanks are due.

In conclusion the editors would like to thank UNESCO for the grant towards the financing of this dictionary, and the publishers for their sympathetic awareness of the difficulties involved.

The Editors

CONTENTS

CONTRIBUTORS

AK	Mrs Adéla Křikavová, Oriental Institute, Prague
AS	Dr Dr Annemarie Schimmel, Professor of Indo-Muslim Culture, Harvard University, Cambridge, Mass.
AT	Dr Andreas Tietze, Professor of Turkish, University of California, Los Angeles
BH	Dr Blahoslav Hruška, Oriental Institute, Prague
DL	Dr Dušanka Lukač, Academy of Science, Belgrade
DNM	Dr David Neil MacKenzie, Reader in Iranian Languages, University of London
EP	Mrs Eva Pantůčková, Oriental Institute, Prague
FT	Dr Felix Tauer, Emeritus Professor of Islamic History and Culture, Charles University, Prague
HB	Dr Henri Broms, Finnish Oriental Society, Helsinki
HS	Miss Helena Spurná, Oriental Institute, Prague
HT	Dr Helena Turková, Prague
IH	Dr Ivan Hrbek, Oriental Institute, Prague
JB	Dr Dr Jiří Bečka, Oriental Institute, Prague
JČ	Dr Jaroslav Černý, Professor of Egyptology, Oxford University
JML	Dr Jacob M. Landau, Associate Professor at the Hebrew University, Jerusalem and at Bar-Ilan University, Ramat-gan
JO	Dr Jaroslav Oliverius, Charles University, Prague
JOS	Dr Jiří Osvald, Charles University, Prague
JPA	Dr Jes P. Asmussen, Professor of Iranian Philology, University of Copenhagen
JS	Dr Jana Sieglová, Náprstek Museum of African and American Cultures, Prague
JZ	Dr Jitka Zamrazilová-Weltmanová, Oriental Institute, Prague
KB	Kamil Baňák, Oriental Institute, Bratislava
KP	Dr Karel Petráček, Professor of Arabic, Charles University, Prague
LH	Dr Luděk Hřebíček, Oriental Institute, Prague
LJKČ	Dr Ladislav J. Krušina-Černý, Oriental Institute, Prague
LK	Dr Lubor Kropáček, Charles University, Prague
LM	Mrs Ludmila Motalová, Oriental Institute, Prague
LMa	Dr Lubor Matouš, Professor of Cuneiform Studies, Charles University, Prague
ML	Dr Manfred Lorenz, Humboldt University, Berlin
MM	Miloslav Manoušek, Oriental Institute, Prague
NN	Dr Nea Nováková, Charles University, Prague

CONTRIBUTORS

NZ	Mrs Naděžda Zimová, Oriental Institute, Prague
OK	Dr Otakar Klíma, Oriental Institute, Prague
OS	Dr Dr Otto Spies, Professor of Arabic and Islamic Studies, University of Bonn
PC	Dr Pierre Cachia, Reader in Arabic, University of Edinburgh
RCO	Dr R. C. Ostle, Lecturer in Arabic, School of Oriental and African Studies, University of London
RS	Dr Rudolf Sellheim, Professor of Arabic and Islamic Studies, Johann Wolfgang Goethe University, Frankfurt am Main
SP	Dr Svetozár Pantůček, Oriental Institute, Prague
SZS	Dr Stanislav Segert, Professor of Semitic Philology, University of California, Los Angeles
VAČ	Dr Václav A. Černý, Oriental Institute, Prague
VK	Dr Věra Kubíčková, Prague
VS	Dr Vladimír Souček, Professor-Assistant of Cuneiform Studies, Charles University, Prague
VSD	Dr Vladimír Sadek, State Jewish Museum, Prague
WH	Dr Wilhelm Heinz, Professor-Assistant of Islamology and Iranian, Würzburg University
ZV	Dr Zdenka Veselá, Oriental Institute, Prague

NOTE ON VOLUME III

This volume contains entries dealing with the literatures of West Asia, the region most bound up with the Islamic cultural tradition, to which of course the countries of North Africa also naturally belong; the Arabic literature of Egypt and the Sudan has been included, as has the literature of Algeria and Morocco, whether in Arabic or French. Similar reasons guided the inclusion of the literature of Ancient Egypt alongside the other Near Eastern literatures of antiquity. Of authors whom readers may seek in this volume, Baluchi writers, and Indo-Persian poets and writers of local importance only, will be found in Volume II; the more important authors such as Amīr Khosrou or Bēdil, known throughout the area of Persian literature and influential on it to a considerable degree, are included here.

It is sometimes extremely difficult to determine the nationality of classic writers, even the greatest, like Ibn Sīnā (Avicenna), al-Bīrūnī, al-Fārābī, or Bābur. The first three, like many others, were simply described as Arabic writers because they wrote only or mainly in Arabic. In modern times, naturally enough, they are claimed by individual literatures which have grown up in the regions where the great men were born or where they lived and wrote. Bābur is claimed by the Uzbeks, the Indians and the Afghans; Alīshēr Navā'ī should belong to Uzbek literature, by right of most of his work, yet he also wrote Persian poetry and spent the whole of his life in Herat, in what is now Afghanistan; his work is rooted in the region. The same can be said of writers like 'Onsorī, Farrokhī, or Jāmī, all of whom wrote and lived in what is now Afghanistan or Tajikistan, and who are naturally claimed today by the literary historians of both. Some of the most important writers have therefore been listed in more than one national literature.

It has been the Editors' aim to make the transcription of names and terms (usually transliteration) as simple as possible, and the use of special letters has therefore been avoided. At the same time the transcription conforms to the requirements of scholarship, so that names and titles can be transcribed back into the script of the original. Diacritic signs have therefore been used, in accordance with scholarly usage; they can simply be ignored by the layman. A horizontal line above the vowel indicates its length.

The authors are given under the name (or pseudonym) which is best known, with cross-reference where the case is not clear (eg Khayyām, see 'Omar Khayyām). In the attempt to indicate adequately the pronunciation of the same name (usually Arabic) in different though related languages, various transcriptions have been used according to the vowel system of each language. Thus the name 'Uthmān (Arabic) will appear as 'Os̱mān (Iranian), Osman (Turkish), or Usmōn (Tajik), rather as the European John may be Jean, Johann, Giovanni or Ivan.

There is a different problem when we come to writers from lands in the USSR, who are sometimes better known under the Russianized form of their name (eg the Ossetian writer Khetagurov or the Uzbek Aybek). The original form of the name (Khetaegkaty, Oybek) is nevertheless used for the entries in the dictionary, and the Russianized form provided with cross-reference.

ABBREVIATIONS USED FOR BOOKS AND JOURNALS

AAP	A. J. Arberry, *Arabic Poetry; A Primer for Students* (Cambridge 1965)
AAPJ	N. Adontz, *Armenia in the Period of Justinian* (Lisbon 1970)
AB	Armenische Bibliothek, Leipzig 1880–90
AGEMPL	Bozorg Alavi, *Geschichte und Entwicklung der modernen persischen Literatur* (Berlin 1964)
ALAC	*Anthologie de la littérature arabe contemporaine. La poésie par L. Norin et E. Tarabay* (Paris 1967)
AMAP	A. J. Arberry, *Modern Arabic Poetry. An Anthology with English Verse Translations* (Cambridge 1967^2)
ANET	*Ancient Near Eastern Texts Relating to the Old Testament* (ed J. B. Pritchard) (Princeton 1955^2)
BGAL	C. Brockelmann, *Geschichte der arabischen Literatur, I–II, Weimar, 1898–1902*, S I–III (Leyden 1937–42)
BHLA	R. Blachère, *Histoire de la littérature arabe des origines à la fin de XVe siècle de J.—C.* (Paris 1952)
BLHP	E. G. Browne, *A Literary History of Persia*, 4 Vols. (London 1902–24 (several reprints))
CAR	*Contemporary Arabic Readers V, Modern Arabic Poetry*, Part 1, texts; Part 2, notes and glossaries. Ed. by J. A. Bellamy, E. N. McCarus, A. I. Jacoub (Ann Arbor 1966)
DCML	G. R. Driver, *Canaanite Myths and Legends* (Edinburgh 1956)
DLMEF	J. Déjeux, *La littérature maghrébine d'expression française* (Algiers 1970)
EI	*Encyclopaedia of Islam* (Leiden–London 1913–34, and reprints)
ELA	*Europe, Littérature arménienne* (Paris, février–mars 1961, No. 382–3)
FGAL	F. N. Finck, *Geschichte der armenischen Literatur in Geschichte der christlichen Literaturen der Orients, II* (Leipzig 1909)
FKH	K. Fuad, *Kurdische Handschriften* (Wiesbaden 1970)
GAL	H. A. R. Gibb, *Arabic Literature* (2nd ed.), (Oxford 1963)
GHOP	E. J. W. Gibb, *A History of Ottoman Poetry* t. I–VI (London 1900, repr. 1958)
GIPh	*Grundriss der iranischen Philologie*, ed. W. Geiger and E. Kuhn, 3 vols. (Stuttgart 1891–1904)
GLALG	H. A. R. Gibb and J. M. Landau, *Arabische Literaturgeschichte* (Zurich 1968)
GSLA	F. Gabrieli, *Storia della letteratura araba* (Milano 1956^2)
GT	Th. H. Gaster, *Thespis: Ritual, Myth and Drama in the Ancient Near East*, 316–76 (New York, 1966)

HBKJV	*The Holy Bible: King James Version* 1611, and reprints
HBRSV	*The Holy Bible: Revised Standard Version* 1952–57, and reprints
HGTM	P. Horn, *Geschichte der türkischen Moderne* (Leipzig 1909)
HO	Handbuch der Orientalistik, Bd. 7, 1. Abt. (Leiden 1963)
KLL	*Kindlers Literatur Lexikon I–VII* (Zurich 1964–72)
KMPPL	*H. Kamshad, Modern Persian Prose Literature* (Cambridge 1966)
LACC	D. M. Lang, *Armenia: Cradle of Civilization* (London 1970)
LCHAMA	V. Langlois, *Collection d'historiens anciens et modernes de l'Arménie* (Vol. 1, Paris 1867; Vol. II, 1869)
LLS	A. Leist, *Literarische Skizzen* (Leipzig 1886)
LRPOA	R. Labat et al., *Les religions du Proche-Orient asiatique* (Paris 1970)
LSATC	J. M. Landau, *Studies in the Arab Theater and Cinema* (Philadelphia 1958)
MLIC	F. Machalski, *La littérature de l'Iran contemporain, II* (Kraków 1967)
MRPRPV	R. Munibar, *Post-revolution Persian Verse* (Aligarh 1955)
NACL	F. Nève, *L'Arménie chrétienne et sa littérature* (Louvain 1886)
NEB	*The New English Bible* (Oxford–Cambridge 1970)
NLHA	R. A. Nicholson, *A Literary History of the Arabs* (London 1907, 1969)
OeO	*Oeuvres et opinions*, Numéro special: *La littérature arménienne* (Moscou, Mars 1969)
PHTF	*Philologiae Turcicae Fundamenta* (Wiesbaden 1964)
RHIL	Jan Rypka, *History of Iranian Literature* (Dordrecht 1968)
RSPA	H. G. Raverty, *Selections from the Poetry of the Afghans* (London 1862)
SLM	*Soviet Literature Monthly*, devoted to Literature and Art of Soviet Armenia, No. 3 (Moscow 1966)
SPL	C. A. Storey, *Persian Literature. A Bio-Biographical Survey* (London 1927–58)
THLA	H. Thorossian, *Histoire de la littérature arménienne* (*des origines jusqu'à nos jours*) (Paris 1951)
TMSLZ	J. C. Tavadia, *Die mittelpersische Sprache und Literatur der Zoroastrier* (Leipzig 1956)
TPAAM	*Poèmes arméniens anciens et modernes. Traduits par Archag Tchobanian* (Paris 1902)
TRA	*La Roseraie d'Arménie. Traduction precédé d'une étude et suivie de notes par Archag Tchobanian* (Vol. I, Paris 1918, Vol. II, 1923; Vol. III, 1929)
TTA	*Les trouvères arméniens. Traduction française avec une introduction par Archag Tchobanian* (Paris 1906)
UAGP	Venera Urushadze, *An Anthology of Georgian Poetry* (Tbilisi 1948)
WILA	G. Wiet, *Introduction à la littérature arabe* (Paris 1966)
WSHSL	W. Wright, *A Short History of Syriac Literature* (London 1894)

VOLUME III

WEST ASIA AND NORTH AFRICA

A

Abasheli, Aleksandre (b 1884 Sachochio, near Kutaisi, d 1954), Georgian poet. After leaving secondary school in Kutaisi he studied in Russia, and was arrested and exiled for several years for his part in the 1905 revolution. He began writing in Russian, but his first published volume (*Mzis sitsili*—Laughter of the Sun, 1913) was already written in Georgian. After the establishment of a Soviet government in Georgia he joined the Academic Association, a literary group led by K. Gamsakhurdia (qv). His meditative and nature lyrics are in the spirit of Pasternak and N. Zabolotskiy, who both translated his verse into Russian. With G. Abashidze (qv) he wrote the words for the Georgian national anthem. Further works: *Antebuli kheivani* (Burning Avenue, 1923); *Mze da samshoblo* (Sun and Country, 1939); *Gmiruli dgheebi* (Heroic Days, 1942).

Trans.: UAGP VAČ

Abashidze, Grigol (b 1913 Chiatura), Georgian poet and writer. He studied philology at Tbilisi University, completing his studies in Moscow. He is editor-in-chief of the journal *Mnatobi* and works in the executive of the Union of Georgian Writers. He is very interested in Georgian history, and the glorious moments of his country's past inspired much of his meditative and nature poetry: *Aspurtsela* (Millefoil, 1940); *Samkhretis sazghvarze* (On the Southern Frontier, 1949). He also writes film scripts, poetry for children, historical novels (*Lasharela*, 1957; *Didi ghame*, The Great Night, 1962) and plays. His drama *Mogzauroba sam droshi* (Wandering through Three Times, 1961) gives a lyrical view of the past, present and future of Georgia, and makes use of some science-fiction techniques. In addition to his own work, he has translated Neẓāmī (qv), Petöfi and others. With Aleksandre Abasheli (qv) he wrote the words for the Georgian national anthem. VAČ

Abasıyanık, Sait Faik (b 1906 Adapazarı, d 1954 Istanbul), Turkish writer. He studied at the Arts Faculty, Istanbul, and later travelled in Switzerland and France. Abasıyanık was a teacher, who also tried trade and journalism before finally devoting himself to literature. The form in which he excels in modern Turkish literature is the short story. Two novels *Birtakım İnsanlar* (A Few People, 1944) and *Kayıp Aranıyor* (Wanted—a Loss, 1953), are less famous than his stories, at first emphasizing the plot, later a free lyrical prose form often philosophical in mood. He was the poet of Istanbul and her people, of the sea and its fishermen. In his last volumes the human mind and what goes on hidden in it became his theme, presented in modern prose techniques. Volumes of stories: *Semaver* (Samovar, 1936); *Sarnıç* (Reservoir, 1939); *Lüzumsuz Adam* (A Useless Man, 1948); *Son Kuşlar* (The Last Birds, 1952); *Alemdağda Var Bir Yılan* (The Snake above the World, 1954).

Trans.: Sabri Esat Siyavuşgil, *Un point sur la carte* (Leyden 1962). LH

Abay, Kazakh folk poet, see **Kunanbayev,** Abay.

'Abdalmu'ti Ḥijāzi, Aḥmad (b 1935), Egyptian poet. Coming from a small village in the Nile delta, he has a sympathetic eye for poverty. After leaving his village school he went to the teachers' training college in Manūfiya, graduating in 1944. He then went to Cairo, where he has been on the editorial staff of *Rose al-Yūsuf* since 1956 and is now head of the cultural section. His own experience has made him a socialist. He is one of the finest modern poets in Egypt, together with the older Ṣalāḥ 'Abdaṣṣabūr (qv) and the younger Amal Dunqul. Besides writing poetry he is the author of critical studies and introductions to anthologies, giving preference to the forgotten 'damned' poets like Ibrāhīm Nājī, who foreshadowed the likeness of the present generation of Egyptian poets. He has published three volumes of verse to date: *Madīna bilā qalb* (City without a Heart, ^{2}1968), *Lam yabqa illā i'tirāf* (There Was Nothing for it but to Confess, 1965),

Marthiyat lā'ib sirk (Lament for a Circus Clown), which contains verse written since 1965. He has also published a long *qaṣīda* (qv), *Aurès* (1956).

ALAC–P 235. KP

'Abdaṣṣabūr, Ṣalāḥ (b 1931), Egyptian writer. Born in the country, he came to Cairo early; he has travelled widely and met intellectuals of both West and East. A journalist and essay-writer, he has recently turned to play-writing; his play *The Tragedy of al-Ḥallāj* (1965) gives an interpretation of this Islamic mystic who died on the cross. 'Abdaṣṣabūr translates plays into Arabic (Ibsen, 1961) as well as articles and essays on many subjects, ranging from the works of John Osborne and D. H. Lawrence to books on British politics or atomic submarines. A calm, thoughtful man, he is determined to struggle against all obstacles. Starting as a poet in the traditional manner, he has progressed to the trend now uppermost in Arabic poetry, namely free verse. Unlike the Syrian-Palestinian poets of purely intellectual interests like Adonis (qv) who are reviving old myths and applying their philosophical learning to poetry, 'Abduṣṣabūr draws his subjects and his symbolism from contemporary life. His work reflects the feelings and desires of the individual. His latest volume (*The Old Rider's Dreams*) stresses the search for new values, disillusionment with the old, and longing for youth and for rural life.

He is rightly claimed as the leader of the modernists in Egypt, by the followers of the movement themselves as well as by the critics. He dealt with the theory of literature in *Qirā'a jadīda li-shi'rinā'l-qadīm* (Re-reading Our Old Poetry, 1968) and *Ḥattā naqhur al-mawt* (When We Overcome Death, 1966), and above all in the autobiographical *Ḥayati fi'sh-shi'r* (My Life in Poetry, 1969). His volumes of poetry include *an-Nās fī bilādī* (The People of My Region, 1957), *Aqūlu lakum* (I Tell you, 1961), *Aḥlām al-fāris al-qadīm* (The Old Rider's Dreams, 1964); earlier poems appeared (with a translation) in *Riḥla fi 'l-layl* (Way through the Night, 1970). Besides the play given above he has published *Laylā wa Majnūn* (1970) and *Musāfir layl* (Nocturnal Pilgrim, 1969). Other writings have appeared in periodicals. The *dīwān* (qv) *Intiẓār al-layl wa 'n-nahār* (Waiting for Night and Day) is perhaps still being printed; his last *dīwān* is *Ta'ammulāt fī zamān jarīḥ* (Meditations on Time Wounded).

CAR 287; ALAC–P 211. KP

'Abduh (Abdo), Muḥammad (1849–1905), Egyptian Muslim theologian and thinker, founder of the Islamic modernism in Egypt. He was educated at the Tanta theological school, the Muslim university al-Azhar, and then became a teacher at the *Dār al-'ulūm* (House of Science) college in Cairo. In 1880 he became editor-in-chief of the official paper *al-Waqā'i'al-miṣrīya* (Egyptian Facts). After the 'Urābī rising of 1882 he went into exile in Beirut, Paris and Tunis. He returned to Egypt in 1889 and held high office in the judiciary, devoting himself also to journalism. In his early public life he was profoundly influenced by Islamic mysticism and tended to asceticism. His ideas were radically changed in 1872 after he met the greatest pioneer of pan-Islamic ideas, Jamāluddīn Afghānī (qv), who turned his attention to the contemporary problems of Islam, and its attitude towards the science and civilization of Europe. In his journalistic writings as in his philosophical and theological works, he called for a renaissance of Islam, drawing on its original impulses. He pondered over the position of Islam in the modern world and stressed the essential need for cultural, social and political reform. He believed that it was possible to combine the fundamental ideas of Islam with modern European thinking and scientific advance, and to achieve a harmonious synthesis of religious and scientific thought. His extensive journalistic and popularizing writings played a very important part in the struggle for social, cultural and political reform in Egypt and thus helped the renaissance in Egyptian literature. Main works: *Risālat at-tawḥīd* (Treatise on the Oneness of God, 1897); *Al-Islām wa'n-naṣrānīya* *ma'a'l-'ilm wa'l-madanīya* (Islam and Christianity in Science and Civilization, 1902).

Trans.: Kenneth Cragg and Isḥāq Musa'ad, *The Theology of Unity* (London 1966).
Charles C. Adams, *Islam and Modernism in Egypt* (London 1933). JO

'Abdulḥamīd Mōhmand (d c1732), Pashtun romantic poet. Ḥamīd was born in the village of Māshō Khēl, south of Peshawar, and was a near contemporary of 'Abdurraḥmān Mōhmand (qv). Less popular than Raḥmān, his verse is more polished, learned and subtle. He translated two romances from Persian and wrote a *dīwān* (qv), also known as *Durr o marjān* (Pearls and Coral).

RSPA 85–141. DNM

'Abdulqādir Khān Khaṯak, (b 1650, d c1720), Pashtun poet and translator. A younger son of Khushḥāl Khān (qv), 'Abdulqādir translated both Jāmī's (qv) *Yūsof Zoleykhā* and Sa'dī's (qv) *Golestān* from Persian into excellent Pashto. His own *dīwān* (qv) has a strongly mystical flavour.

RSPA 268–86. DNM

'Abdurraḥmān Mōhmand, (b c1650 Peshawar, d c1720), Pashtun religious poet. '*Raḥmān Bābā*', as he is known, was born and died in the village Bahādur Kelay, south of Peshawar. Little is known of his life except that it was one of extreme piety and love of God, which is fully reflected in his poetry. This has been published both as two separate *dīwāns* (qv) and collected into one. He is probably the most popular of all Pashto poets.

RSPA 1–50. DNM

Abovian, Khatchatur (b 1809 Kanaker near Yerevan, d 1848 Yerevan), Armenian writer active in national revival, of poor peasant family. As a monk of Etchmiatsin he climbed Ararat with Prof. F. Parrot, and with his help then studied theology, philology, philosophy, ethnography and pedagogy in Dorpat (Tartu). As a teacher, inspector, and head of a private boarding school he was victimized by the Church and Tsarist authorities for progressive ideas. The circumstances of his death are obscure. Abovian wrote poems, stories, fables, plays, travel books, pedagogical and ethnographical studies, at first in German and Russian. His goal was to combat ignorance and backwardness and revive consciousness, reforming education. After writing in the classical Armenian (*grabar*) he was one of the first to write in colloquial Armenian (*ashkharhabar*), adapting the Yerevan dialect for his greatest work *Verk' Hayastani* (The Wounds of Armenia, 1858), the moving story of the peasant rising against Persia in 1826. The heavy style is encumbered with Turkish and Persian elements, and crude in composition, but the description of popular customs, though slow, is informative, and the heroes are well drawn. The satirical treatment of foreigners and priests, the patriotic humanism and the lyrical descriptions of nature are highly convincing. The tendency to hyperbole, holding back the action, and the emotional intensity recall the folk heroic epos. This most important of the literary efforts of the first phase of the modernization in East Armenian literature stimulated the development of the historical and social novel.

Trans.: TPAAM; AB II.
HO7; THLA. LM

Abū 'Alī Ebne Sīnā, Central Asian scholar, see **Ibn Sīnā.**

Abubacer or **Abū Bakr,** Arab philosopher, see **Ibn Ṭufayl.**

Abul'alā Ganjevī, Persian poet, see **Khāqānī,** Afẕaloddīn Badīl b 'Alī.

Abu 'l-Atāhiya, (b 748 Kufa, d 825 or 826), Arabian poet of the Abbasid era who lived in Iraq. From a modest home he brought a hatred of wealth and power, but nevertheless he lived at court as the panegyrist of the Caliph al-Mahdī. Under the Caliph Hārūn ar-Rashīd he was imprisoned, but later freed and pardoned. About the year 178 of the Hijra, however, he gave up the gay life reflected in his love poetry and turned exclusively to ascetic verse on a decidedly pessimistic note. He wrote many prayers in verse, in which he criticized those who held power in this world, not excepting the Caliph. He was even accused of heresy. His social origins and lack of a traditional education are reflected in his light, comprehensible style, full of fresh figures of speech. He also introduced new elements into his metres.

BGAL I, 74–6, S I 114–8; NLHA 296–303; GSLA 157–9; WILA 94; AAP 46; EI² 107–8. KP

Abu 'l-Fidā', Ismā'īl b. 'Alī al-Ayyūbī (b 1273 Damascus, d 1331 Hamāt), Arab historian and geographer. Of the family of the Ayyubid sultans, he took part in the fighting against the Crusaders under the Egyptian Mamluks, and was rewarded with the governorship of Hamāt, which became his hereditary fief. His historical work *Mukhtaṣar tārīkh al-bashār* (Epitome of the History of Mankind) narrates the history of the world from the creation to the year 1329. Except for the account of his own times, it is not an original work but one based on older chronicles, summarizing the most important events recorded there. His geographical work *Taqwīm al-buldān* (Gazetteer of the Countries), also based on earlier sources, presents the geographical picture of the world as the Arabs of his times knew it.

Trans. into Latin: J. J. Reiske, *Abilfedae Annales Moslemici*, 5 vols. (Leipzig 1778); into French: M. Reinaud and St Guiyard, *Géographie d'Abulféda*, 2 vols. (Paris 1848–83). IH

Abū Mādi, Īlīyā (b 1890(?) Muḥayditha, d 1957), one of the most prolific and gifted of the Syro-American or *Mahjar* poets, he first emigrated from Lebanon to Egypt where he lived in Alexandria in comparative poverty before emigrating again to the USA in 1911–12. From 1916 he co-operated closely with the other *Mahjar* writers in New York, both as a poet and a journalist. He contributed frequently to their periodical *al-Funūn* (The Arts). From 1929, he began to edit a periodical of his own called *al-Samīr* (The Companion). Two of his best-known *dīwāns* (qv) are *al-Jadāwil* (Streams, New York, 1927) and *al-Khamā'il* (Thickets (NY?) 1940), containing deeply introspective poems which dwell on individual episodes of romantic malaise, doubt, and confusion, often combined with that strong sense of nostalgia which is the hallmark of much of the *mahjar* poetry. RCO

Abū Nuwās (b between 747 and 762, d between 813 and 815), Arabian poet. His mother was a Persian. He was the best writer in the modernizing school of Abbasid poetry. After studying in Basra and Kufa he lived in Baghdad as a panegyrist at court. He led a wild life, denying himself none of the world's pleasures, and his free-thinking verse even attacked Islam. He was said to have made up for it by asceticism at the end of his life, but his work does not suggest that this was so. His poetry is among the finest written in the Abbasid period, comprising mainly drinking and love songs, the latter mostly in praise of youths, only occasionally of women. His love poetry echoes with sincere emotion, harsh cynicism, humour and irony at his own expense. He also wrote panegyrics (in the spirit of the old forms), satires, moralizing verse, and poems on obscene, ascetic and hunting themes. In the latter he took a traditional subject but established it as a new genre. His language and style show the influence of the vernacular and of the strophe forms of folk poetry, but in other respects his language is correct, with many words of Persian origin. His contacts with the court and especially with the Caliph Hārūn ar-Rashīd became a popular theme in folk literature (see *Alf layla wa layla*). His *dīwān* (qv) has been published (Ahlwardt 1861, E. Wagner 1958).

BGAL I, 74–60, S I 114–8; AAP 46; WILA; EI ²I 143. KP

Abū Sa'id b. Abī'l-Khayr (b Mehne 968, d 1049), Persian poet, he is widely known as a mystical leader who formulated the first rule for Sufi communities. He was for a long time considered the first author to use the *robā'ī* (qv), a quatrain-form to express mystical ideas. However, recent research has proved that none of the quatrains ascribed to Abū Sa'īd were composed by him, although at approximately the same time Persian writers began using in their quatrains the imagery of wine, love, and beauty which could convey an ambiguous, or purely mystical, meaning. It is worth mentioning that the biography of Abū Sa'īd by his grandson Moḥammad b. al-Monavvar, *Asrār at-tauḥīd* (Secrets of Divine Unification) is an important example of early Persian literary prose.

R. A. Nicholson, *Studies in Islamic Mysticism* (Cambridge 1921). AS

Abū Shādī, Aḥmad Zakī (b 1892 Cairo, d 1955 Washington, USA), Egyptian poet. He studied medicine, and completed his student career in England between the years 1912–22, after which he returned to Egypt accompanied by his English wife. He showed precocious talent as a poet, and his first *dīwān* (qv) was published in 1910 when he was only eighteen. As a young man he was a friend and admirer of Khalīl Muṭrān. His career was one of great variety and output: in addition to his numerous *dīwāns* of poetry, he wrote scripts for operas, produced translations, and was an accomplished painter. In 1932 he founded the Apollo Club and edited its periodical *Apollo*. This review was of the greatest significance as a source of encouragement to young writers and poets, and as a forum for literary and artistic discussion. Abū Shādī's ideas and activities created hostility in certain sections of Egyptian society, and after the closure of *Apollo* he withdrew from the public eye. After the death of his wife he emigrated to America in 1946. RCO

Abū Tammām (b early 9th century, d about mid-century), Arab poet. He came from Syria and travelled through a number of countries (Egypt, Syria, Iraq) trying to establish himself at their courts, as a panegyrist. He succeeded in Baghdad. He is more of a rhetorician than a poet, and critics accuse him of artificiality. He became famous as a collector of ancient poetry with his anthology *al-Ḥamāsa* (Fortitude). Arabic literary criticism constantly compares this with the one of the same name by al-Buḥturī (qv).

BGAL 84, S I 134; NLHA 324; WILA 92; EI. KP

Achchygyya, Amma (real name Nyukulay Jögyörebis Muordinap, in Russian Nikolay Yegorovich Mordinov, b 1906 Nizhniy Amginsk, Alexeyev district), the greatest contemporary Yakut writer, dramatist and poet. He studied at the literary and language faculty of the Pedagogical Institute, Moscow University, graduating in 1931. Like all the first generation of Yakut writers he began in the literary groups *Saydyy* (Self-Education) and *Kyhyl sulus* (Red Star) which were formed in the 1920's in Yakutsk. His first short stories were written under the influence of the classical Russian writers and of critical realism. His stories written 1927–8 and published in the papers *Cholbon* and *Khotugu ychchat* were published in book form 1937 and 1940, entitled *Kepseenner* (Tales). They describe the changes in Yakut life after the Revolution, and the growing political consciousness of the working people. Later, particularly in the novel *Saasky kem* (Springtime, 1944, [2]1952) which made a name for him abroad, he portrayed the new Soviet citizen with his moral and spiritual qualities. He translated Tolstoy's *Anna Karenina* into Yakut. MM

Achikar (Achiacharus), an Aramaic narrative from the 5th century BC, discovered on papyri from the island of Elephantine (Jeb) in the Nile near Aswan, where a garrison of Aramaic-speaking Jewish mercenaries guarded the southern frontiers of Egypt for the Persians. The fragments comprise the story of how Achikar, wise advisor to the Assyrian king Esarchadon, was almost killed by his ungrateful nephew. They also include a collection of proverbs and short fables. The Aramaic text seems to be based on an Akkadian original; it was not discovered until 1906–7, but Achiacharus was known from the *Book of Tobit*, and from Syrian and Arabic versions, and appeared in the *Arabian Nights* (see *Alf layla wa layla*) under the name *Haikar*.

Trans.: H. L. Ginsberg, in ANET, 427–30
SZS

Acts of Thomas (perhaps early 3rd century AD), a Syriac romance describing the (legendary) work of the Apostle Thomas in Southern India. Besides the narrative, which also draws on Indian Buddhist motifs, the book includes two songs of poetic value probably of gnostic origin (also preserved in a Greek version), the *Song of the Church* and the *Song of the Prince and the Pearl*, known also as the *Song of the Soul.*

Trans.: A. F. J. Klijn, *The Acts of Thomas* (Leiden 1962).
SZS

adab, a term used in Arabic literature. At the present time it means literature in general, especially *belles-lettres*, but in the Middle Ages it signified didactic literature, that sum of knowledge which makes a man courteous and urbane. It comprised particularly poetry, the art of oratory, historical and tribal traditions, grammar, lexicography and other diciplines. *Adab* in time broadened its base from the original Arabic nucleus and embraced the knowledge of other nations as well, the Iranians, Hellenes and Indians. It thus came to include Iranian epic, gnomic and narrative material, Indian fables, and the philosophy, ethics and economic knowledge of Greece. There were different forms of *adab* according to the audience for whom it was intended (scribes, viziers, judges, court circles and even rulers, for whom the form was that of the *Fürstenspiegel*). Although *adab* played a progressive role in furthering the development of Islamic and Arabic secular learning, and its propagation, it nevertheless strengthened those trends which led to sterility in Arabic thought.

J.-C. Vadet, *L'esprit courtois en Orient dans les cinq premiers siècles de l'Hégire* (Paris 1968); EI[2] 175–6; NLHA 283; 346; GSLA 196; WILA 55–6.
KP

Addarian, Garnik (b 1925 Beirut), Armenian poet. Since the forties he has been one of the most popular personalities in the public and literary life of Armenians in the Near East. He is a poet, editor, and organizer of the Armenian Club in Beirut. He has published verse in the Lebanon, France, USA and USSR (*Patneshin vra*, At the Barrier, 1954; *Aprim-meŕnim*, Chrysanthemum). He was awarded the prize offered on the fiftieth anniversary of the Armenian genocide, for his poem *Matean cawi ew narucman* (Book of Pain and Recompense, 1965). He has written articles on many West Armenian writers (V. Shushanian, etc) and on contemporary Armenian culture in the diaspora.

Trans.: P. Loriol, in ELA.
LM

Adıvar, Halide Edib (b 1884 Istanbul, d 1964 Istanbul), Turkish writer. After finishing at the American High School in Üsküdar in 1901, she started her career as teacher, school inspector, and writer. In 1917, she married the physician, writer, and politician Abdülhak Adnan Adıvar, whom she followed in his exile to France and England in 1926–39. Returning with him, she became professor of English Literature at the University of Istanbul in 1940, and deputy in the National Assembly, 1950–54. By her English, not French, education she differed from the majority of the intelligentsia of her time. Halide Adıvar was the first woman to actively participate in the public life of writers and journalists, riding on horseback through war-torn Anatolia talking to the leaders of the Nationalist revolution. In World War I and the post-War years she was the idol of emancipated womanhood.

Determined and enthusiastic, her novels combine a personal, womanly emotionalism with a universal message. There is much Western sophistication in her thought and in her psychological approach while the language itself is rather conservative. In her later novels the portraiture of the individual is overshadowed by an attempt to bring to life a milieu or period, for example in *Sinekli Bakkal* (The Clown and his Daughter, 1936, in English 1935). In this particular novel, which was written in exile and first published abroad (1935), a concern to interpret the traditional Turkish background to the Western reader is noticeable. Similar concerns loom in her political essays. Occasionally she also produced very fine essays in literary criticism. But her main work, especially in her early and middle years, were her 19 novels. She can be termed one of the most representative Turkish writers of fiction at the end of World War I and in the early period of the Turkish Republic. Her best known novels: *Yeni Turan* (New Turan, 1912); *Atesten Gömlek*, 1922 (Shirt of Flame, New York, 1924); *Vurun Kapiye* (Knock on the Door, 1926); *Zeyno'nun Oğlu* (The Son of Zeyno, 1928). Short stories in two volumes (1911, 1922). Two plays (one also in English: *Masks and Souls* (1953)). Mémoirs and essays: *Mémoirs of Halide Edib* (New York–London, 1926); *The Turkish Ordeal, Being the Further Memoirs of Halide Edib* (New York–London, 1928); *Turkey Faces West, a Turkish View of Recent Changes and their Origin* (New Haven–London, 1930); *Conflict of East and West in Turkey* (Lahore, [2]1936); *Inside India* (London, 1937). AT

Adonis ('Alī Aḥmad Sa'īd, b 1930), Syrian poet, one of the leading modern Arabic poets. After studying philosophy in Damascus (1954), he worked as a journalist and edited the periodical *ash-Shi'r* (Poetry). From 1956 he has lived in the Lebanon, working as a journalist (cultural column) and critic, writing poetry and publishing anthologies of Arabic poetry (*Dīwān ash-shi'r al-'arabī*, Dīwān of Arabic Poetry, 1964). In his poetry his philosophical knowledge combines with the esoteric doctrines of Islam (Sufism, qv), ancient myths of the Near East, and the influence of modern trends in French poetry, including existentialism. It is highly intellectual poetry, but extends beyond the sphere of metaphysics to touch on social conflict, too. He is the poet of rebirth, the rebirth of Arab society as well as the rebirth of the human individuality, as suggested in his poem *al-Ba'th wa-r-ramād* (Resurrection and Ashes). He is more inclined to state problems than to solve them. His constant search for new forms has led him to write poems in prose. He has published volumes of verse: *Qālat al-arḍ* (The Earth Said, 1954); *Qaṣā'id ūlā* (First Poems, 1957); *Awrāq fi'r-rīḥ* (Leaves in Wind, 1958); *Aghānī Mihyār ad-Dimashqī* (Songs of Mihyar the Damascene, 1962); *Kitāb at-taḥawwulāt wa'l-hijra fī aqālīm an-nahār wa'l-layl* (Book of Translations and Journey into the Regions of Day and Night).

CAR 249; NTALAC 165. KP

Afghānī, 'Alī Moḥammad (b 1925 Kermānshāh), Iranian writer. He graduated from the Military Academy in Teheran, and after study in the USA (1950) served as an officer in the Iranian army. In 1954 he was sentenced to death for political reasons; the sentence was later commuted to life imprisonment and in 1960 he was pardoned. He was completely unknown in the literary world when in 1961 he published at his own expense his first novel *Shouhare Āhū khānom* (The Husband of Ahu khanom), almost 900 pages long. It created a sensation and was quickly sold out. One of the largest Iranian publishers issued a second edition in 1962 at which time the book was proclaimed the best novel of the year. Some Iranian critics consider it the best Persian novel ever written. It is a social novel, dealing with family relationships, and particularly polygamy among the craftsmen of an Iranian town. The critics stress its encyclopaedic character which, however, is rather to the detriment of the literary quality of the book. In 1967 Afghānī published his second novel, again 900 pages long, *Shādkāmāne darreye Qarasū* (The Happy People from the Qarasu Valley) which deals with relations between young people in the social context of the author's native Kurdistan.

RHIL 410; KMPPL 131–4. JOS

Afghānī Jamāluddīn, Islamic reformer, see **Jamāluddīn** Afghānī.

Afrem (b c315 Nisibis, d 373 Edessa, ie Urhoy, now Urfa, south-east Turkey), Syrian poet and theologian. Baptized at 18, he became a monk and seems to have been ordained deacon; from 365 he served as a teacher in Edessa. He was canonized as a saint and teacher of the Church. A prolific writer, he left over 3 million lines. His prose includes Biblical exegesis, theological writings and sermons, and epistles. In his poetry, Afrem adopted the forms developed by the poet Bardaisan (Bardesanes, 154–222) for the propagation of Gnostic teaching, and used them to refute it. These polemic poems (*madrasha*) were directed against heretical trends in Christianity and against the emperor Julian. His ballads of the siege of Nisibis by the Persians are of historic interest. Besides these epic and sometimes dramatic songs in simple metres, intended to be sung, Afrem also wrote poems (*memra*) not meant for singing. His liturgical songs are still sung in the Syrian church and were translated into Greek, Old Slavonic and many other languages. Works by other authors are found under his name.

Trans.: J. B. Morris, *Selected Works of S. Ephrem the Syrian* (Oxford–London 1847). A. Voobus, *Literary, Critical and Historical Studies in Ephrem the Syrian* (Stockholm 1958); WSHSL. SZS

Afẕal Khān Khaṯak (b 1661, d ?1747), Pashtun historian and translator. Son of Ashraf Khān and grandson of Khushḥāl Khān (qqv), Afẕal succeeded to the chieftaincy of his tribe in 1689. His most important work is a history of the Pashtuns, in part a translation, called *Tārikhi muraṣṣa'* (Bejewelled History). He also translated Abu'l-Fazl's version of the fables of Bīdpāy, the *'Iyāri dānish.*

T. C. Plowden, *Translation of the Kalid-i-Afghání* (Lahore 1875), 167–208. DNM

Agathangeghos (5th century AD), Armenian historian. Possibly Roman in origin, author of one of the earliest Armenian texts. His *Patmut'yun Hayoc* (History of Armenia) deals with the life and work of the first Armenian Patriarch, St Gregory the Illuminator, the first Armenian martyrs and the conversion of Tiridates III and of the whole Armenian nation to Christianity. From a literary point of view, his most effective works are the lyric prayers of the Holy Virgin Hripsimé and the Vision of St Gregory; descriptions of miracles point to folk sources. The language is influenced by Greek, and may be a translation from a Greek of heterogeneous original. The treatise is known also in old Arabic and Greek versions.

R. W. Thomson, *The Teaching of Saint Gregory: an Early Armenian Catechism* (Cambridge 1970); French trans. in LCHAMA I. A. Gutschmid, *Agathangelos, Kleine Schriften III* (Leipzig 1892); A. Meillet, *Remarques sur le texte de l'historien arménien Agathange* (Paris 1911); G. Garitte, *Documents pour l'étude du livre d'Agathange* (Vatican 1946); *Une version arabe de l'Agathange* (Louvain 1950); AAPJ; HO 7; THLA. LM

aggadah, religious stories, see **haggadah.**

Ahmedî, Taceddin Ibrahim (b before 1335 Kütahya, d 1413 Amasya), Turkish poet. He is known to have studied in Cairo. Returning to Anatolia he entered the service of the Germiyan Prince Süleyman and in 1380 went to the Ottoman court. His most fertile period was the time he spent at the court of Süleyman, the son of Bayezid. He wrote both epic and lyric poetry, surpassing his contemporaries in both, and introduced the legend of Alexander the Great to Turkish literature, in a *mesnevi* (see *maṣnavī*) of 8250 couplets entitled *İskendernâme.* An important part of this work is the *Dâsitâni Tevârihi Âli Osman,* which is the earliest work dealing with the history of the Ottoman dynasty to survive. Ahmedî's best epic is the *mesnevi Cemşîd ve Hurşîd* (5,000 couplets), a free version of the work of the same name by Salmāne Sāvejī (qv). His talent found its finest expression in his lyrics, however, in his *dīwān* (qv). In his time he represented the peak of the poet's art; his epic *mesnevi* inaugurated a period of literature in which longer poems predominated, while the language of his *dīwān* became the standard Turkish of poetry for centuries.

GHOP I, 260–98; PHTF II 417 ff. EP

al-Aḥwaṣ al-Anṣāri (b c655, d c723–8), Arab poet of the Umayyad dynasty and the finest of the love poets of Medina. Most of his life was passed in the cultured ambiance of Medina, as the poet of both the Anṣār and the Umayyads. His wild life and his satirical pen aroused the enmity of many powerful men, bringing him various punishments including a period of exile on the island of Dahlak in the Red Sea. He wrote love poetry, satire and panegyrics; his famous counterpart in Mecca was 'Umar b. Abī Rabī'a (qv).

BGAL I 48–9, S I 80; EI[2] 305; GSLA 133. KP

Akhnaton, Egyptian king of the 18th dynasty (c1367–50 BC). He changed from his original name Amenhotep, when he decided to introduce the worship of one god, Aton—the sun-disk, in the place of the multitude of local gods. Having moved his court from Thebes to a new site (near el-Amarna) which he named Akhetaton, he lived there in seclusion till his death. The essence of the new religion can be found in the *Hymn to Aton*, probably composed by the king himself. This explains that the only creator and giver of life is Aton, who cares for the people of Egypt and other lands. Akhnaton's death was followed by a sharp reaction; his successor Tutankhaton took the court back to Thebes, restored the old gods (including Amen), and changed his name to Tutankhamen.

Sir Alan Gardiner, *Egypt of the Pharaohs* (Oxford 1961), ch. IX; J. Černý, *Ancient Egyptian Religion* (London 1952), 61–66. JČ

al-Akhṭal (d 710), Arab poet at the court of the ruling Umayyads. He belonged to the Taghlib tribe, which had been converted to Christianity; his adherence to the faith of his fathers gave him the chance of referring to subjects no Muslim would have dared to. In this he was often able to perform valuable service for his Umayyad masters, although more than once at the risk of his life. New trends appeared in Umayyad poetry, the love poetry of the Hijaz towns (see 'Umar b. Abī Rabī'a), and political poetry. The former meant a departure from the earlier Arabian tradition (see *qaṣīda*), while the latter carried on traditional forms, although the content had changed, reflecting the struggle for power between different religio-political sects. The work of al-Akhṭal belongs to this last trend. In his poetry satire and panegyric predominate over drinking and love songs and poems of the traditional type. Basically he followed the old models, but the social and political context of his poems is new. The sharp polemics in which the poets al-Farazdaq and al-Jarīr and the elderly al-Akhṭal took part made all three famous. The purity and polish of al-Akhtal's style is highly appreciated. His *dīwān* (qv) has survived in manuscript and was published in Beirut 1891–2.

NLHA 238 ff. WILA 45; BGAL I 49–52, S I 83–4; EI[2] 33. KP

Akhtoy (formerly called Khety by scholars), son of Duauf, ancient Egyptian writer under King Senwosret (c1971–1828 BC) who commissioned him to write the *Instruction of Amenemhēt* (see Amenemhēt I); he also wrote *Instruction of Duauf*, the so-called *Satire on the Trades*, enjoining the young man to study and to become a scribe; he describes the disadvantages of other occupations, and praises that of the scribe, who acquires authority over others. This work was very popular reading in the New Kingdom (16th–11th century BC), frequently copied; all extant manuscripts date from this period.

Trans.: J. Wilson, in ANET, pp. 432–4. JČ

Ākhund Darwēza (b c1540, d 1639 Peshawar), Pashtun divine. Darwēza was a champion of Muslim orthodoxy and a virulent opponent of the heretical teachings of Bāyazīd Anṣārī (qv). He wrote a large number of works in Persian and compiled the *Makhzan ul-islām* in Persian and Pashto, a compendium of the faith later augmented by his son Karīmdād.

J. F. Blumhardt and D. N. MacKenzie, *Catalogue of Pashto Manuscripts in the British Isles* (London 1965), 1–28. DNM

Akhundzadä (Akhundov), Mirzä Fäthäli (b 1812 Nuha, d 1878 Tiflis), Azerbaijan writer, poet, dramatist, philosopher and leading figure in the national awakening, best known by the Russian form of his name Akhundov. He studied at the Muslim university of Ganje and spent his life as official interpreter in Tiflis (Tbilisi). Through Russian, he came to know other European literatures; in addition he had a profound knowledge of the languages and literatures of the Muslim Near East. He introduced to Azerbaijan prose the genres of short story and comedy, and the conception of literature as a force for national awakening. His philosophical story *Aldanmysh kävakib* (Disillusioned Stars) is the first literary prose work in the language. His comedies, depicting contemporary life, inaugurated literary drama; they drew on the folk dramatic tradition, and were in fact among the first plays in the Islamic cultural sphere; translated into Persian they laid the foundations of Persian and Afghan drama. A linguistic reformer, Akhundzadä tried to introduce the Latin instead of the Arabic script for the West Turkish languages (Azerbaijan and Turkish, which he wanted to form into a single literary language). In his philosophical and critical writings he fought against religious fanaticism and other forms of social backwardness. His disciples revived Azerbaijan literature, with new ideas influenced from Europe, but in the national spirit.

Trans.: Louis Bazin, *Comédies* (Paris 1967).

ZV

Akyn, Kazakh folk singer, see **Zhabayev,** Zhambyl.

'Alavī, Bozorg (b 1904 Teheran), Iranian writer, one of the pioneers of modern prose. Of an old merchant family, he was educated first at home and then in Berlin, where he learned German. On his return he taught at the Teheran Technical School, at the same time beginning to write. Together with other young modernists he founded the famous literary circle *Rab'e* (see S. Hedāyat). His interest in politics brought him into an illegal Marxist group in Teheran, led by Dr Erānī. Along with 52 other members of the group he was arrested in 1937 and imprisoned until the amnesty of 1941, when a new ruler came to the throne. 'Alavī's first book was a collection of six stories, *Chamadān* (The Suitcase, 1934), with unusual, perverted characters for its heroes, such as the impotent opium addict of *Sarbāze sorbī* (The Lead Soldier), or people in abnormal situations like the father and son in love with the same girl, in the title story. Underlying the text is the author's Freudian interpretation of character and human relations.

In his later work, marked by the years in jail and his critical approach to the problems of contemporary life in Iran, he tried to drop this attitude. *Varaqpārehāye zendān* (Prison Scrap-papers, 1941), a collection of short stories published soon after his release, presents the tragic fates of some of his fellow-prisoners. *Panjāh o se nafar* (The Fifty-three Men, 1942), a description of his arrest, interrogation and imprisonment, is of documentary rather than literary worth. The nine prose pieces in *Nāmehā va dāstānhāye dīgar* (Letters and Other Stories, 1952) include a suggestive lyrical picture of provincial life, *Gīlamard* (The Man from Gilan), and the moving title story about the conflict between a Teheran judge devoted to the regime and his daughter, member of an illegal political group. His only novel, *Chashmhāyesh* (Her Eyes, 1952), is the story of a group of intellectuals working against the Iranian dictatorship at home and abroad, to bring about a more just social order; the two main characters, Mākān the painter of genius, uncompromising in his devotion to the cause, and the lovely Farangīs, wealthy and aristocratic but unreliable, bring the novel into the category of literature for the broad public. It has been translated into Polish and German. Since settling in Germany 'Alavī has changed his literary language and written several stories in German. Some of his books however are not in the field of *belles-lettres*, such as *Kampfendes Iran* (1955); an imaginary trip to the Iran of today and its traditional past, entitled *Das Land der Rosen und Nachtigallen* (1957); *Geschichte und Entwicklung der modernen Persischen Literatur* (1964); and a compendious dictionary, in collaboration with H. Junker, *Persisch-deutsches Wörterbuch* (1965). In addition to his own writing, 'Alavī translates Chekhov, Priestley, Schiller, Shaw and others.

Trans.: H. Melzig, *Ihre Augen* (Berlin 1959). RHIL 414–5; KMPPL 113–24. VK

Āle Aḥmad, Jalāl (b c1920, d 1971), Iranian writer. A teacher by profession, he is of the generation which made its literary debut after the Second World War. His first stories were contributed to the modernist review *Sokhan* (The Word); they deal with the contemporary problems of Iranian rural and urban life, and especially those which stem from religious prejudice and conventions. His attitude is critical, perhaps because of his strict family upbringing, but is enhanced by his experience as a teacher. Between 1945 and 1954 a number of volumes of these short prose pieces appeared; *Dīd o bāzīde 'eyd* (Exchange of Festive Visits, 1945), *Az ranj ke mībārīm* (From Our Suffering, 1947), *Story of the Beehives* (1954), while *Modīre madrase* (The Headmaster, 1958), with a dry humour, is a longer work about an Iranian village school. Āle Aḥmad was also an authority on folklore and dialects and published several valuable studies in this field. This gave an additional dimension to his literary work. He also made a useful contribution to Persian literature by his translations from the French, especially the work of Jean-Paul Sartre and Albert Camus.

RHIL 416; KMPPL 125–6; AGEMPL 221–2. VK

Alf layla wa layla (known as the **Arabian Nights' Entertainments,** or A Thousand and One Nights), an Arabic collection of folk tales within a fictitious framework, gathered over a period of centuries; it never reached its final form. The prototype was the Pahlavī work *Hazār afsāna* (Thousand Tales) that had its origin in Iran during the Sasanian period (224–642). Besides Persian it also included the Indian elements then permeating Iran, treated in a Persian spirit. The framework, containing about 200 stories, tells a story of a king who took a new wife every night and had her executed the next morning, until he got a clever princess, Shahrazād, who, with the aid of her maid Dīnāzād told the king such fascinating stories night after night, for a thousand nights, that he let her live. This book in the early years of the Abbasid Caliphate (in the second half of the 8th century) was translated into Arabic in Iraq, under the title *Alf layla* (1,000 Nights). It gave an Arabic character to the stories taken over from the original model and added new ones in which important personages appear, like Hārūn ar-Rashīd (786–809), his vizier Ja'far Barmacid and the black eunuch, Masrūr; these figures, recalling the great times of the Islamic Empire, were often inserted even into the older stories. Besides anecdotes, these collections also included new tales of love, humorous stories and attractive descriptions of travels and sea voyages to far lands.

Arab sources rarely mention this type of literature, for official writers looked down on it. The economic decadence in Iraq after the Seljuq and Mongol invasions in the 11th to the 13th centuries did not favour this type of literature which appealed chiefly to the middle classes and was handed down, mainly orally, by professional narrators. In Egypt, on the other hand, from the 11th century onwards there are certain hints of a collection of stories entitled *Alf layla wa layla* (1001 Nights), the change in the title reflecting the Turkish liking for the numeral *bin bir* (1001) which suggests a large number of anything. It would appear that one of the Iraqi collections had somehow reached Egypt, thus starting a new phase of development on new soil. The framework was kept, except that Shahrazād now appeared as the vizier's daughter and Dīnāzād as her sister. Under the Mamluk Sultans (1250–1517) and then during the Ottoman Government the older stories were adapted to fit the social and economical conditions of the time, giving a faithful picture of contemporary society; new Egyptian stories were also added.

From the manuscripts which have survived (or rather from the professional narrators' manuals) it seems that the title was never taken literally. Attempts to spread the contents over 1001 nights involved adding other elements that had previously existed separately, like the long story of King 'Umar (ibn) an-Nu'mān, which appears in a redaction of the late 18th century and is found in printed versions in India, Egypt (Bulak, Cairo) and Beirut from the first quarter of the 19th century onwards. Of other redactions divided up into 1001 nights, a particularly interesting one exists in the Wortley Montague manuscript of the Bodleian Library, Oxford, dating from the middle of the 18th century. It differs from the previously mentioned redaction in that it includes stories not found in any other version, but which were probably a part of the repertoire of contemporary story-tellers.

In Europe in 1825, M. Habicht began publishing the Arabic text, completed after his death by H. L. Fleischer, in 1843. The Tunis manuscript which this edition, according to its sub-title, is supposed to follow, is nevertheless mysterious; it seems never to have existed in fact and it seems that Habicht's own fictitious arrangement was followed.

The most delightful and, from the literary point of view, the most valuable element in the text are the fairy-tales; there is a large group of stories and romances of love, and of adventurous and humorous themes. These latter are longer narratives; they have for the most part a framework of their own, and they are interspersed with short legends and anecdotes in which some of the characters are historical personages. The relatively few fables, and the longer cycles of didactic and cautionary tales, are meant to be instructive. Even Islamic science appears occasionally, in the form of a learned disputation. In most of them the prose is decorated with passages of verse, chanted by the narrator to the accompaniment of a stringed instrument.

A French paraphrase by Antoine Galland (1701–17) introduced a selection of the stories to Europe; it was then translated into many languages. Relatively complete translations from the original Arabic were made during the 19th and 20th centuries of which the most important are: into English by E. W. Lane (1838–41, with omissions), J. Payne (1882) and F. R. Burton (1885); in German by G. Weil (1837–41, re-edited 1866–7), M. Henning (1895–9) and E. Littmann (1921–8); in Danish by J. L. Rasmussen (1824) and I. Oestrup (1927–8); in French by J. C. Mardrus (1899–1906); in Czech by F. Tauer (1928–55); in Russian by M. A. Sal'e (1929–36) and in Italian, by a team led by F. Gabrieli (1948–9). Most of these translations are based on the printed Egyptian version and appeared in several editions, some of them being provided with more or less detailed annotations. Translations of some parts of other redactions were published, either separately or as supplements to the above-mentioned editions. Galland's paraphrase differs considerably from these translations, and includes some very popular tales, of considerable literary value, whose Arabic original is unknown today.

A detailed bibliography up to 1885, V. Chauvin, *Bibliographie des ouvrages arabes*, IV–VII, XI, (1900–5); F. Tauer, *Tausendundeine Nacht im Weltschrifttum als Gegenstand der Lektüre und der Forschung*, *Auslese der Bibliographie*, in *Insel Almanach auf das Jahr 1969* (Frankfurt a.M.), 121–47. FT

Ali, Sabahattin (b 1906 Gümülcine Komotini, d 1948 Kirklareli), Turkish writer. Son of an army officer, he spent his childhood and youth in various provincial towns. He was then a boarding student at the Teachers College in Istanbul. Winning a scholarship, he went to Germany where he stayed from 1928 to 1930. After returning Ali became a teacher of German at schools in Aydın, Konya, Ankara, and later a drama teacher at the State Conservatory, Ankara. Because of an offensive political poem, he served one year in jail (1932–33). In 1944 he was dismissed from his teaching job. Subsequently he tried various kinds of work, among others in the publishing business. In 1948 he was found slain near the Bulgarian border.

The years the intelligent, gifted young man spent in Germany had a lasting effect on him. Returning with a thorough knowledge of German and a familiarity with Western culture and literature, he became painfully aware of the social woes of his country, and especially of its rural areas. After a short period of romantic self-expression, in which he cast poetry into the mould of the folk song, he turned to prose and discovered in provincial Anatolia an inexhaustible source of material for a novel (*Kuyucaklı Yusuf*, 1937) and a large number of brilliant short stories, which opened a new era in Turkish literature. Their limpid style, penetrating psychological observation, and pitiless exposure of social evils at once brought the author into the limelight. In another remarkable novel, *İcimizdeki şeytan* (The Devil Within, 1940) he described the problematic position of the Turkish intellectual between East and West, old and new. The post-World War II years pushed the fighter Sabahattin Ali deeper into politics. He founded a bold and dangerously successful satirical paper (*Marko Paşa*, 1946) which constantly fought with the censor for its life. Bitterness and frustration tinge the last batch of his short stories (1947). A treacherous murder interrupted his creative and successful life. Works: poetry, *Dağlar ve rüzgar* (Mountains and Wind, 1934); 5 volumes of short stories, *Değirmen* (The Mill, 1935); *Kağnı* (The Cart, 1936); *Ses* (The Voice, 1937); *Yeni Dünya* (1943); *Sirça Köşk*, (The Glass Pavilion, 1947); and the novel, *Kürk Mantolu Madona* (Madonna in a Fur-Coat, 1943). AT

Alimjan, Hamid, Uzbek poet, see **Olimjon,** Hamid.

Alishan, Ghevond (b 1820 Constantinople, d 1901 Venice), Armenian poet and scholar. Son of an archaeologist and numismatist, he was educated at the Mkhitarist seminary in Venice (see Sebastatsi Mkhithar) and universities in Europe; as a priest, he taught at Paris and Venice Armenian schools, and collaborated in editing the journal *Bazmavep*. Alishan soon deserted classical Armenian for a new literary language which would be intelligible to the common reader. This he did independently of Abovian (qv) and even earlier, on the basis of his mother-tongue learned in childhood, and later philological studies. His lyric poetry and epic poems on historical themes (eg *Plpuln Avarayri*, The Avarayr Nightingale, the hero of which is the historian Yeghishe, qv) were among the first signs of Romanticism in Armenian poetry; immediately popular, they influenced the younger generation of poets. They were patriotic in mood, and gloried in the past and in the beauty of the homeland. The growth of national consciousness and the movement for national liberation in the 40's and 50's gave rise to his studies of the philology, history, ethnography, geography and botany of Armenia; his editions of old manuscripts and mediaeval folk poems are still a mine of information for students of Armenia. Works: poetry *Nuagk'* (Songs, Vols. I–V, 1857–8); *Nahapeti erger* (A Patriarch's Songs, 1847–50); history-ethnography, *Yushikk' Hayreneac Hayoc* (Reminiscences of the Armenian Fatherland, 1869–70); *Hayapatum* (Armeniaca, 1901–2); *Shirak* (1881); *Sisuan* (1885, in French *Sissouan ou l'Arméno-Cilicie*, Venice 1889); *Ayrarat* (1890); *Sisakan* (1893).

A. Leist, *Literarische Skizzen* (Leipzig 1886); HO 7. LM

Amenemhēt I, (c1991–62 BC), Egyptian king, founder of the XIIth dynasty, into whose mouth the scribe Akhtoy placed the *Instruction*, written after the king's assassination at the orders of his son and successor Senwosret I. Amenemhēt is represented as speaking to his son to remind him of his father's merits; the purpose of the work was to strengthen Senwosret's position as well as the authority of the monarch in general. This work was one of the most widely used in the schools of the New Kingdom (16th–11th century BC); several complete texts and many fragments have survived, but the text is corrupt and difficult to understand.

Trans.: J. Wilson, in ANET, 418–9. JČ

Amenemope, son of Kanaht, a scribe in Akhmin, Upper Egypt, c11th century BC. He was the author of a body of teachings preserved in their entirety in a British Museum hieratic papyrus (No. 10474 from the 7th century). The work comprises brief instructions and advice to his son Harmakhor, in thirty chapters. The links between this and the Hebrew *Book of Proverbs* (qv, especially 22, 7–23, 14) are undeniable, but the question is which borrowed from which, or whether both had the same Semitic source at their disposal. Since the *Instructions of Amenemope* is in fact a late example of a genre popular in Egypt since at least the middle third millennium BC, there can be no doubt that the *Book of Proverbs* is indebted to it.

Trans.: J. Wilson, in ANET, 421–5; F. L. Griffith, in *Journal of Egyptian Archaeology*, XII (1926), 191–231. JČ

Amenhotep, see **Akhnaton.**

Amir Khosrou (b 1253 Patiala, d 1325), Persian poet, the son of a Turkish officer and an Indian lady. He took part in the activities of the court from early life, attached himself to Sultan Balaban's son, and was ever willing to win the patronage of a new ruler by praising him. He was a disciple of Neẓāmoddīn Ouliyā, the Chishti saint of Delhi, from whose sayings he collected the *Afẓal al-favā'ed*. Amīr Khosrou excelled in all fields of literature. His collection of five lyrical *dīwāns* (qv), highly appreciated by Ḥāfeẓ (qv), shows an amazing technical perfection, a dazzling display of puns and rhetorical refinement, but is, on the whole, somewhat cerebral. His epic poetry consists of an imitation of Neẓāmī's (qv) *Khamse*, widely read in India and Iran, and a novel creation, a number of *maṣnavīs* (qv) about contemporary Indian subjects, like *Doval Rānī Kheẓr Khān* (1316), the love story of a prince of the ruling house, *Qerān as-sa'deyn* and *Noh sepehr*. They are partly interspersed with *ghazals* (qv), and sometimes the headings of the chapters again form a *qaṣīda* (qv). These epics, as his whole work, are of special interest because of their local colour and the information they give about Indian customs, festivals, fruits etc. Amīr Khosrou also composed books on style in letter writing (*E'jāze Khosravī*) which also are interesting sources for mediaeval Indian culture. His treatises on riddles show his amazing mastery of languages, not only Persian, but also Arabic and Hindi (though the Hindi verses attributed to him cannot be verified). Amīr Khosrou, surnamed 'the Parrot of India', was a master of music. None of his works in this field seems to have survived, but his theories are well known to Indian musicians. He was probably the most versatile Indian writer during the whole Muslim period. His works have often been illustrated by miniaturists since Baysonghur Mīrzā's time.

Elliot, *History of India* III 524–66; Wahid Mirza, *The Life and Works of Amir Khusraw* (Calcutta 1935); M. Habib, *Life of Amir Khusro* (Delhi, s.d.). AS

Amirani, the hero of a Georgian folk epic, was the son of a hunter, Sulkalmakh and the goddess of the forest animals, Dali; as a punishment for rebellion against the gods he was chained in a cave. It is a very old epic, with pre-Christian features even in the form in which it is known today. The figure of Amirani also appears in Ossetian and Abkhazian folklore under the name Abrskil. The oldest of the literary versions which point to the popularity of the theme is *Amirandarejaniani* (The Romance of Amirani and Darejan), attributed to Mose Khoneli and written in the 12th century, according to references in Rustaveli (qv). It presents Amirani as a mediaeval knight. This version was put into verse in the 17th century by the otherwise unknown Sulkhan Taniashvili. The folk version presents the separate episodes in prose with verse interludes, which has led some scholars to attempt the reconstruction of an 'original' versified text (*Nutsubidze*). The epic has been systematically collected and studied since the middle of the 19th century.

Trans.: UAGP; R. H. Stevenson, *Amiran-Darejaniani.* VAČ

Amīrī, Mīrzā Ṣādeq Khān, known as Adīb ol-mamālek, Scholar of the Kingdom, (b 1860/1, d 1917), Persian poet and journalist, a leading figure in the Iranian national revival. Of a noble family, he received a traditional education; as a young man he met the important official and scholar 'Alī Khān Garrūsī in Tabriz, and became his panegyrist. The desolate state of Iran drove Amīrī, like many other educated and patriotic men, to endeavour to spread enlightenment. They believed that one of the major causes of the sad plight of Iran was its backwardness and the lack of modern education, and endeavoured by their own efforts to make good some of these shortcomings. Amīrī established and directed an educational weekly in Tabriz, *Adab*, and later did the same in Mashhad and Teheran. He worked in the official translation institute, founded a modern secondary school in Tabriz, travelled widely, and collaborated with the enlightened Persian emigrés in Baku. All his life his verse appeared in papers and magazines, and in occasional educational publications; it was not collected for publication in book form until 1933, when it was edited by Vaḥīd Dastgardī. In form his verse presents the classical style of the *qaṣīda*, *qiṭ'a* (qqv) and *mosammaṭ*; in subject it is a chronicle of the national revival, in which men of letters, traditional *adībs* in modern guise, played such an important role. Amīrī was a pure poet of this new type, and was deeply committed.

BLHP IV 346–8; AGEMPL 35–6; MLIC I 48–56, with examples in French translation.

VK

Amos (8th century BC), Israelite prophet, originally a shepherd from Judaea. He preached against the idolatry practised in the shrines of Israel, against the wrongs committed by other nations, and against oppression. His prophecies are made more vivid by the use of imagery from nature and the life of the shepherds, and by his lyrical style. See also Bible.

J. L. Mays, *Amos* (London–Philadelphia 1970).

SZS

Anday, Melih Cevdet (b 1915 Istanbul), Turkish poet and writer. He studied sociology and worked as a journalist; later he became a professor at the Istanbul Conservatoire and a member of the administrative board of the Turkish Radio and Television. Together with Orhan Veli Kanık and Oktay Rifat (qqv) he founded a group of poets *Yeni Şiir* (New Poetry) and contributed to the anthology *Garip* (Strange, 1951). His later work is a reaction to the abstract principles of the poetic movement *İkinci yeni* (Second New). In his novels *Aylaklar* (Tramps, 1965) and *Gizli Emir* (Secret Orders, 1970) he attempted an exact picture of the social background of his story. He has published a travel book on the Soviet Union, Bulgaria and Hungary (1965), a number of books of essays and volumes of poetry, for example *Rahatı Kaçan Ağaç* (The Tree that Lost its Peace of Mind, 1946), *Yan Yana* (Together, 1956), *Göçebe Denizin Üstünde* (Nomad at Sea, 1970); and the plays *İçerdekiler* (Those Who Are Within, 1965) and *Mikado'nun Çöpleri* (Mikado Sticks, 1967). LH

Anṣārī, Bāyazīd, Pashto writer, see **Bāyazīd** Anṣārī.

Anṣārī, Sheykh ol-Eslām Abū Esmā'īl 'Abdollāh b. Moḥammad b. 'Alī al-Haravī (b 1006 Kuhandiz, d 1089 Herat), Persian poet and mystic. Anṣārī claimed descent in the sixth generation from one of the companions (*anṣār*) of the prophet, Zayd, called Abū Ayyūb. A famous traditionalist and Sufi (see Sufism), he received his spiritual education from Sheykh Abu'l-Ḥasan al-Kharaqānī. An outspoken supporter of the Hanbalite school, Anṣārī was repeatedly persecuted for his convictions. Attracted to mysticism and poetry early in life, his first poetical attempts are said to have been made at the age of nine. He studied in Bistam, Tus, Nishapur, and travelled widely to gather traditions about the prophet (*ḥadīth*, qv), claiming to have collected about 300,000. Three times expelled from Herat by the jealousy of the Sufis of Herat, who charged him with heresy, he was recalled each time. His last banishment was by the famous Seljuq minister Neẓāmolmolk (qv), who soon recalled and honoured him. Anṣārī received acknowledgement of his merits as an orthodox thinker by the Caliph Qā'imbillāh in 1069, and by the Caliph Muqtadirbillāh in 1081, who gave him the title of *Sheykh ol-Eslām*.

Anṣārī wrote in Persian and Arabic. His most famous work in Persian is the *Monājāt* (Prayers), monologues of the soul, composed in verses and rhymed prose *saj'* (qv). Other works are the *Anīs al-morīdīn va shams al-majāles* (The Comrade of the Adepts and the Sun of the Sessions), perhaps the first account of the famous Qur'ān story of Joseph and Zoleykhā in Persian prose, and *Manāzel as-sāyeirīn* (Stations of the Travellers) in prose and verse (*Pseudo-Manāzil*, not to be confused with the work in Arabic of the same title). Leaving out the verses as later interpolations, the remaining rhymed prose is one of the earliest specimens of Sufi didactic literature and one of the sources leading to later Sufi didactic verse initiated by Sanā'ī (qv) in the 11th century. The other source is the didactic poem of the traditional ethical type (*pandnāme*), eg Nāṣere Khosrou's (qv) *Sa'ādatnāme* and *Roushanā'ināme*. Anṣārī is one of the earliest composers of versified *resālas* (tracts) and mystical *robā'īs* (qv). Of the former *Ṣad meydān*, *Naṣīheye Neẓāmolmolk*, *Vāredāt*, *Qalandarnāme*, *Asrārnāme*, *Maḥabbatnāme* should be mentioned. Anṣārī is reported to have composed a commentary to the *Qur'ān* (qv) and three collections of poems in Persian, which are all lost. Important for later hagiographers were his *Ṭabaqāt aṣ-ṣūfiyya* (Biographies of Sufi Saints), a Persian version of the Arabic hagiographical work *Ṭabaqāt aṣ-ṣūfiyīn* by Sulamī. These contained biographies of 120 Sufi mystics in the old language of Herat, most probably written after his death by a disciple using notes dictated by Anṣārī.

BLHP II 269–70; RHIL 234–5; SPL 924–7.
WH

'Antara b. Shaddad (6th century), Ancient Arabian poet of the pre-Islamic era, hero of the later 'romance' *sīrat 'Antar* (qv), and one of the most famous poets of his time. His father was an Arab and his mother a Negress; it was his deeds that gained him his renown. His verse is decidedly heroic in character, as can be seen too in his famous long poem (*mu 'allaqat*, qv). His *dīwān* (qv) has survived.

Trans.: AAP 148–84.
BGAL I 22, S I 45; NLHA 111; WILA 31; EI²I, 521.
KP

antuni, Armenian folk form, see **Mkrtitch** Naghash.

Anvari, Ouḥadoddīn 'Alī b. Vaḥīdoddīn Moḥammad (b c1126 Abivard, d 1189/90 Balkh), Persian poet. Together with Mo'ezzī (qv) the greatest in the circle of poets around Sanjar at the Seljuq court. Equally versed in poetry and the sciences of the time, mathematics, astrology and philosophy, he preferred the career of court poet to science, for financial reasons. That he suddenly turned to poetry, as suggested by Doulatshāh and handed down by tradition, is refuted by the poem quoted, in which Anvarī speaks of years of previous poetic activity. He was a convinced Sunnite. A false prediction of a gale, for which he was ridiculed for years, and a satire on Balkh wrongly ascribed to him, were some cause of trouble. He detested his work as propagandist of the Seljuqs and eulogist of Sanjar, and ended his life in retirement and scholarly pursuits.

Anvarī is regarded as the greatest poet on the *qaṣīda* (qv). His poetry is in the Azerbaijan style, the second of the five main styles of Persian poetry, and has much in common with the works of Khāqānī (qv). It is characterized by excessive use of rhetoric figures, scientific terminology and learned allusions to the sciences of the day, and to music. This is most evident in his *qaṣīdas*, while his *ghazals* (qv) are more conventional, showing less of his individual manner. In the *qaṣīda* an element of humour and dialogues give a personal touch and somewhat relieve the rigidity of highflown learned rhetoric. The introducton of the *qaṣīda*, the *naṣīb*, is loosely connected with the main eulogistic part or sometimes absent. Thematically the subject of nature is less represented in his poems. The influence of Arabic poetry is still obvious. There is a moralistic strain in Anvarī's satire. *Robā'īs* (qv) and *moqaṭṭa'āt* of a more personal character, reflecting actual life, complete his poetical works. Of his scientific works a commentary to Ibn Sīnā (qv) may be mentioned. Anvarī's *dīwān* (qv) was widely commented, the best known commentary being the one by Abo'lḥasan Farahānī.

BLHP II 365–91; RHIL 197–9. WH

Apocalypse (in Greek: revelation, vision), Jewish writings revealing the future, especially the end of the world. Apocalyptic features can already be seen in the Old Testament (*Isaiah*, c24–27, and the vision in the latter part of *Daniel*, qqv), but most writings of this kind date from about the birth of Christ; some are placed among the Old Testament pseudepigrapha (qv), like the *Book of Enoch* and *IV. Esdras*. The canon of the New Testament (qv) includes the *Revelation* of St John the Divine, written in Greek, but betraying Semitic origin in the style and language. The images are drawn from the Old Testament and early Oriental sources. The impassioned visionary style of the apocalyptic writings was taken over by the religious movements of the late Middle Ages and early Reformation, and by Romantic and Symbolist writers later.

D. S. Russell, *The Method and Message of Jewish Apocalyptic* (London-Philadelphia 1962). SZS

Apandi, the hero of Central Asian folk tales, see **Nasreddin Hoca** and **Mushfiqī.**

Apocrypha (in Greek: 'hidden' books), according to the Jewish manner of expression, books not suitable for public worship. The Old Testament *Apocrypha* are either additions to canonical Books (*Daniel*, *Esther*, qqv) or variants (*III. Esdras*), or they may be books dating from too late to be accepted in the canon. The *Book of Tobit* includes motifs from the Aramaic story of *Achikar* (qv); the heroine of the *Book of Judith* is a courageous woman. The *Book of Ecclesiasticus* (qv), modelled on wisdom literature, is partially preserved in its Hebrew original, but the *Wisdom of Solomon* is more Greek in both spirit and language. The first two *Books of the Maccabees* give a fairly accurate historical account of the Jews' fight for religious freedom and national independence against their Seleucid rulers in the middle of the 2nd century BC. The New Testament *Apocrypha* comprise sayings attributed to Jesus and later imitations of the Gospels and Epistles, as well as various Apocalypses.

R. H. Charles (ed.), *The Apocrypha and Pseudepigrapha of the Old Testament* (Oxford 1913 and reprints); B. M. Metzger, *An Introduction to the Apocrypha* (New York 1957). SZS

Aqhat, Ugaritic legend incompletely preserved on cuneiform alphabetic tablets of the 14th century BC (see Baal). King Daniel was grieved because he had no son, but the god El granted his wish. When his son Aqhat grew to manhood he received a magic bow from his father, which he refused to give up to the goddess of war and love, Anath, even in return for her love and immortality. She had the youth killed by a mercenary, but the murder of the innocent made the fields barren. The sorrowing father cursed the place of the murder and Aqhat's sister Pughat was preparing to avenge his death. The continuation has been lost, but it may have described Aqhat being brought to life again. The tradition of the wise Daniel is found in the Apocryphal story of Susanna (see Daniel), the motif of the birth of the son in the *Arabian Nights*.

Trans.: L. H. Ginsberg, in ANET, 149–155.
C. H. Gordon, *Ugarit and Minoan Crete* (New York 1966); DCML; GT 316–76. SZS

'Aql, Sa'īd (b 1912 Beirut), Lebanese poet, dramatist and short story writer. His greatest success was the play *Kadmūs* (1944), re-telling themes from classical antiquity and introducing into Arabic literature tragedy of the classical type. His most successful volume of verse was *Rindalā* (1950). He has been called the poet of joy. He has published a volume of lyrics written in the Lebanon dialect, *Yara*, in an attempt to develop the colloquial language and raise it to a literary level. This raised controversy, because he abandoned the Arabic orthography for one based on the Roman alphabet. On his initiative, and financed by him, the Sa'īd 'Aql Prize is awarded every month. Further works: poems, *Ajmal minka? Lā!* (Lovelier than You? No!, 1960); stories, *Lubnān in ḥakā* (If Lebanon were to speak, 1960); play, *al-Majdalīya* (The Magdalene, 1937); essays, *Ka's li -Khamr* (A Wine Glass, 1960). SP

al-'Aqqād, 'Abbās Maḥmūd (b 1889 Assuan, d 1964 Cairo), Egyptian writer, critic, literary historian and poet. From a conservative Muslim family, he was educated in Assuan and became a civil servant, later a journalist and writer. He was imprisoned in 1930–31 for attacking the royal court in Parliament. Al-'Aqqād, who was greatly influenced by English literature, exerted a marked influence on modern Egyptian literature and criticism. Lyrical (nature, love) and contemplative poetry predominates in his verse, written in classical forms; it is fundamentally pessimist in mood. His only novel *Sāra* (1938) is a subtle psychology study of a young love-affair wrecked by jealousy and doubts. In 1921, with al-Māzinī (qv), he published *ad-Dīwān*, critical studies attacking the blind imitation of traditional forms in modern Arabic poetry, particularly in the poetry of Aḥmad Shawqī (qv). He stressed the essential unity of a work of literature, and regarded poetry as the expression of the poet's personality, and his personal feelings and experience. He used the psychological approach (Freud) in his studies of Arabic poets (Abū Nuwās, qv, Ibn ar-Rūmī, etc), convinced that a writer's work is the mirror of his soul and the history of his inner life. Al-'Aqqād also wrote studies of great religious personalities (eg *'Abqariyyat Muḥammad*—The Genius of Muhammad), many essays on literature, art and philosophy, and several autobiographical works. Other verses, *Dīwān* (1916), *Waḥj al-arba'īn* (The Inspiration of Forty Years) 1933; essays, *Sā'āt bayn al-kutub* (Moments Spent with Books); *'Abqarīyyat al-Masīḥ* (The Genius of Jesus); autobiography, *Anā* (I). JO

Aragvispireli, Shio (b 1867 Kariskhevi, d 1926), Georgian writer. The son of a village priest, he went to the Tiflis seminary and then to the Veterinary Faculty in Warsaw, where, with other Georgian students, he founded a secret 'league for the liberation of Georgia'. He was arrested and imprisoned in Warsaw and Kutaisi. After graduation he worked as a veterinary surgeon. He was one of the leading 'populists'; his stories of village life described the misery and helpless despair of simple honest people, contrasted with the debauched and morally decadent rich and those who lived 'on the upper floors', the urban upper class. VAČ

Archil (1647–1713 Moscow), Georgian king and poet, of the Mukhran branch of the Bagration family, which ruled over two east Georgian kingdoms. Politically he followed unreal chimeras, and was not very successful as a ruler. For a short time (1664–75) he resigned over Kakheti, and when driven out he made several attempts to hold down the throne of Imereti, until he was finally forced to live in exile in Russia, where he died. A follower of the national trend in Georgian literature, his poem *Dabaaseba Teimurazisa da Rustavelisa* (Argument between Teimuraz and Rustaveli) sharply attacked the former for his imitation of Persian models. *Dabaaseba katsisa da soplisa* (Argument between Man and the World) deals with the vanity of human endeavour, inspired by the author's successive misfortunes. In his didactic *Sakartvelos zneobani* (Georgian Customs) Archil described the ideal Georgian nobleman; not only physical valour, but real learning was to be cultivated. Although as a young man he had been forced to accept Islam, he later translated the Bible and theological works from Russian and church Slavonic. It was also due to his efforts that a Georgian printing-shop was established in Moscow.

W. E. D. Allen, *A History of the Georgian People* (London 1932); D. M. Lang, *The Last Years of the Georgian Monarchy* (New York 1957).

VAČ

'Āref, Mīrzā Abolqāsem Qazvīnī (b c1882 Qazvin, d 1934), Iranian poet, the bard of the national revival. Besides a traditional Islamic education he was trained in music, for his father realized the unusual musical talent he possessed and intended him to follow the pious profession of singer and reciter of the lives of the Shiite saints. After his father's death 'Āref changed; he led a gay life with the youth of wealthy families, and his attractive appearance, good voice and talent for improvisation brought him to court as a musician in the service of the Qajar rulers. Then his life again took a sudden turn; his mind and heart were captured by the ideas of the patriots fighting for national revival, for a political constitution, and for the modernization of Iran, still deep in mediaeval backwardness. He gave his wholehearted service to the movement, as a poet, musician and reciter. His poetry was inspired by the incidents of the struggle in those stormy times, calling upon his hearers to fight, encouraging friends and condemning enemies. In a sense it is a chronicle of the times, but the quality of the verse, and the uncompromising fervour of the message, give it a timeless appeal. His favourite poetic form was the *taṣnīf*, the ballad, of folk origin. He himself recited his verses to his own accompaniment and his concerts were enthusiastically greeted wherever he made one of his unexpected appearances. He was in every respect a real *shā'ere vatanī*, bard of his country. His work was collected and published by the literary historian R. Shafaq.

RHIL 372 ff; AGEMPL 36–44; MRPRPV 52–6. VK

Arif, Ahmed, Turkish poet, see **Özel,** Ismet.

Artāk Virāz Nāmak (Book of the Holy Virāz), a Middle Persian ethicodidactic work in the form of a pilgrimage through supernatural worlds. The date and author are unknown. The hero is a wise man, Virāz, chosen from a great assembly of wise men to fetch from the next world the knowledge of good and evil, which deeds merit reward and which earn punishment in Hell. Virāz reaches these other worlds in a hashish dream and on his return seven days and nights later, dictates his visions. The first three and the last chapters form the framework, while the text proper is a very stereotyped description of a pilgrimage through Heaven, Purgatory and Hell. The work is informed with a spirit of unbounded cruelty combined with the most scrupulous justice. It is a theme which had already been elaborated in some oriental literatures.

RHIL 36 f; TMSLZ. OK

'arūḍ (Persian 'arūẓ) is the system of Arabic verse which dominated the poetic theory of Persian, Turkish (Ottoman and eastern Turkic), Pashto, Urdu and other literatures. According to Arabic writers *'arūḍ* is the science of the rules by means of which one distinguishes the correct metres from the faulty in Arabic poetry. It is based on the alternation of long and short syllables (quantitative verse), but was also used in Turkish verse, for instance, where there is no difference in the length of the syllables. Open syllables with a short vowel are short (∪); long syllables (–) may be open with a long vowel or closed with a short vowel. A closed syllable with a long vowel is considered as a syllable and a half (– ∪). Certain syllables may be either long or short as required; they occur mostly in grammatical words (eg in Persian poetry, the negative *ne*, *eẓāfe*, the conjunctions *ke* and *che*).

A group of syllables forms a foot (*rukn*), of varying composition. There are eight basic feet (some authorities give ten; but only the first six are used in Persian poetry); they are known by forms of the Arabic verb *fa'ala*, for mnemotechnical purposes: *mafā'ilun* (∪ – – –), *fā'ilātun* (– ∪ – –), *fā'ilun* (– ∪ –), *maf'ūlātu* (– – – ∪), *fa'ūlun* (∪ – –), *mustaf'ilun* (– – ∪ –), *mutafā'ilun* (∪ ∪ – ∪ –), and *mafā'ilatun* (∪ – ∪ ∪ –).

Various combinations of these feet form the 19 basic metres (*baḥr*). Seven of these which in the verse (*bayt*) use only one kind of foot are called *sālim*: *hazaj* (∪ – – –), *rajaz* (– – ∪ –), *ramal* (– ∪ – –), *mutaqārib* (∪ – –), *mutadārik* (– ∪ –), the *wāfir* (∪ – ∪ ∪ –) and *kāmil* (∪ ∪ – ∪ –) and are used only in Arabic verse. The other 12 metres use various feet in the *bayt*: *ṭawīl*, *basīṭ*, *sarī'*, *muḍāri'*, *mujtathth* (*P. mujtass*), *khafīf*, *munsariḥ*, *muktaḍab*, *gharīb* (or *jadīd*), *qarīb*, *mushakil*; the first three are exclusively Arabic, the last three exclusively Persian.

The basic unit of the line (*bayt*) formed by two hemistichs (*miṣrā'*) may be of eight feet, and is then known as *musamman*, of six feet (*musaddas*) or four feet (*muravva'*). *Ramal*, for example, follows the arrangement: – ∪ – – | – ∪ – – | – ∪ – – | – ∪ – – or four identical feet in the distich, while *mushākil* has two different feet: – ∪ – – | ∪ – – – | ∪ – – –. In all metres any foot may be changed in certain conventionally acceptable ways, which gives the opportunity for great variations. These new metres have then their own names, as in Persian poetry where shortening of the last foot in the *mutaqārib* metre ∪ – – | ∪ – – | ∪ – – gives

∪ – – | ∪ – – | ∪ – – | –, *mutaqāribi musammami abtār*. Different again is the *mutaqāribi musammani maḥrūf* used by Ferdousī (qv): *tavānā bovad har ke dānā bovad* ∪ – – | ∪ – – | ∪ – – | ∪ –.

'*Arūḍ* as a system dominated the literatures of the Middle East, with the exception of Turkish, until modern time, and it was not until the 1920's that innovators in some of these national literatures began to use syllabic or free verse.

G. Weil, *Grundriss und System der altarabischen Metren* (Wiesbaden 1958); H. Blochmann, *The Prosody of the Persians according to Saifī, Jāmī and other writers* (Calcutta 1872); GHOP I, ch. 3 and 4; EI² 1, 667–77. JB

Asadī, Abū Manṣūr 'Alī b. Aḥmad (b c1010 Tus, d c1080), Persian poet. During the disturbances when Khorasan passed from the Ghaznavid dynasty to the Seljuks he left Khorasan for Nakhchavan in Transcaucasia, for the court of Prince Abū Dolaf. His heroic epic *Garshāspnāme* (Book of Garshāsp) makes him chronologically the first of the generations of epigones of the epic poet Ferdousī (qv). In the introduction, Asadī said that it was his intention to fill in what the great author of the *Shāhnāme* had deliberately left out in certain passages. Like Ferdousī he drew the material for his epic from the oral traditions of Sistan, and also from the *Ketābe Garshāsp* of Abolmo'ayyad Balkhī. He was the first Persian poet to revive the *monāẓere* (tenzon), an autochthonous genre which had been very popular in pre-Islamic Iran. He was also the author of the first Persian dictionary *Loghate Fors* (Dictionary of the Persians), in which examples are taken from the earliest Persian poets; thus fragments of works long since lost have been preserved. Some scholars (J. E. Bertel's) consider this dictionary as a preparatory study for the *Garshāspnāme*. Asadī was a man of great learning, with leanings towards the didactic, which his verse often reflected; he was therefore called Asad ol-ḥokamā (Lion of the Wise). The older generation of Iranologists made two men out of him: the father, author of the tenzon, and son, author of the epic poet and lexicographist, but modern Iranian scholars (B. Forūzānfar and Ẕ. Ṣafā) have rejected this division as unfounded.

Trans.: H. Massé, *Le Livre de Gerchâsp, poème persan d'Asadî de Toûs* (Paris 1951).
RHIL 164–5; BLHP II 148–52, 272–4. JOS

al-A'shā Maymūn (6th–7th century), Ancient Arabian poet born before the Islamic era, who probably survived to the time of Muhammad although the tradition that he accepted Islam cannot be taken as conclusively proved. He was an itinerant poet who travelled from one court and tribe to another, writing panegyrics on prominent people. Although the state in which his poems have survived is not very reliable, the author is revealed as a satirist as well as a panegyrist, and also as a writer of Bacchic verse. There are hints of monotheistic thought. His style is remarkably simple, although those who criticize his language stress his frequent use of Persianisms. His *dīwān* (qv) has survived and was first published in 1928 (R. Heyer).

BGAL I 36, S I 65; NLHA 123; BHLA 321. KP

al-Ash'arī, Abu'l-Ḥasan (b 873/4 Basra, d 935/6 Baghdad), Arab Islamic theologian. At first a follower of the *Mu'tazilite* school, which used rational arguments influenced by Greek philosophy in the formulation and interpretation of religious doctrine, he deserted it about the year 912/3 and founded a school of his own. It tended towards the traditionalist conception, but al-Ash'arī used more exact formulations; the arguments by which he defended them were of the same rationalist type as those of the *Mu'tazilite* school. Although it was sometimes criticized, his school was recognized as orthodox, and has remained so practically to the present day. The leading Sunnite theologians followed it, particularly at the peak period (roughly to the fifteenth century). Unlike theologians of the *Mu'tazilite* school, al-Ash'arī held that corporeal terms did apply to God (like hand, face, speech, sight, sitting on the throne) and were real attributes, but were not to be taken either literally (ie as anthropomorphisms) or metaphorically (as 'grace', 'power' etc), but 'without knowing how'. Similarly man cannot understand the true nature of the eschatological features of the traditional beliefs, but they have to be taken as really existing. The *Qur'ān* (qv) is an eternal attribute of God and therefore uncreated. There is nothing to limit the omnipotence of God, and man is therefore not free to create his acts; God creates them and man only acquires them. Al-Ash'arī's theological conclusions are often called the *via media* between extreme views.

LK

Ashraf Khān Khaṯak, 'Hijrī' (b 1635, d 1693 Bijapur, Deccan), Pashtun poet. Eldest son and worthy successor of Khushḥāl Khān (qv), Ashraf suffered a harsher fate. He succeeded his father in the chieftaincy of his tribe in 1672, but in 1681 was arrested and imprisoned in India by the Moghuls till his death. His pen-name means 'the exile' and suits the melancholy strain of his verse.

RSPA 249–67. DNM

ashugh, Armenian folk singer, see **Frik, Nahapet** Khutchak, **Sayath-Nova** and **âşık.**

âşık, Turkish folk singer, poet, composer and interpreter of his melodies, who accompanied himself on a stringed instrument. There is a reference to *âşiks* in one of the oldest Turkish folk epics, *Kitâbi Dede Korkut* (The Book of Dede Korkut, qv). Some of the classical Ottoman poets (see Abu Bekir Kanî, Nedim) came close to the *âşıks* in their style and in their performance of their own works, but nevertheless there remained a considerable difference between them in formal structure, metre and language, as well as in their subjects. The *âşıks* recited long epic *destans* (see *dāstān*) about the great warriors of the past, about heroic outlaws, the stories of great lovers; their shorter songs praised justice and love, gave good counsel for life, criticized and rebelled against injustice, or were simply folk songs. Many of their songs became folk songs, and it is very difficult to determine the boundary between the two. In the 17th and 18th centuries, in particular, they helped popular rebellions and even sometimes became the heroes of such rebellions, so that further songs were made up about them. Their verses represented the real culture of strata of the people to whom classical poetry was quite alien. From the last century onwards the songs were recorded, and exercised a strong influence on modern poetry (see Kemal Namık) right up to the present day (see Hikmet Ran Nazım). There are folk singers of this type in the Caucasus still, where they reflect the new life of the region in their songs. Of the Turkish *âşıks* A. Ömer, A. Garip and A. Kerem were outstanding.

GHOP V 46, 51 ff; PHTF II 2 ff. ZV

Atra-hasis ('the exceedingly wise'), sobriquet of Ziusudra or Uttanapishtim (Greek Xisuthros), the hero of the Babylonian myth of the Flood. The myth of Atra-hasīs was known in ancient Mesopotamia by the opening words: 'When the gods were as men'. Of the various versions and copies of different date, the most complete is on three tablets of the 17th century BC; there are also later versions. The myth begins when there were only gods in the world, who had to work for their living as man did not yet exist. The three most powerful gods ruled the world between them: Anu, Enlil and Enki; the younger gods were forced to work hard, and after forty years of this they determined to rebel. At this critical moment Enki suggested they create man; with the help of the mother-goddess he made seven men and seven women from clay mixed with the flesh and blood of a dead god.

In the course of the centuries man multiplied so much that the noise disturbed the god Enlil in his sleep. The gods decided to destroy man, but Enki, who had created man and felt sorry for him, twice helped mankind to survive. For the third time Enlil insisted on sending a great Flood, and so Enki told King Atra-hasīs to pull down his house of reeds and build a boat instead, in seven days. Atra-hasīs and his family and animals took refuge in their boat, and the Flood destroyed mankind. The gods were distressed at the destruction of mankind, especially Enki and the mother-goddess, who reproached Enlil for having destroyed the rivers, channels and fertile land as well.

When the waters subsided and Atra-hasīs landed, he sacrificed to the gods to win their favour. Enlil was angry to see that mankind had again survived, but apparently allowed himself to be persuaded by the other gods that men are useful to them, especially if the gods govern them better (the text is very defective here). The myth of Atra-hasīs seems to have been intended for public recitation, like that of the creation of the world, *Enuma elish* (qv). The Flood is also the subject of Tablet 11 of the epic of Gilgamesh (qv).

W. G. Lambert–A. R. Millard, *Atra-hasis, The Babylonian Story of the Flood* (Oxford 1969); LRPOA 26–35. NN

'Aṭṭār, Farīdoddīn (d c1220, extreme dates 1193 and 1235), Persian poet. Lived in Nishapur, the old centre of Sufism (qv). As his name indicates, he practised as a pharmacist, but also treated the sick, after having wandered through the lands of Islam in his youth. Besides his profession he produced a large number of mystical books in both poetry and prose which belong to the classics of Persian language. Hellmut Ritter sees in his literary products a development from more orthodox mysticism to a pantheistic feeling. Some epics without higher poetical value which show Shi'a tendencies should be attributed to another 'Aṭṭār. The poet's collection of 97 biographies of Sufi saints, *Taẓkerat ol-ouliyā'*, deeply influenced later hagiographers; it shows his tendency to story-telling and romantic description and does not always conform to the original sources. The biography of Ḥallāj in this book forms the basis for the later romantic stories about the fate of this martyr-mystic to whom 'Aṭṭār felt spiritually attached.

His most famous epic poem is the comparatively short *Manṭeq oṭ-ṭayr* (The Birds' Conversation) in the *ramal musaddas* metre. It describes the pilgrimage of thirty birds, led by the hoopoe, towards the heavenly Phoenix, Sīmorgh; after traversing seven valleys, searching with love, they reach their goal and find, at the end, annihilated and illuminated, that they themselves (*sī morgh* ie thirty birds) are identical with the Sīmorgh. The subject of the birds' journey is taken from the tradition of the Ikhvān aṣ-Ṣafā and from Ghazzālī's (qv) *Resālat oṭ-ṭayr*; Sanā'ī's (qv) *Sayr ol-'ebād* has also influenced it; but the conclusion is 'Aṭṭār's invention. The book furnished Persian poets with a large number of symbolical stories and metaphors.

The *Manṭeq oṭ-ṭayr*, *Moṣībatnāme* and *Elāhīnāme* (the story of the conversations of the king and his six sons) have the same literary form; out of one frame story the parables and tales develop logically. The *Moṣībatnāme* (Book of Affliction), tells how the wayfarer asks for forty mystical, cosmic, historical and psychological powers for help in his affliction and is informed by his master about the meaning of their refusal, until he casts himself into the Ocean of the Soul and recognizes himself as the essence of everything, and begins the journey *in* God. The subject of suffering and dying for the sake of love is best expressed in the *Oshtornāme* (Book of the Camel) whose hero even commits suicide. Among 'Aṭṭār's epics the didactic *Pandnāme* (Book of Counsel), was formerly often read in schools. His lyrical poetry is full of ecstatic verses, but even in his lyrics the narrative element is visible. Hymn-like praise, with long chains of anaphors, is found both in his lyrics and epics. The *Mokhtārnāme*, his *robā'iyāt*, together form a kind of developing discourse.

Trans.: Silvestre de Sacy, *Pandnāme* (Paris 1819); Garcin de Tassy, *Mantic ut tair ou le Language des Oiseaux* (Paris 1857); R. P. Masani, *The Conference of the Birds: a Sufi Allegory* (London 1924); Fuad Rouhani, *Le livre divin* (*Elahi-nameh*) (Paris 1961); A. J. Arberry, *Muslim Saints and Mystics: Episodes from the Tadhkirat al-Auliyā* (London 1965, Chicago 1966).
RHIL 237–40. AS

Äuezov, Mukhtar (b 1897 Semipalatinsk district, d 1961 Alma-Ata), Kazakh writer and dramatist; of a nomad family closely related to Abay Kunanbayev (qv), his interest naturally turned to literature. He studied at Leningrad and Tashkent University, became professor of literature and a member of the Kazakh Academy. In 1917 he wrote the first Kazakh play *Englik-Kebek*, inspired by a folk story. Besides contemporary themes his plays and prose writings draw on the Kazakh past, the most popular work being the biographical novel about the founder of Kazakh literature *Abay zholy* (Abay's Road, 1946, 1955) translated into many languages. Other works: stories, *Karash-Karash* (1960); plays, *Abay* (1939); *Bes dos* (Five Friends, 1952, with L. Sobolev); novel, *Ösken örken* (Reborn Generation, 1962).

Trans.: Antoine Vitez, *Abaï* (Paris 1960); Léonide Sobolev et A. Vitez, *La jeunesse d'Abaï* (Paris 1959). LH

'Aufī (Awfī), Muḥammad, Persian poet, see **'Oufi**, Moḥammad.

Averroes, Arab philosopher, see **Ibn Rushd.**

Avesta (pre-Christian era), the modern Persian name for the canon of the religious texts of Zoroastrianism (Parsi religion, Mazdaism). It consists of the *Yasna*, invocations and prayers. The most important part of the *Yasna* are the *gathas* of Zarathushtra (qv), and the *Hōm-yasht*, a remnant of ancient epic poetry dating from the warrior-peasant society before the time of Zarathushtra. The *Visprat* contains monotonous invocations of divine beings. The *Little Avesta* comprises shorter prayers and also important long hymns to the foremost gods of the old faith, *yashts* (song of praise).

From the poetical point of view the *gathas* and the *yashts* are the most noteworthy parts of the *Avesta*. Their form is metrical and rhythmical, and for the most part they represent the ancient heroic epic verse, composed by bards in the times before Zarathushtra. They praise the ancient heroic ideals, with allusions to or longer treatment of, the oldest Iranian legends. Later the *yashts* were revised by priest-poets, so that the religious and ethical aspect, and the emotional element, predominate, and the manner is that of invocation. A minority of the *yashts* are of later date and of little significance. The *Little Avesta* also includes two prayers in calendar form (*Sī rōchak*, Thirty Days). The last part of the *Avesta*, *Vidēvdāt* (Law against Evil Spirits), is a religious codex, dealing with ritual purity, man's duties towards his fellows and towards animals and the rewards for purification and cure. It tells us of types of agreements and how they were made, family life, relationships, punishments and fines, the care of domestic animals, and the remuneration of farming. There are also several legends of the earliest times in Iran. There are several shorter writings, in a later language, which stand outside the canon.

According to Parthian tradition the *Avesta* was originally much longer, the present version being only about one third of the original. For a long time it was handed down orally, and written down only much later, and not very well. It is of immense significance for research not only in philology, but in the history of civilization, folk lore, literary history, law and religion. It provides a wealth of material on the state of Iranian society in ancient times, in all aspects, although often only in hints. The value of the text varies, and from the literary point of view it rarely reaches a high level, particularly when compared with the Jewish and Christian canons.

Trans. F. Wolf, *Avesta, die heiligen Bücher der Parsen* (Berlin–Leipzig 1924).
I. Gershevitch in E. B. Ceadel, *Literatures of the East* (London 1953); J. Duchesne-Guillemin, *La religion de l'Iran ancien* (Paris 1962); RHIL 3–17. OK

Avicenna, Central Asian philosopher, see **Ibn Sinā.**

Avyātkār I Zarērān (Memoir of Zarer), a Middle Persian tale, c500, by an unknown author, based on a Parthian poem. It describes the fighting between the Iranians and the Huns, whose king did not approve the conversion of the Iranian king Vishtāsp to the faith of Zarathushtra. In this 'holy war' Vishtāsp's brother Zarēr won great fame, until he was treacherously killed by a magic weapon. His young son Bastvar became his avenger, and together with the king's son, Isfandyār, he destroyed the whole of the enemy army. Bastvar's lament for his dead father is a beautiful lyric embodied in the text. Hyperbole in the descriptions and in the numbers involved are frequent, while repetition of the principal parts of the story points to folk narrative and fairy-tales. The last author to touch up the work regarded it as a treatise on the ideal of feudal chivalry nurtured in the aristocratic religious Sasanian civilization. Zarēr fell fighting for his king, his brother, his religion and Iran, the highest values of ancient Persian society.

RHIL 43 f; TMSLZ. OK

Aybek, Uzbek writer, see **Oybek.**

Aynī, Sadriddin (b 1878 Soktare, d 1954), Tajik writer and outstanding personality in modern cultural affairs in Central Asia, whose direct influence has profoundly affected the formation of the Tajik nation of today in many important aspects. At first a poet, after the Revolution he became primarily the founder of modern Soviet Tajik prose as well as making important contributions to philology and historiography. His father, a peasant and stonemason, saw to it that Aynī acquired some education, but both his parents died when he was 12, and he went to join his elder brother in Bukhara and study at the *madrasa*, the Islamic high school. At first he waited on his wealthier fellow-students and in later years took jobs; his studies lasted 15 years. From the Turkish press he learned of the movement for progress and reform, Jadidism, and began to set up school in which the syllabus was of the Western type. Aynī was forced to leave Bukhara on account of his opinions, and on his return in 1917, he was flogged in the palace prison, sentenced to 75 strokes as an enemy of the Amir's regime. He was freed by Russian soldiers and went to Samarkand to recover; from here he organized a press campaign against the autocracy in Bukhara. After the Bukhara revolution in 1920 and finally after the Tajik Republic was established, he held various leading positions in cultural affairs. He was the first President of the Tajik Academy of Sciences, holding this post until his death.

Before the Revolution, Aynī's work took the form of traditional poetry, although social questions and reform ideas appeared in it (*Zaminro boyad nafurūshed*, You Must not Sell the Land, 1917; *Mozī va hol*, Then and Now, 1913). The text-books he wrote for Tajik elementary schools at this time were adapted to the child's mind both in content and language, and drawn up according to very progressive ideas on education. During the Revolution his verse was directed against the Amir and called for freedom for the people (*Surūdi ozodī*, Song of Freedom, 1919). Afterwards it served the people's government. The novella *Jallodoni Bukhoro* (The Bukhara Executioners, 1920), like his historical works *Ta'rikhi amironi manghitiyai Bukhoro* (History of the Bukhara Manghit Amirs, 1920) and *Bukhoro inqilobi ta'rikhi uchun materiallar* (Material on the History of the Bukhara Revolution, in Uzbek, 1926), exposed the inhumanity of the old regime. *Namunai adabiyoti tojik* (Pictures of Tajik Literature, 1926), showing the thousand-year-old tradition of the national literature, was aimed against pan-Turkish tendencies in Central Asia. This was followed by a number of historical and autobiographical stories, novellas and novels which initiated a new era in Tajik literature and became the model for most of the prose writers who followed. *Dokhunda* (The Mountain Villager, 1927) was the first novel to be written in Tajik; it tells of the hard life led by people in the mountains and their struggle to change it. The epic *Ghulomon* (Slaves, 1935) is a mature work, an allegory of Tajik over three generations.

His autobiographical writing forms an important part of Aynī's work: *Maktabi kūhna* (The Old School, 1935), describing his childhood and the village religious school; *Ahmadi devband* (Ahmad the Magician, 1936), about his boyhood; the social satire *Margi sudkhūr* (Death of a Usurer, 1937) reflects his student years; and his unfinished memoirs *Yoddoshtho* (four vols. 1949–54) represent the highest point of his work. They give a truthful and fascinating picture of life in the country and in Bukhara, showing the corruption rife in officialdom, the law courts and the army, describing the pleasures and the troubles of student life in Bukhara, and introducing the reader to the most important writers of his youth. Some of his books appeared simultaneously, or even earlier, in Uzbek, and he is therefore also held to be the founder of modern Uzbek prose. Most of his works have been translated into other languages in the Soviet Union and elsewhere.

His works on literature are important for Tajik literary history: on the tomb and birthplace of Rūdakī (qv), which finally determined that the founder of Persian and Tajik poetry lived near Samarkand; and works on Firdavsī (Ferdousī), Abū'Alī Ibn Sīnā, Sa'dī, Navā'ī, Vāsifī, and Bēdil (qqv). The number of works written about Aynī's life and writings is considerable. Other works: *Mukhtasari tarjimai holi khudam* (My Life in Brief, 1955), a volume of verse *Akhgari inqilob* (Sparks of Revolution, 1923), the novels *Odina* (1927) and *Yatim* (The Orphan, 1940), a historical work on heroes of Central Asia, published during the war, *Kahramoni khalqi tojik Temurmalik* (Temurmalik, Hero of the Tajik People, 1944), and *Is'yoni Muqanna'* (Muqanna's Rebellion, 1944). In 1940 he went back to poetry in the ballad *Jangi odamu ob* (Man against Water).

Trans.: J. Hann, *Pages from My Own Story* (Moscow 1958); S. Borodine and V. Voinet, *La mort de l'usurier* (Paris 1957); S. Borodine and P. Korotkine, *Boukhara* (Paris 1956); H. Brunschwitz, *Buchara. Errinerungen* (Leipzig 1953); T. U. G. Stein, revised by M. Lorenz, *Der Tod des Wucherers* (Berlin 1966).
RHIL 559–564. JB

Aytmanov, Chingiz (b 1928, Sheker), Kirghiz writer. He worked in an agricultural research institute and translated Kirghiz writers into Russian. Trained at the Gorkiy Literary Institute, he attracted attention with a love story of World War II, *Zhamiyla* (Jamila, 1958), whose popularity later ensured wide publicity for his realistic works in the Soviet Union and abroad. His other works are the stories *Obon* (Melody, 1959) and *Samanchy zholu* (Milky Way, 1963) and the novel *Birinchi mughalim* (First Teacher, 1963). LH

ayyām al-'Arab, narratives of pre-Islamic tribal fighting and raids in Arabia (6th–7th centuries AD) handed down orally and written up by Arab philologists and historians in the first centuries of the Hijra. These books have survived only in later literary anthologies and historical works. Some of the narratives are fairly true to historical fact; in others, especially the longer epic cycles, historical reality is lost. Many of the popular *ayyām* became the basis or a part of the heroic Arabic epics (eg the epic *sīrat 'Antar*, qv, and that of Zīr Sālim, etc). The *ayyām* are an important source of information on social and political relations in Arabia, and the only examples of pre-Islamic Arabic prose.

EI² I 793–4. JO

Ayyūb, Dhū'n-Nūn (b 1908 Mosul), Iraqi Arabic prose writer. Educated in Mosul, he studied in a teachers' college and worked as a teacher, but left to devote himself to writing. He was a Member of Parliament, but had to leave Iraq in 1954 because of his non-conformism. He went to Vienna and returned to Iraq only after the 1958 Revolution, to direct Baghdad radio. Ayyūb started writing in 1933, and has since published many articles and short stories, as well as several full-length novels. His writing is sympathetic to the simple folk, in a realistic way, often tinged with pessimism. Ayyūb's first novel was *Doktōr Ibrāhīm* (1940). He tells the story of corruption in the upper echelons of the Iraqi civil service at the time. Its hero, Dr Ibrāhīm, an agronomist, enriches himself in Iraq, by devious ways, then emigrates to England. His second novel, *al-Yad wa'l-arḍ wa'l-mā'* (The Hand, the Earth and the Water, 1948), tells the fictitious but poignant story of Iraqis led by a lawyer, who attempt to assist landless peasants by renting uncultivated land and forming a cooperative there. They fail, but receive the full sympathy of the author.

GLAIG 265–7; WILA 304. JML

Azerbaijan style, see **Anvari,** Ouḥadoddīn.

B

Baal, a collection of Ugaritic myths called after the god Baal, the central figure. The text cannot be reconstructed without lacunae, because of the fragmentary state of the alphabetic cuneiform tablets on which it is preserved. Written in an ancient Semitic language (14th and 13th century BC), the tablets are from the city of Ugarit (now Ras Shamra) on the northern Syrian coast near Ladikiya. Baal, the god of storm and of fertility, conquers the god of the sea (Yamm) in battle, while the issue with the god of death (Mot) is fought with varying success. Baal is helped by his sister and lover, Anath, who persuades the old god El to allow Baal to build a palace. Some features of these myths, and the fragment describing El's feast, are reminiscent of the ancient Greek myths of the life of the gods on Olympus. Mythological motifs and charms against snake-bite are also found in the hymns and ritual texts. The verse used, with its constant number of words and frequent parallel repetitions, gives these myths epic breadth and poetic effectiveness. The motifs, images and forms of these mythical poems of Canaan can also be found in the earliest Hebrew poetry.

Trans.: ANET 129–42.
DCML; GT 114–244. SZS

Ba'albakki, Laylā (b 1937), a Lebanese Shiite Muslim prose writer in Arabic, who has produced several remarkable novels. Born in Lebanon to a wealthy and conservative family, she had to fight them to be allowed to study, earn a living, and write fiction. A marked autobiographical element pervades her first two novels, published in Beirut. The first, *Anā aḥyā* (I live, 1958) is written with fine sensitivity and throbs with revolt against the obstacles of tradition and conservatism, speaking of the desire felt by the young for a life of their own. Arab critics were shocked both by the heroine's hatred of her mother and by her unabashedly sensual thoughts. French critics, however, were impressed by the quality of her writing and called her 'Colette from Beirut' or 'the Lebanese Sagan'. Her second novel, *al-Āliha al-mamsūkha* (The Monster Gods, 1960) is shorter than the first. It pushes the rebellion against the conventions still further in respect of the sensual feelings and illicit love of 21-year-old Myra, for 40-year-old married Nadīm. Without surrendering to the erotic element, Laylā Ba'albakkī succeeds in drawing an arresting picture of Lebanese society in change, and of Arab youth in particular.

Trans.: Michel Barbot, *Je vis* (Paris 1961).
WILA 301; GLALG 242–3. JML

Bābā Ṭāher 'Oryān (b c1000, d after 1055/6 Hamadan), Persian poet popular among the broad masses of the Iranian people, who still sing his songs today. Only sporadic and contradictory reports about his life and personality have come down to us. He is said to have been an important member of an order of mystics, a dervish, and the author of works in Arabic. There is even a tradition which describes him as a fool made wise by a miracle. His poetry places him among the early practitioners of the *robā'ī* (qv), but the form he used had a different metre (the curtailed hexameter) and is usually called the *dobeytī*. The popular character of his verse (which, according to some scholars, is connected with pre-Islamic Persian poetry) is enhanced by the dialect in which he wrote. His fragmentary poetical legacy consists mainly of love lyrics, interpreted in terms of Sufi mysticism (see Sufism), which are striking in their ardour, passion and melancholy.

Trans.: E. Heron-Allen and E. C. Brenton, *The Lament of Bābā Tāhir* (London 1902); A. J. Arberry, *Poems of a Persian Sufi* (Cambridge 1937); Mehdi Nakosteen, *The Rubáiyyát of Bábá Táhir Oryán of Hamadán* (Boulder 1967). HS

Bābur, Muḥammad Ẓahīruddīn (1483–1530), of the family of Tamerlane, became ruler of Farghana at the age of fourteen. He was involved in warfare against his neighbours and relations, and conquered parts of Afghanistan. Like his cousin Ḥusayn Bāyqarā, he was defeated by the Uzbeks in 1501, and conquered Kabul in 1504. Eventually he reached north-west

India where in 1526 he won the decisive battle of Panipat north of Delhi, and laid the foundations of the Mughal Empire in India. Bābur was not only a very successful warrior and ruler but also a man of letters, a poet and calligrapher. His autobiography, *Tuzuki Bāburī* or *Bāburnāma*, belongs to the most attractive products of Turki (Chaghatāy) prose. His descriptions of landscapes, human beings and animals are always to the point. Of special interest are his remarks about the poetical activities of his numerous relations, most of whom tried to write Turki or Persian verses (thus Ḥusayn Bāyqarā is blamed for writing poetry exclusively in one metre). Bābur's language is chaste, simple and fresh, and his presentation sometimes shows touches of humour. In the present form of the book some years are lacking. It was translated twice into Persian, Zaynuddīn Vafā'i Khāfī wrote a paraphrase of parts of it, Khānkhānān 'Abdurraḥīm translated it in full in 1590. The book contains a number of Turki poems by Bābur which were collected in his partly edited *dīwān* (qv); they, too, are written in unaffected language.

Bābur also translated a mystical-didactic work by the Naqshbandī saint 'Ubaydullāh Aḥrār into Turki poetry, as well as some smaller treatises, not all of them very poetical. He composed a long, unpublished, treatise, *Risālayi 'arūẓ*, about the art of poetry in Turki. The poetical talent of Bābur was inherited by most of his children (Humāyūn, Hindāl, Kāmrān, Muḥammad 'Askarī and especially his daughter Gulbadan). The later rulers of the Mughal dynasty also composed autobiographies, and others excelled as mystical writers (Dārā Shikōh, Jihānārā) and poets Zēbu'n-nisā (see Vol. II), Aftāb, Bahādur Shāh Ẓafar.

Trans.: A. S. Beveridge, *The Bābarnāme* (Leiden–London 1905, 1922); J. Leyden and W. Erskine (London 1826). AS

Babylonian Job, see **Ludlul Bel Nemeqi.**

Babylonian Theodicy, acrostic poem in the form of a dialogue, discussing divine justice. It derives from the Sumerian Disputes (qv), in which two friends vie with each other in boasting of their abilities and achievements. In the Babylonian Theodicy the discussion is between two friends, one, the sufferer, stressing the evils of human society while the other defends the justice of the divine ordering of the world. The 300 lines of the poem form 27 strophes of eleven lines; each of the eleven lines begins with the same syllable, while the 27 initial syllables reveal the name of the author, a priest called Saggil-kinam-ubbib who seems to have lived at the end of the second millennium BC. The original text has not come down to us; there is a late copy from Assurbanipal's library and fragments of others which are even later.

The friends argue by turns, a strophe each, beginning with the pessimistic sufferer, who doubts the justice of the order appointed by the gods; he himself has known nothing but poverty and misfortune since his youth, while the just around him are oppressed and only the wicked prosper. All moral values are topsy-turvy; how can man believe in divine justice? His friend tries to convince him that sooner or later evil will be punished and good rewarded; the human view of the divine order is one-sided; how can we understand the intentions of the gods? In the end the pessimist admits that the ways of the gods are hidden from man, and that his friend is right. The Babylonian Theodicy is important literary evidence of the religious thought of the Babylonians in the middle of the second millennium BC.

W. G. Lambert, *Babylonian Wisdom Literature* (Oxford 1960), 63–91; LRPOA 320–7. LMa

Baghdādi, Shawqī (b 1928), Syrian short-story writer; he graduated as a teacher in 1951 and worked in Lādhiqīya and Damascus. He was one of the founders of the Union of Syrian Writers. He wrote poetry and stories as a student, winning many honorary awards. Most of his stories analyse the psychological state of lonely people, alienated from their environment, work, or family, and endeavouring to renew contact. The style, frequently employing a stream of consciousness technique, uses hints and stylistic short-cuts and forces the reader to supply the framework of facts and circumstances himself. Principal work: *Ḥayyunā yabṣuq dam* (Our Quarter is Spitting Blood, 1954). JO

Bahār, Moḥammad Taqī (b 1886 Mashhad, d 1950 Teheran), Iranian poet and writer. His life was typical for the intellectual and writer in Iran of the first half of this century. His father was a poet and miniature painter (M. K. Sabūrī). Bahār studied poetry and the classical Islamic sciences under the greatest literary figures of the time (eg A. Nīshābūrī), revealing an individual poetic talent in early childhood. Although his father intended him for a business career, he remained faithful to literature and soon earned fame. He was awarded the title *malek osh-sho'arā* (King of Poets) for the *qaṣīda* he composed in honour of the coronation of Mozaffaroddīn Shāh; he was the last poet in Iran to hold the title. A member of the Democratic Party during the stormy years of revolutionary ferment, Bahār decided against a career as court poet and wrote tendentious political poetry (1907–8), editing the Democratic Party papers *Noubahār* and *Tāzebahār* (Early Spring). In 1917 he founded a literary society *Dāneshkade* (Place of Knowledge) and published a literary review of the same name. He was a member of Parliament several times, but after Reẕā Shāh Pahlavī came to the throne he retired from public life and devoted himself entirely to literary research and university lecturing.

Bahār is considered the last great poet of the classical forms, and indeed the greatest poetic talent to appear in Iran in the last few centuries. He revived the Khorāsān style, and laid unusual stress on the strophe forms (*mosammaṭ, tarkībband, tarjī'band*). The importance of Bahār lay in his ability to give the classical forms a new political and social content. His lyrics were also very popular. A two-volume *dīwān* (qv) of his verse appeared in Teheran in 1956–7. He has a considerable reputation as a scholar; the outcome of his studies of Middle Persian under Professor Herzfeld was the translation of two Middle Persian texts into New Persian. He is the author of an unsurpassed three-volume work on style: *Sabkshenāsī yā tārīkhe tatavvore naṣre fārsī* (Stylistic, or the History of the Development of Persian Prose, 1941). He has also published critical editions of mediaeval Persian texts, of which the most important is *Tārīkhe Sīstān* (History of Sistan, 1935–6). His historical study, in the nature of a memoir, *Tārīkhe mokhtasare aḥzābe siyāsī—enqerāẕe Qājārīye* (Brief History of Political Parties—the Fall of the Qajars, 1945), is an important source of material for the study of modern Iranian history.

RHIL 373; BLHP IV 223, 345; E. G. Browne, *Press and Poetry of Modern Persia* (Cambridge 1914), 260–89; AGEMPL 56–64, 175, 201–4; M. Rahman, *Postrevolutionary Persian Verse* (Aligarh 1955), 25–42; F. Machalski, *La littérature de l'Iran contemporain* (Wroclaw-Warszawa-Krakow 1965), 89–105. JOS

Bahrām Gūr (ruled 420–438), Persian king, according to eastern tradition the author of the first Persian verses. Persian is said to have acquired the rhythm of connected speech for the first time in a passionate conversation between Bahrām Gūr and his beloved, Dilārām. Later writings have preserved several poems attributed to him.

BLHP I. OK

Bakî, Mahmud Abdülbakî (b 1526/7, d 1600 Istanbul), famous Turkish poet. The son of a poor *muezzin*, he was apprenticed to a saddler but, being unusually gifted, he was able to attend the Muslim university and soon became well-known as a poet. He served as a judge in the provinces and later in Istanbul, finally becoming the supreme judge of the Ottoman Empire, whose decisions fixed the legal norm. He translated Arabic treatises into Turkish and himself wrote on law. All his life he wrote lyric poetry, and although his work is not voluminous, it is considered the peak of classical Turkish literature. Most of it has been translated. His *dīwān* (qv) was read and copied while he was still alive, not only throughout the Ottoman Empire, but at the Turkish-speaking court of the Persian Shahs. His *dīwān* comprises *ghazals* (qv) in praise of wine and beauty, meditations on life, and panegyric *qaṣīdas* (qv).

While the thought is interesting, it is primarily the perfect form which is admired, the masterly use of involved poetic ornament to stress the thought and express ideas he did not wish to make explicit. Much of his verse can be interpreted in two or even more ways according to which of the basic ideas of the whole poem the reader is following. The classical Turkish distich is a complete unit loosely subordinate to the general idea of the poem; Bakî observed this autonomy while achieving marked gradation in the treatment of the central theme. He succeeded in utilizing all the possibilities offered by intricate rhetoric, without detracting from the poetic effectiveness of his language. A number of elegies revealing deep sorrow were inspired by the deaths of his father and of Sultan Süleyman, his protector and friend, who was himself a poet.

Throughout the classical period, perfection of form in treating an often-used subject was the decisive criterion in poetry, particularly successful imitation of a great model in *nazire* form. Bakî tried to break with this convention of admiring imitation (which even led to original verse being declared an imitation of Persian, or more rarely of Arabic, models). His work marks the final maturity of Turkish poetry, freed from dependence on other literatures. His followers developed two trends: one (Nef'î, qv) stressed formal perfection to the point of artificiality, and overburdened its verse with Persianisms unintelligible to Turkish readers. The other trend (followed by, eg, Yusuf Nabî, Alauddin Sabit, qqv) took up Bakî's observation of reality, shifted the stress from form to content, and wrote of everyday life. Copies of Bakî's *dīwān* found their way throughout the Turkish cultural sphere, and his verse was translated into European languages at the beginning of the last century.

J. Rypka, *Báqí als Ghazeldichter* (Prague 1926); GHOP III, 133 ff; PHTF II, 431 ff. ZV

al-Bakrī, Abu'l-Ḥasan, Arabian writer, see **maghāzī.**

Bakunts, Aksel (real name Aleksandr Tevosian, b 1899 Goris, executed 1937), Soviet Armenian novelist. An agriculturalist, teacher and journalist, he wrote sketches, stories, novels, film scripts and literary criticisms, and translated from Russian and French. Bakunts described Armenian country people in patriarchal and post-revolutionary times, and the conflicts arising from the changing way of life both in the individual and the social sphere. His short psychological stories penetrating lyrical and epical elements founded the tradition of this form in Soviet Armenian literature; (collections: *Mt'nadzor*, The Dark Valley, 1927; *Sev çeleri sermnaçan*, The Sower of Black Fields, 1933).

Trans.: A. Wixley, *The Mountain Cyclamen*, in SLM. LM

Bal'amī, Abū 'Alī Moḥammad b. Moḥammad, vizier of the Samanids in Central Asia in the tenth century. He translated into Persian (*Tarjomeye tārīkhe Ṭabarī*) a shorter version of the chronicle written by aṭ-Ṭabarī (qv) which is important as the earliest historical work in modern Persian.

Trans.: L. Dubeux and H. Zotenberg (Paris 1836, reprint 1967).
SPL I 101 and P. 1229. FT

Bannā'ī, Persian poet, see **Benā'ī.**

Baratashvili, Nikoloz (b 1817 Tbilisi, d 1845 Ganja), Georgian poet. Of an ancient noble family, he was intended for a military career, but an injury forced him to take an administrative post; for some years he served in the district commander's office in Nakchichevan and Ganja, where he fell seriously ill and died. In his life as well as his work he was a Romantic, influenced by German and Russian Romantic poets in his student verse, but unlike other Georgian Romantic poets (G. Orbeliani, A. Chavchavadze, qqv), he did not sing the praise of old-time Georgia. He was concerned to interpret the feelings and inspirations characteristic of his own day. His long epic *Kartlis bedi* (The Fate of Georgia) treats of the union of Georgia with Russia, coming to the conclusion that although King Erekle II put an end to the glory of ancient Georgia when in 1773 he signed the Treaty of Georgievsk making the country a Russian protectorate, he nevertheless opened up new opportunities for it. The political aspects of the poem took second place, however, while the lyrical passages of natural description, still sung today as folk songs, won it wide popularity. Some of Baratashvili's love lyrics have also been adopted as folk songs.

Trans.: G. Boitchidzé, M. P. Fouchet, B. Gamarra, J. Gaucheron and Guillevic, *Le Destin de la Géorgie, poèmes* (Paris 1968); UAGP. VAČ

Bārbud (c600), the court minstrel of the Persian king Husrav II. His birthplace is not certain; none of his poems have survived and there are only brief references to his life by later Persian and Arabic writers. He was believed to have created the Persian musical system and to have composed seven 'royal songs', 30 tunes and 360 songs, one sung each day at the royal table. The titles of the songs give some indication of the type of inspiration and the literary genre. They include songs on such subjects as myths and legends; historical and religious subjects; odes; the celebration of the seasons and holy days; nature lyrics in praise of the sun, the stars, parks; panegyrics on the wonders of the royal court, on the queen, Shiren, on the royal treasures, throne, horses, white elephant and other wonders.

BLHP I OK

Barhebraeus, Grigor Abu'l-Faraj bar Evraya (b 1225/6 Melitene, d 1286 Maragha, Azerbaijan), Syrian writer and polyhistorian. Son of a Jewish doctor (his name means 'son of the Jew'), he was consecrated Bishop of the Jacobite Church (1246) and elected *mafreyan* (1264), filling this high office in Mosul, Tabriz and elsewhere. He was a prolific writer, if often only compilatory. Works include Biblical interpretation (*Ausar raze*, Treasury of Secrets), dogmatics (*Ktava d-nughe*, Book of Lightnings), summaries of philosophical and theological knowledge (*Mnarat kudshe*, Sanctuary Candlestick), ethics, a summary of Aristotle's philosophy (*Khemta d-hokhmta*, Cream of Wisdom), treatises and translations of Arabic philosophical works, a Syriac grammar in prose (*Book of Rays*) and in verse, and medical treatises. His long chronicle of secular and ecclesiastical history is of great value. He also wrote down *The Laughable Stories.* His poems are in part didactical. His extensive work was an attempt to collect and preserve knowledge that it might not be lost in troublous times.

Trans.: E. A. Wallis Budge, *The Chronology of Gregory Abū'l Faraj* (London 1932); *The Laughable Stories* (London 1897 and 1899). WSHSL. SZS

Bardaisan (Bardesanes), Syrian poet and theologian, see **Afrem.**

al-Bārūdī, Maḥmūd Sāmī (b 1839 Itāy al-Bārūd, d 1904 Cairo), Egyptian poet from an aristocratic family of Circassian origin. After studying in Cairo at the Military Training School, he embarked on a career of great distinction in the army, politics and in literature. During the reign of the Khedive Tawfīq, he was Minister of Awqāf, and then Minister of War. He figured prominently in the 'Urābī Revolt, and was head of the short-lived nationalist government. When the revolt failed, he was exiled to Ceylon where he remained for 17 years until his return to Cairo in 1900. Much of his most impressive poetry dates from this period of exile, when he wrote verse full of nostalgia for Egypt and his former acquaintances. His sustained expressions of deep feeling are often complemented by detailed descriptions of appropriate mood and atmosphere. His *dīwān* (qv) in two volumes was published posthumously in Cairo. Al-Bārūdī compiled an anthology of the 'Abbāsid *muḥdathūn* poets, many of whose works had previously existed only in manuscript form. This was the *Mukhtārāt al-Bārūdī* (Selections by al-Bārūdī, Cairo 1909–11).

M. A. Khouri, *Poetry and the Making of Modern Egypt* (Leiden 1971). RCO

Bashirov, Gomar (b 1901 Archa), Tatar writer and public figure. He studied at a *madrasa* in the traditional manner, where he became interested in classical literature. He fought with the Red Army and later studied at the Marx–Leninist Institute. He became editor of the paper *Kyzyl Tatarstan* (Red Tartary) and was chairman of the Union of Tartar Writers 1956–9. He takes his themes from the war years and from the period when kolkhoz farms were being established. His most important works are the story *Sivash* (1937) and the novel *Namus* (Honour, 1947). NZ

Bashshār b. Burd (d 783), Arab poet of Persian origin who lived in Basra in the late Umayyad period and at the beginning of the rule of the Abbasids. He was blind from birth, but displayed poetic talent at an early age and turned it to good use as a panegyrist praising various governors, princes and caliphs. This did not prevent him from strong support of Iranian nationalism. His religious views were not entirely orthodox, either, and although his creed is not absolutely clear, he was counted among the heretics. He was persecuted for his views. He was famous as an orator and a writer of prose, but best known as a poet whose innovations influenced the further development of poetry, especially love poetry. He was also a pioneer in satire and in the traditional *qaṣīda* (qv) form; in addition he wrote elegies and Bacchic verse.

J.-C. Vadet, *L'esprit courtois en Orient dans les cinq premiers siècles de l'Hégire* (Paris 1968), 159 ff.; BGAL I 73–4, S I 108–9; EI² 1080; NLHA 373–4; WILA 86; GSLA 156–6; AAP 42. KP

Batiray (b 1832 Urakhi, d 1910 Urakhi), Darg lyric poet. Little is known of his life, which seems to have been that of a wandering singer and musician. His verse is on the borderline between folk and art poetry; his cycles of love poems and heroic ballads use all the folk-lore formal means including poetic cliché. In his autobiographical lyrics, apparently written late in life, Batiray at last became factual and personal. He himself did not write his verse down; it has survived as recorded by his contemporaries and as handed down by word of mouth. He used the Dagestan folk-song form of seven-syllable syllabic lines, unrhymed, but with an intricate pattern of alliteration and assonance. This simplicity, harmony and formal perfection of his poetry made it the model not only for Darga poets but for all those of Dagestan. It has been translated into many languages of the Soviet Union, the Russian translations by Effendi Kapiyev being masterly. VAČ

Bayalinov, Kasymaly (b 1902 Kok Moynok), Kirghiz writer. Trained as a teacher, he then studied at the Institute of Journalism and worked as a journalist. His first poems were not remarkable, but he later developed a striking epic talent especially in the novella *Azhar* (1926), describing the moving fate of Kirghiz women before the Revolution. His most important novel is *Köl boyunda*, an epic in two parts: *Bakyt* (Happiness, 1947) and *Köl boyunda* (On the Shores of the Lake, 1950), dealing with Kirghiz life during World War II. LH

al-Bayātī, 'Abdalwahhāb (b 1926 Baghdad), Iraqi poet. Trained at the Baghdad Teachers' College, he taught Arabic until he was dismissed in 1954 for his patriotic verses. Victimized for his revolutionary activities, he was sentenced to prison and concentration camp. After making his escape he emigrated to Beirut and Cairo, travelling through Europe; in 1958 he returned home after Qāsim revolution, and was appointed Cultural Attaché in Moscow. Later he lectured at the University there. As a result of the turn taken by the Qāsim regime he again became an exile, living in Moscow until 1965, and then in Cairo and Baghdad.

His first volume of verse, *Malā'ika wa shayāṭīn* (Angels and Devils, 1950), was followed by '*Abārīq muhashshama* (Broken Pitchers, 1954), the protest of a humanist against the oppression in Iraq. In *al-Majd li'l-aṭfāl wa z-zaytūn* (Glory to Children and Olives, 1956) the poet deserted the traditional forms for new ones, influenced by European poetic theory. *Ash'ār fī al-manfā* (Verses from Exile, 1957) marks the culmination of his development away from the old forms, and combines personal and political motifs, using free verse. *Khamsat 'ashara qaṣīda min Viyana* (Fifteen *qaṣīdas* from Vienna, 1958) reveals the unhappiness of exile, while *'Ishrūn qaṣīda min Barlin* (Twenty *qaṣīdas* from Berlin) returns to political themes. Among the poems in *Kalimāt lā tamūt* (Words which do not Die, 1960) are twelve *qaṣīdas* (qv) for Iraq, the poet's message to his country, and his creed, as well as the *Letter to Nazım Hikmet* (qv), in which he calls the Qāsim regime to account.

The poet's attitude to the world, in the spirit of humanism and progress, shorn of illusions, gradually crystallized in the volumes which followed: *an-Nār wa 'l-kalimāt* (Fire and Words, 1964), *Safar al-faqr wa-th-thawra* (The Road of Poverty and Revolution, 1965), *Alladhī ya'tī wa lā ya'tī* (He who Comes and Does not Come, 1966), *al-Mawt fī al-ḥayāt* (Death in Life, 1968), *al-Kitāba 'alā'ṭ-ṭīn* (Letters in Clay, 1970), *'Uyūn al-kilāb al-mayyita* (Dead Dogs' Eyes, 1969). Al-Bayātī also published a poetic drama *Muḥākama fī Nīsābūr* (Trial in Nishapur, 1963), giving to an incident in the life of 'Omar Khayyām (qv) a topical political undertone. He has collected his thoughts on writing poetry in the autobiographical *Tajribatī ash-shi'rīya* (My Poetic Experiment, 1968), and has translated the work of French poets (Paul Eluard, 1957; Louis Aragon, 1959). His style is simple and his verse-forms free, rarely arranged in strophes. His metaphors and symbols (less evident in his recent work) are taken from the context of European poetry. From the idea of patriotism, seen from the standpoint of class, and some traces of existentialism, he has gradually worked out a progressive poetic humanism. His chief subjects are ordinary people, especially children, his beloved Baghdad, exile and the feeling of loneliness it brings, friendship, universal love, and man's search for his own heart.

CAR 236; ALAC 189. KP

Bāyazīd Anṣārī, 'Pīr Rōshān' (b 1524 Jullundur, d c1580 Hashtnagar), Pashtun heresiarch. Bāyazīd was the son of 'Abdullāh Anṣārī, an Ormuṛ of Kāṇīguṛam in Wazīristān, where he spent his youth. He gave evidence of piety at an early age and, after his marriage, came under the influence of a sect believing in metempsychosis. This seemed apostasy to his father, who tried to kill him. Bāyazīd fled to Nangrahār and declared himself the 'Saint of Light', preaching an extreme pantheistic Sufism (qv). Attacked by the Mughal governor of Kabul, he took refuge in the Tirah, later driving out or killing the inhabitants. Throughout his lifetime he had a considerable following and was constantly at war with the Mughals, dying in flight after a battle. He had five sons who continued his rebellion, as did his grandsons till their final defeat. Of Bāyazīd's writings the *Khayr ul-bayān* (Best Exposition), preserved in a unique manuscript, is best known. It purports to be a new revelation, in Arabic, Persian, Hindustani, but mainly Pashto. Ostensibly it is simply an exposition of Sufi doctrine. Bāyazīd's teachings provoked an orthodox response from Ākhund Darwēza (qv).

D. N. MacKenzie, '*The Xayr ul-bayán*', in *Mélanges . . . Morgensterne* (Wiesbaden 1964), 134–40. DNM

bayt, verse couplet, see **masnavī** and **'arūḍ.**

bāzgasht, literary movement in Iran, see **Khorasani style** and **Ṣā'eb,** Mīrzā.

Bēdil or **Bidel,** 'Abdulqādir (b 1644 Azimabad, now Patna, d 1721 Delhi), Persian poet in India towards the end of the Mughal era. His family may have come from Central Asia; his father was a military official. He did not learn Persian until he went to school. Sufism (qv) engaged his interest early in life. He served some time in the army but refused to be court poet; he preferred to devote himself to literature, supported by his pupils. Bēdil is a philosopher-poet, whose verse deals with fundamental human problems and is not content with the answers provided by Islam. He also drew on Hindu philosophy, and expressed his non-conformist views quite openly. He believed that all that is material is the product of the natural world, which is eternal and in constant motion.

His *kolliyāt* (collected works) comprises sixteen books of 147 thousand lines, as well as prose works. Of his more important writings the *masnavī* (qv) *Tilismi hayrat* (Talisman of Surprise, 1669) deals with human existence. The *masnavī Muhīti a'zam* (The Great Ocean, 1681) is written in the spirit of Sufism; *Tūri ma'rifat* (Sinai of Learning, 1684) describes travel and nature. *Nukāt* (Tracts) is a discussion in prose and verse of religious questions, Sufism, etc, while the prose and verse *Chahār 'unsur* (Four Elements, 1680–94) contains much of the poet's philosophical ideas, autobiographical details, and reminiscences. Probably his most important work is one of the last, *'Irfān* (Knowledge), which includes the romantic ballad *Komde va Mudan*, about the love of a dancing girl and a singer, as well as the poet's thoughts about life and the world. Bēdil brought the 'Indian Style' (qv) to its culmination in poetry, and this style is often linked with his name. It is a difficult style, with involved metaphors and intricate syntax, although the language itself is simple. Bēdil exercised an immense influence on the further development of poetry in Central Asia, in what is now Afghanistan and the Central Asian Soviet Republics; to a greater or lesser degree his style still prevails in this area. He was less known however in India itself, and even less in Iran.

RHIL 515–20. JB

Behāzin, Moḥammad E'temādzāde (b 1915 Rasht), Iranian writer. He attended a French naval academy and was for a time a naval officer; later he became a librarian and devoted himself to writing and translation. In the latter field he is very skilled. His own work is mainly in the short prose forms, on social themes, and he stresses man's efforts to progress. The heroes of his stories and sketches are mostly simple country folk trying to make their dreams of fortune come true. He writes in a skilful poetic style and cultivated idiom, in keeping with the general lyrical mood of the stories and with his own humanist outlook. Works: novel *Dokhtare ra'iyyat* (The Peasant's Daughter, 1954); stories, *Besūye mardom* (Towards People, 1948) and *Naqshe parand* (Silk Design, 1955).

RHIL 413–14; AGEMPL 218–19. KB

Behramoğlu, Ataol (b 1942 Çatalca), Turkish poet. He graduated in 1966 in Russian Language and Literature at the University of Ankara. Since 1960 he has published poems in literary journals. In 1970 he and Ismet Özel (qv) founded the journal *Halkın Dostları* (Friends of the People). He has also made some fine translations of Russian literature. He is an outstanding member of the youngest generation of writers which has successfully combined the aesthetic values and achievements of the New School and the Second New School of Literature, with what is best in the tradition of socially-engaged poetry. Of his poems, those that reflect his dissatisfaction with himself and his milieu, and his restless and desperate search for solutions are the most successful. His militant poems, although sincere, are marred by a certain complacency and an excess of rhetoric. Collections of poems: *Bir Ermeni General* (One Armenian General 1965), *Bir Gün Mutlaka* (Certainly one Day, 1970). DL

Beliashvili, A., Georgian poet, see **Chikovani,** Svimon.

Benā'ī, also Bannā'ī, Binoī, Kamāloddīn Shīr 'Alī (b 1453 Herat, d 1512 Qarshī), Central Asian Persian poet. His father was a Herat craftsman. Later he lived in Tabriz and finally moved to Samarkand and to Qarshī, where he lost his life when an Iranian army captured the town. Only a part of his considerable body of work has been preserved, principally his didactic *masnavī* (qv) *Behrūz o Bahrām*, also known under the title *Bāghe Eram* (The Garden of Eram), condemning immorality and evil natures, pointing out the importance of training and education and stressing that human experience is essential in upbringing. Part of the *qasīda* (qv) *Majma' ol-gharā'eb* (Collected Wonders) has also survived; the introduction is written in the Herat dialect and includes lines dedicated to A. Navā'ī (qv). He also wrote historical works in combined prose and verse, like *Sheybānīnāme* (Book of the Sheybanids) and its enlarged version *Fotūḥāte khānī* (The Victorious Campaigns of the Khans) about the reign of the Central Asian Sheybanids up to the disintegration of the Timurid empire.

RHIL 497–500. JB

Bēnawā, 'Abdurraūf (b 1913 Kandahar), Afghan poet, playwright and historian, writing mainly in Pashto. He lived in Kandahar until 1936, publishing his first poems in the paper *Ṭulūi afghān.* He went to Kabul to take up journalism, and from 1940 worked in the Pashto Academy (*Pashto ṭolena*), then in diplomatic service and later in high office, eg Minister of Culture in 1967. One of the leading poets in the endeavour to give poetry a new content in keeping with modern life, he is a fertile writer. One of the first reformers of Afghan prosody, he tries out new metres and has experimented with free verse. At the turn of the 40's–50's he was one of the leaders of the progressive literary and political movement *Wesh zalmiyān* (Youth Awakes). Important themes in his poetry are love of country, of the simple people, wrath at the sight of oppression, defence of the rights of the Pashtuns, praise of work (eg *Pe speka ma rā ta gōra!*, Do Not Look Down On Me!), the struggle against backwardness and such outworn customs as bride-buying (eg in the play *Zōr gunahgār*, The Old Sinner), and disgust at the oppression of women. In 1947 Bēnawā called on other poets to write for the paper *Kābul*, on new, modern subjects. His poems appeared in periodicals, and in two collections, *Prēshānē afkār* (Troubled Thoughts, 1957) and *De ghanemē wezhay* (Heads of Wheat). He published editions of such classical Pashto writers as Raḥmān Bābā (see 'Abdurraḥmān Mōhmand), 'Shaydā' Kāẓim (qv); literary studies like *Khushḥāl Khaṭak tse wāyi* (What Khushḥāl Khatak says, 1948); historical works like *Mīrways Nīke* (Grandfather Mirways, 1946) and *Hōtakīhā* (The Hotaks, 1956); and journalistic writing such as *Pashtanē mērmanē* (Pashto Women, 1944) and *Pashtunistān* (1951). Together with I. Sharīfī he published a collection of folk verses, with English translations, *Landey* (1958). JB

Berk, İlhan (b 1916 Manisa), Turkish poet. He studied Romance languages in Ankara and taught in Anatolia. His isolation from the centres of cultural life has meant that this original poet comes to public notice only sporadically. After his first lyric volumes he joined the *İkinci Yeni* movement (Second New, see Oktay Rifat). Several years of silence were followed by a return to his early style. Poems: *Güneşi Yakanların Selim* (Greetings from Those who Fire the Sun, 1935); *İstanbul* (1947); *Aşıkane* (With Love, 1968); translations of the poetry of Rimbaud (1962). LH

Besiki (real name Besarion Gabashvili, b 1750 Tiflis, d 1791 Yassy, Rumania), Georgian poet and diplomat. His father was a priest, who was excommunicated and exiled with his family from the kingdom of Kartli. He found refuge with King Solomon of Imereti, who made Besiki a counsellor and sent him to Russia at the head of a diplomatic mission, to arrange the terms of a protectorate. The poet died in Moldavia, as an attaché of Field-Marshal Potemkin. He wrote love poetry and lyrical meditations, and later composed eulogies on the victories of Georgian kings, and intricate punning poems. In form and style his verse is close to that of the folk poets, but his vocabulary is richer and his diction more involved. His chief popularity came from the polished and varied form and playful mood of his love poetry. His meditative verse and a few epics follow classical models.

Trans.: UAGP. VAČ

Bētāb, Ṣūfī 'Abdulhaqq Khān (b 1888 Kabul, d 1969 Kabul), Afghan poet writing in Dari. He was Professor of Persian Literature at Kabul University, and was awarded the traditional title *Malik ush-shu'arā* (King of Poets) by the monarch in 1952. He was a connoisseur and consistent practitioner of the 'Indian style' (qv), writing *ghazals* (qv) on traditional subjects. He also introduced new themes, such as patriotism (in *Qaṣīdai istiqlāl*, Ode on Independence, 1957), electricity, etc, and introduced technical terms into his verse. His poems appeared in periodical publications; he published a number of works on Dari poetry and prosody, eg *'Ilmi badī'* (Rhetorics, 1951), *Qāfīye wa 'arūẓ* (Rhyme and Metre, 1954). JB

Beyatlı, Yahya Kemal (b 1884 Üsküp/Skopje, d 1958), Turkish poet. He lived in Paris between 1903 and 1912, and was Turkish ambassador to Spain and to Poland. He is considered the last classical poet of Turkey, using the traditional metres; he was not infrequently attacked by progressive Turkish writers. However, his poems reveal great sensitivity, influenced by both 18th century Turkish literature (Ahmed Nedim, qv) and modern French poetry (the symbolists). There are poems telling of the old glory of the Ottoman Empire, of Turkish music, sometimes touching a mystical chord, but the masterpieces in his very small collection of poems are those written about Istanbul. Yahya Kemal elaborated his poems very carefully, and they excel in their inimitable music and harmony. AS

Beyhaqi, Abu'l-Faẓl Moḥammad (996–1077), Persian historian in Central Asia. For many years he was an official and finally vizier to several sultans of Ghazna from Maḥmūd (999–1030) to Farrokhzād (1053–59). He wrote an extensive history of the sultans of Ghazna, with long excursions into the history of earlier times. His work was planned in 30 volumes, of which only a small part exists which deals with the reign of Sultan Mas'ūd (ie with the years 1030–42), and is therefore called *Tārīkhe Mas'ūdī* (History of Mas'ūd). The work is more in the nature of memoirs, and beside historical facts presents detailed descriptions of court customs and state administration.

SPL I, No. 334, and P 1271. FT

beyt, verse couplet, see **maṡnavi** and **'arūḍ.**

Bible (Greek [*ta*]*biblia*, books), comprises the Old Testament and the New (qqv), the terms being of Christian origin; the Jews also use the word *Bible* for their canon, but in Hebrew prefer the word *Tenach*, formed from the initials of each of the three groups of books of the Bible: *Law*, *Prophets*, and *Writings*. The Old Testament books, in Hebrew with short passages in Aramaic, were chosen from Ancient Hebrew literature (12th–2nd centuries BC) for use in the Jewish religious community. Taken over by the Christian church, they were supplemented (2nd–3rd centuries AD) by a set of Greek books, the New Testament. Those books composed at various times and places but not ultimately accepted into the canon are called *Apocrypha* (qv). The connection of the pseudepigrapha (qv) with the books of the Bible is even looser.

The Jews always preferred to use the Hebrew original of the Bible texts, but after they were scattered among different peoples in the Diaspora, and knowledge of Hebrew was no longer general, translations were made. The earliest and most significant was that into Greek, known as the *Septuagint* from the number of scholars working on it (Alexandria, 3rd century BC); it was later revised by the Christians for their own use. The most notable modern Jewish translation is that into German, by Martin Buber and Franz Rosenzweig, because it attempts to present the characteristic language and style of the Hebrew in a totally different language. Greek-speaking Christians used the Old Testament in the *Septuagint* version and the New Testament in the original.

The Bible spread eastwards in various Syriac translations which, together with the original, formed the basis for translations into other languages. The first Latin translations were made in North Africa, but the one by Jerome (c400 AD) later known as the Vulgate, became the most important. In the Middle Ages translations of the Bible laid the foundations of national literatures: the Gothic translation was by Ulfilas (4th century), and the Old Slavonic by Cyril and his disciples (9th century). In central and northern Europe the translations of the Bible made in the course of the Reformation helped to form the literary languages of the new age: Luther in German, John Hus in Czech, and the Authorised Version completed under James I in 1611, in English. Biblical themes and ways of expression are found in secular as well as religious poetry in mediaeval Europe, and in the 16th century Luther, the Czech Brethren, and others, wrote hymns and metrical versions of the Psalms (qv). Many great works of 17th century literature were inspired by the Bible (Milton's *Paradise Lost*, Bunyan's *Pilgrim's Progress*, Commenius' *Labyrinth of the World*); Biblical themes appear in the classical and romantic periods (Victor Hugo), while modern European literature has produced such works as Thomas Mann's tetralogy on Joseph and his brethren.

Modern Biblical research rightly pays attention to style, composition and literary genres, bearing in mind the tradition of the original text and of the translation. A comparison of New Testament Greek with the vernacular as preserved on papyri found in Egypt has made it possible to determine the literary character of the different books of the New Testament, while the discovery of works of Ancient Egyptian, Sumerian, Babylonian and Assyrian literatures and more recently of Ancient Aramaic and Canaanite works (the Ugaritic epics), has substantiated the historical and literary study of Old Testament Hebrew literature. The study of the Bible from the literary standpoint, now developing in volume, can provide valuable information for the analysis of the historical background and the religious concepts of the Bible.

Trans.: HBKJV; HBRSV; NEB.
C. M. Laymon (ed.), *The Interpreter's One Volume Commentary on the Bible* (Nashville 1971); R. E. Brown, J. A. Fitzmyer, R. R. Murphy, eds., *The Jerome Biblical Commentary* (New York 1968); *The Cambridge History of the Bible I–III* (Cambridge 1970, 1969, 1963).
SZS

Bidel or **Bidil,** Persian poet in India, see **Bēdil,** 'Abdulqādir.

Bilbaşar Kemal (b 1910 Çanakkale), Turkish writer. After attending training college (Edirne) and Education Institute (Ankara), he became a secondary school teacher working in the west Anatolian countryside. He draws his short story themes from country life, belonging to the school of writers who see in this sphere the specific quality of Turkish literature. He is a realist with a strong sense of local colour; his plots are well-developed. His popular novel *Cemo* (1966) received the *Türk Dil Kurumu* prize. Other works: short stories, *Anadoludan Hikâyeler* (Tales from Anatolia 1939), *Pembe Kurt* (Pink Wolf, 1953); novels, *Denizin Cağrısı* (Call of the Sea, 1943), *Ay Tutulduğu Gece* (The Night the Moon was Caught, 1961). LH

Binawā, Afghan poet, see **Bēnawā,** 'Abdur-raūf.

Binoi, Central Asian Persian poet, see **Benā'ī,** Kamāloddīn Shīr 'Alī.

al-Birūni, Abu'r-Rayḥān Muḥammad b. Muḥammad (b 973 Khwarezm, d 1048 Ghazna), Central Asian polyhistorian, one of the greatest Islamic scholars. Although he wrote mainly in Arabic, he remained attached to Iranian culture. When about 20 years old he went to Rayy and then to Jurjān, where he wrote his chronological work for Qābus b. Vushmgīr (976–1012). The rest of his life was spent at the court of Sultan Maḥmūd and his successors (999–1048) in Ghazna. From there he made several travels to India, which Sultan Maḥmūd had opened up to Islam. He learned Sanskrit and was able to write about this country from Indian sources. His most important works are: *al-Athār al-bāqiya 'an al-qurūn al-khāliya* (Traces left by Past Eras, English translation by E. Sachau, London 1879); *Tārīkh al-Hind* (History of India), a general description of the country, (English translation by E. Sachau, London 1888, new ed. London 1910); *al-Qānūn al-Mas'ūdī* (The Canon of Mas'ūd), a general astronomical work; *at-Tafhīm li-awā'il ṣinā'at at-tanjīm* (Instruction on the Basic Principles of the Art of Astrology), a popular compendium of astrology, translated into English by R. Wright, London 1934.

BGAL I, 475–6, Suppl. I, 870–5. FT

Book of the Dead, title given in 1842 by R. Lepsius to an Egyptian religious work.

In origin, the Book is a collection of magic spells, some of which are to be found in texts from the royal pyramids of the Old Kingdom (c2350–2150 BC), and on the inner walls of private coffins dating from the Middle Kingdom (c2150–1800 BC). By reciting these spells the soul would be better able to overcome the difficulties and dangers of the other world. The Book of the Dead contains in addition the verses in which the soul declares its innocence before the court of Osiris.

The Book of the Dead was compiled under the XVIIIth dynasty (after 1570 BC), when it became customary to write these texts in hieratic script on papyrus rolls which were placed in the coffin or in the wrappings of the mummy; this custom survived almost to the disappearance of pagan religion. The selection and order in which the verses (called 'chapters' by scholars) are found varies, the most complete (a Ptolemaic copy in Turin Museum, published by Lepsius) comprising 165; altogether 186 different 'chapters' are known. Throughout its long history, the text has become corrupted to such an extent that it is not always intelligible. Many copies contain appropriate illustrations to the different 'chapters', often in colour, making the Book of the Dead the oldest illustrated book in the world. Although not of great literary value, it is an important source of knowledge about Egyptian ideas on religion. The title in ancient Egyptian was *Per-em-hrow* (literally, going out in daylight), from the title of an important 'chapter' (17), never on any account omitted; it ensured the soul the right to revisit the world and return to the grave.

Trans.: Sir E. A. Wallis Budge, *The Book of the Dead* (London-New York 1928). JČ

Bosḥāq, Aḥmad Abū Esḥāq Ḥallāj (d between 1424–1427, Shiraz), one of the few satirical poets in classical Persian literature. His *dīwān* (qv) is made up entirely of verses about food, and he is therefore also known as Bosḥāq al-Aṭ'eme (Bosḥāq of the Food). He parodied the poetry of the most famous poets, line by line, with exceptional thoroughness and attention to formal detail, but transferring it to the sphere of gastronomy. Whether he wanted to ridicule the rigidity of style and subject in poetry up to that time, or whether he simply wanted to draw attention to himself, his work has remained a rarity. Although his verse is not devoid of wit, he did not succeed in writing really lively satire. His example was followed with even less success by Maḥmūd Qārī of Yazd (d 1461/2), whose parodies of great poets took the form of poems about clothing. HS

al-Buḥturī (b 821, d 897), Arab poet from Syria, where he lived; later he became a panegyric poet in Baghdad. He wrote political verse, panegyrics in the old style, and other traditional genres. His elegant simplicity, successfully combining old forms and new themes, is much admired; he is considered one of the neo-classicists of his time. He also acquired fame as a collector of old poetry, which he published in the anthology *al-Ḥamāsa* (Fortitude).

BGAL I 80, S I 125; NLHA 324; WILA 92; AAP 72; EI²I 1289. KP

Bundahishn (First Creation), the chief religious didactic work of Middle Persian literature, an eleventh century redaction based on an earlier text. It is an interesting combination of religious teaching and rich ethnological material (especially myths about the struggle between good and evil), a traditionally conceived history of Iran, the story of the first human beings, lists of mountains, rivers, lakes, stars, planets, the protectors of animals and plants, a list of evil spirits, a description of the Last Judgment, etc. The value of the text lies in its unusually rich traditional material, not in the dry manner of the writing.

RHIL 41ff; TMSLZ. OK

Burhaneddin, Kadi Ahmed (1344–1398), Turkish poet and jurist who made himself ruler of Sivas, Kaiseri and Erzincan in eastern Anatolia; he was attacked by the Egyptians in 1387 and slain in 1398, by the Akkoyunlu Turkmens. His *dīwān* (qv, the only extant copy is dated 1395) belongs to the oldest monuments of eastern Turkish literature. He wrote in Arabic, Persian and

Turkish, but most of the poems are written in Persian metres and use, for the first time in Turkish, the whole heritage of images and symbols as well as the rhetorical devices developed in Persia. But his *dīwān* also contains a number of *tuyugh* (qv) quatrains in lines of 11 syllables, full of eastern Turkish words, which contrast with the vocabulary of the other poems. Kadi Burhaneddin, considered 'the first love poet of Asia Minor' (Gibb) also wrote a commentary for a law-book and a grammatical treatise.

GHOP I, 204ff. AS

al-Bustāni, Buṭrus (b 1819, d 1883), Lebanese philologist, writer and journalist, pioneer of modern Arabic literature and Arab cultural revival. Of a Christian family, he worked with the American Mission in Beirut, founded a boys school, and then turned to philology and journalism. His great dictionary, still a valuable source of information, laid the foundations of modern Arabic and therefore of modern Arabic literature. Al-Bustānī began to publish an Arabic encyclopaedia, with his son Salīm, as well as several journals. Author of several studies of Classical Arabic literature, he endeavoured to revive the Arab cultural heritage, but helped also to propagate European culture and science. Principal work: *Muḥīṭ al-muḥīṭ* (Complete Dictionary, 1867–69). JO

C

Cansever, Edip (b 1928 Istanbul), Turkish poet. He abandoned his studies at the High School of Commerce to become an antique dealer in Istanbul. He is both a founder and one of the most representative members of the Second New School of poetry, which established its reputation in the period from 1950 to 1960. Cansever is a poet who, since his earliest poems, has constantly changed and renewed both his themes and his stylistic techniques. His latest and finest collection of poems, *Kirli Ağustos* is an achievement that surpasses all other works of Turkish surrealist poetry. The poetry of Cansever describes an alienated, drab, despairing world from which man escapes into his subconscious. His nonconformist view of man sees him in his global context, naked, helpless and incapable of transcending his human limitations. Cansever's poetry has evolved in the direction of a gradual withdrawal into the subconscious where the reader must labour to follow him. Collections of poems: *Ikindi Üstü* (Afternoon, 1947), *Dirlik-Düzenlik* (Getting Along, 1954), *Yerçekimli Karanfil* (Gravitational Carnation, 1967), *Umutsuzlar Parkı* (The Park of the Desperate, 1958), *Petrol* (1959), *Nerde Antigone* (Where is Antigone, 1961), *Tragedyalar* (Tragedies, 1964), *Çağrılmayan Yakup* (Uninvited Jacob, 1966). DL

Çelebi Dede, Süleyman (d 1421/2 Bursa), Turkish poet, son of a high civil dignitary, an *imam* in great mosques. Only one of his works has survived, *Mevlidi Nebî* (Birth of the Prophet), which made him famous with a broad public. It is the oldest Ottoman-Turkish treatment of the popular Islamic literary theme; versified legends of events preceding and accompanying the birth of Muhammad. There are also fairy-tale elements in the work, written in simple language, using the dialect of Bursa (then the capital). For its simplicity and its hold on the popular imagination it was the favourite book from the end of the 14th century to modern times, imitated over 100 times, but with less success.

GHOP I 232ff. ZV

Cem Sultan (b 1459 Edirne, d 1492 Naples), Turkish poet, younger son of Sultan Mehmed II the Conqueror. He fled to Europe after an unsuccessful attempt to seize the throne after his father's death. Interned in France and Italy for several years, he fell a victim to quarrelling between the Pope and the King of France, and seems to have been poisoned. As a young man he was regarded as a capable statesman and a learned and accomplished poet; besides lyrics he wrote a romantic epic, *Cemşid ve Hurşid.* His poems from exile were a faithful account of his feelings and experiences on his travels in Europe, but it has not yet been proved whether he was influenced by European literature. For political reasons, and because it was not understood, his work did not influence the further development of Turkish poetry. His adventurous life has often inspired novels and plays in Turkish and Bulgarian literature.

GHOP II 70ff; PHTF II 428ff. ZV

Cemal Süreya (Seber) (b 1931 Erzincan), Turkish poet. Graduated in 1954 at the Faculty of Political Science. Cemal Süreya was employed in the civil service until his resignation in 1965. He was subsequently for some time in Paris. He has contributed to the following literary journals: *Mülkiye*, *Yeditepe*, *Evrim*, *Pazar Postası*, *Türk Dili* and the newspaper *Vatan*. Besides writing poetry, Cemal Süreya is a poetry critic, translator, and the editor of two anthologies of poetry, *Mülkiyeli Şairler* (Poets from the School of Civil Service, 1966) and *Yüz Aşk Şiiri* (One Hundred Love Poems, 1967). He also edits and publishes an excellent literary review *Papirus* (4 numbers from 1960–61, and from 1966–70 numbers 1–44), which is mainly concerned with a recapitulation and survey of Turkish literature of the Republican period. *Papirus* has made an important contribution to the cultural life of present-day Turkey. Cemal Süreya is both a founder and an outstanding member of the literary school known as the 'Second New Literature' (*İkinci Yeni*). Latterly he has been noted for his championship of personal freedom, the writer's individuality, and the importance of form, whereas he began his literary career as a Marxist and surrealist. He is one of the most popular and most highly respected poets of today. His poetry, whether he is writing of love's passions, or of social themes, is always warm and intimate in tone. Collections of poems: *Üvercinka* (Pigeon, 1958), *Göçebe* (Nomad, 1965); both won awards, and were much praised by Turkish literary critics. DL

Cevdet Anday Melih, see **Anday,** Melih Cevdet.

Chakhrukhadze (end 12th–early 13th centuries), Georgian poet who probably came from Khevi. Legend has it that he was the major-domo and secret lover of Queen Tamara (1184–1213); he died in a Georgian monastery in Sinai or Palestine. The only work of his to survive is *Tamariani*, a collection of 12 odes and one elegy, in honour of Tamara. Written in the 20-syllable lines known as *chakhrukhauli*, after the author, they are an ardent eulogy of the queen's beauty, power and wise rule.

Trans.: UAGP. VAČ

Charasha, Tembot (b 1902 Koshabl, Adyghei), Circassian (Adyghei) writer. Son of a serf, he managed to get to the Tatar Teachers' Training College in Ufa in 1914, but was forced by the war to interrupt his studies. He worked as an apprentice for some years, and it was not until after the Revolution that he attended an ordinary school. He worked as a translator and journalist, studying in Moscow from 1929, and returning home to journalism, publishing and research. He has been writing in Circassian since 1922. His short stories are mainly propagandist. The novel *Road to Happiness* describes changing Circassian village life as Soviet power established itself and agriculture became collectivized. The first volume, *Attack* (1929), has a fresh directness, but the work as a whole is reminiscent of Sholokhov's *Virgin Soil Upturned*. In the post-war *Daughter of the Shapsugha* (1951) he returned to the folklore sources which he has been collecting for years. VAČ

Chavchavadze, Aleksander (b 1786 St Petersburg, d 1846 Tbilisi), Georgian poet of the Romantic period. Of an ancient aristocratic family, his father was ambassador of the Georgian king Erekle II at the Czarist court. After Georgia was annexed by Russia the 18-year old Chavchavadze took part in a rising led by Prince Parnaoz Bagration to restore the independent monarchy. Arrested, he was released at his father's request and made a court page. He took part in the war against Napoleon and was an aide to Barclay de

Tolly, the Allied Commander-in-Chief, during the occupation of Paris. He was again arrested in 1832 for his part in an aristocratic plot, and exiled to Tambov. Again released, he returned to Tiflis. His military career did not suffer, however, for when he died he was a Major-General in the Russian army. His daughter married the Russian writer Griboyedov. Chavchavadze's work marks the transition from traditional Georgian to European poetry. Educated in the classical spirit of French literature, he declared himself a classicist, but his verse in praise of the past glory of monarchical Georgia, banquets, drinking parties and romantic love, is entirely in the Romantic spirit. He introduced Racine to the Georgian reading public, and also translated Pushkin from the Russian.

Trans.: UAGP. VAČ

Chavchavadze, Ilia (b 1837 Kvareli, d 1907 Saguramo), Georgian poet, journalist and politician. Of princely family, he studied at St Petersburg University and after his return home devoted himself to literature and public life. He published several journals, presided over various educational and charitable societies, was a judge, and Chairman of the Tiflis Aristocratic Bank; in 1906, he was elected a member of the State Council. In 1907 he was assassinated on his way to his country home in Saguramo, probably by police spies. He is the most important of the '*tergdaleuli*' generation ('those who have drunk of the Terek', ie the second generation of Georgian intellectuals, educated in Russia, who crossed the boundary river, Terek), also known as the 'sons', who unlike their 'fathers' nostalgically singing of the old Georgia before annexation, were not resigned to the fate of their people and considered their chief duty to be the revival of Georgian national feeling and the formation of a modern Georgian culture. Chavchavadze's prose writings included realistic novels and also critical polemics, stressing the immorality and untenable state of serfdom, the moral degeneration and decadence of the aristocracy. In his lyric poetry, particularly his meditative and nature poems, he deliberately turned away from traditional Georgian poetry and sought European models.

Trans.: Marjory Wardrop, *The Hermit* (London 1895); UAGP.
D. M. Lang, *A Modern History of Georgia* (London 1962). VAČ

Chikovani, Svimon (b 1902 Neyesakovo), Georgian poet. Graduating from Tbilisi (Tiflis) University, in the first years of Soviet rule he worked in educational institutions of the Georgian Red Army. In 1922 he joined A. Beliashvili, D. Gachechiladze, the critic B. Zhghenti and others in the futurist manifesto which marked the establishment of LEF, the Georgian group of the Left Front. In 1924 he worked on the editorial board of H_2SO_4, published by the group. After World War II he became a Soviet Deputy and one of the leaders of the Union of Georgian Writers. In his early work he was strongly influenced by the Russian futurists, particularly Mayakovskiy and Khlebnikov. Later he sought his models in earlier poetry, especially in classical Georgian literature. He is regarded as an accomplished lyric poet, making full use of the beauties of literary Georgian. His style makes his work difficult to translate, but it has been done by B. Pasternak and N. Zabolotskiy. Visits to the German Democratic Republic and Poland after the war led him to write of the essential values in international cultural contacts. Since the futurist group in Georgia fell to pieces, Chikovani has remained an isolated phenomenon in Georgian poetry. Major work: ballad *Simghera Davit Guramishvilze* (The Song of David Guramishvili, 1944).

Trans.: UAGP. VAČ

Chiladze, Tamaz (b 1931 Tbilisi), Georgian writer. He studied at the Faculty of Literature, Tbilisi University, and works in the State Committee for the Rustaveli State Prize, Georgian SSR. He belongs to the generation of writers who grew up during the last war, and shares their tendency to take urban themes, with a preference for the emotional and philosophical problems of young people today. His long stories,

Gaseirnoba ponis e lit (Driving in a Pony-cart, 1961) and *Shuadghe* (Noon, 1962), are primarily concerned with interpersonal relations. VAČ

chistān or **chiston,** literary puzzles, see **Vāṣefī,** Zeynoddīn.

Chraïbi, Driss (b 1926), Moroccan writer, writing in French. He studied chemistry and neuro-psychiatry in Paris, but soon turned to literature and journalism. He worked in French radio, and now lectures in a Canadian university. His first novel *Le passé simple* (The Simple Past, 1954) is a sharp attack on both traditional Islam and colonialism. *Les Boucs* (Billygoats, 1955) describes the impact of France on the North Africans. *Succession Ouverte* (Unclaimed Inheritance, 1962) reveals disillusionment with conditions in independent Morocco, which the hero prefers to leave, and with Moroccan relations with Europe. Recent novels deal with more general subjects. His style is very dynamic. Other works: novels, *L'âne* (The Ass, 1965); *La foule* (The Crowd, 1961); *Un ami viendra vous voir* (A Friend is Coming to See You); stories, *De tous les horizons* (From All Horizons, 1958).

J. Déjeux, *La littérature maghrébine d'expression française* (Algiers 1970), 241–267. SP

Chronicles (*Paralipomenon*), two of the historical books of the Old Testament (qv), chronologically parallel to *Samuel I* and *II* and *Kings I* and *II* (qqv), but paying more attention to ritual matters in the Temple of Jerusalem. See also Bible. SZS

Chūbak, Ṣādeq (b 1916 Bushahr), Iranian writer. He acquired a native education in Shiraz and a western education at the American College in Teheran, completing his studies oriented towards English and American culture by visiting those two countries. He now works for the National Iranian Oil Company, which makes him independent of the financial success of his books and able to follow his own literary goals regardless of the tastes of the time. He was one of the close friends of Ṣādeq Hedāyat (qv), who influenced his early writing. Chūbak has published four volumes of stories: *Kheymeshabbāzī* (The Puppet Show, 1945), *Antarī ke lūṭiyash morde būd* (The Baboon Whose Buffoon Was Dead, 1949–50), *Rūze avvale qabr* (The First Day in the Grave, 1965) and *Cherāghe ākher* (The Last Alms, 1966); two novels: *Tangsīr* (A Native of Tangestan, 1963), the story of a deceived man seeking justice through personal revenge; and *Sange ṣabūr* (The Patient Stone) drawn from the dregs of the society of Shiraz. He also wrote two plays: a political satire *Tūpe lāstikī* (The Rubber Ball) and a drama *Hafkhaṭ* (Slyboots).

The central theme of Chūbak's work is the search for the meaning of human existence reflected in the life of the lowest strata of society or on the plane of animal allegory. The philosophical undertone is complemented in an original manner by a creative picturing of reality with a sense of microscopic detail. Chūbak's writing is marked by the combination of typical Iranian subjects and modern western literary techniques. The most significant of his works from the formal point of view is the novel *Sange ṣabūr*, which was the first Persian novel to drop the classical epic construction and replace the narrator by the 'stream of consciousness' expressed in the characters' inner monologues. Linguistically his work represents an original attempt to create a literary language drawing on all levels of Persian including colloquial and vulgar language, slang, argot and local dialects. His work is not accepted by conservative critics, for it is too unconventional in all respects, but among the modernists he is regarded as the most significant prose writer in post-war Iran. His work has been translated into English, German and Russian. Chūbak also translates western literatures: Edgar Allan Poe, Eugene O'Neill, C. Goldoni, Lewis Carroll.

RHIL 415; KMPPL 127–30; AGEMPL 215–16. JOS

Codex Cumanicus, a manuscript dated 1303, Cuman being an ethonym used by Muslim historians for the Kipchak tribes, belonging to the Turkic peoples. The manuscript is probably the work of some Franciscan monk sent as a missionary to the tribes. In 1362 it came into Petrarch's possession. The manuscript comprises paradigms and vocabularies in Latin, Persian, German and Kipchak, and translations of religious texts from Latin to Kipchak. The only authentic Kipchak text is a set of Kipchak riddles. The Italian texts in a different hand are clearly later additions to the original manuscript.

A. Tietze, *The Koman Riddles and Turkic Folklore* (Berkeley-Los Angeles 1966). LH

D

Dağlarca, Fazil Hüsnü (b 1914 Istanbul), Turkish poet. As the son of an army officer with changing asignments, his youth was spent in a number of provincial towns. He attended military schools as a boarding student, then served as an officer in the army for 15 years. In 1950 he resigned, for some time held a government job, then in 1959 started a bookshop and small publishing business in Istanbul. If poetry means a state of enhanced sensibility combined with an ability to communicate, Dağlarca is the perfect model of a poet. He has no protective layers; his naked skin sucks in the world and beams out his emotion in undiluted personal formulations. He called one of his volumes of poetry *Çocuk ve Allah* (The Child and God, 1940): indeed, the poet experiences the world with the directness of a child's eyes. Therein lie both the strength and the universal appeal of his poetry as well as, at times, its impenetrableness. These features can be seen in all his work, whether it is lyrical or epical, political satire or transcendental inspiration, for Dağlarca has produced poetry in all of these genres. Especially powerful in his protest against social injustice, he is undoubtedly the most successful Turkish poet living today. Dağlarca has published his poems in many volumes since 1935, especially: *Asu* (1955); *Batı Acısı* (Bitterness of the Western Sunset, 1958); *Haydi* (Go, 1968).

Trans.: Talât Sait Halman, *Selected Poems: Seçme şiirler* (Pittsburgh 1969). AT

Dānesh, Aḥmad, Tajik writer, see **Donish,** Ahmad.

dangbezh, Kurdish folk singer, see **Khānī,** Aḥmade.

Daniel, book of the Old Testament (qv), written 166/164 BC, during the Maccabaean Wars. The first part, mostly in Aramaic and based on older traditions, tells of a young Jew whose piety won him the favour of the Babylonian and Persian kings. The second part, mostly in Hebrew, consists of apocalyptic prophecies. The example of faithfulness set by Daniel, and his prophecies encouraging the oppressed, helped to make this book popular in times of trouble (persecution of the Christians, the Hussite Wars, etc). In the Greek translation the apocryphal stories of Susanna and of Bel and the Dragon have been added. See also Bible.

N. Porteous, *Daniel* (London-Philadelphia 1965). SZS

Daqiqi, Abū Manṣūr Moḥammad b. Aḥmad (b c thirties of the 10th century, Tus or Balkh, d c977), Central Asian poet writing in Persian at the Samanid court in Bukhara. He may have been a Zoroastrian; he died a violent death, reputedly at the hand of his slave. His importance for literature lies in his beginning the verse chronicle of Iranian history *Shāhnāme* (Book of Kings) on the orders of the monarch. He had only written a part of it when he died, and Ferdousī (qv) took over about 1,000 lines of it for his epic. Daqīqī's contribution has not the poetic mastery of Ferdousī's great work. *Ghazals* (qv) and other poems by Daqīqī have also survived.

RHIL 153–4. JB

Dashti 'Ali (b 1901), Iranian writer, called by Kamshad 'one of the most controversial figures in the letters and politics of modern Iran'. He began his career as a journalist, on the editorial board of the radical paper *Shafaqe sorkh* (see 'Eshqī), whose high standard of writing and unconventional approach roused considerable interest among the public between 1921 and 1930, while earning its editor several prison terms. Later his views became more moderate. He was member of Parliament for many years, ambassador in Egypt and the Lebanon, and finally a senator. His essays and articles present original views on modern life and its problems, on relations between East and West, on autocracy and democracy, etc. A number of these pieces appeared in book form in an enlarged edition of his first book *Ayyāme maḥbas* (Days in Prison) and *Sāye* (Shadow, 1946). Equally unconventional in approach are his essays on classical Persian poetry and poets (Ḥāfeẓ, Sa'dī, Khāqānī, Nāṣere Khosrou, Rūmī, qqv); instead of a learned commentary, these essays give the writer's personal impressions of the poetry. Dashtī's novels and stories deal exclusively with the elegant society of Teheran, lovely capricious women and men who are the ideal companions and seducers; these love stories all follow a similar pattern. The style is appropriate to the subject matter, elegant and decorative, with a refined vocabulary over-inclined to French and other European expressions. Dashtī's work as a translator of Arabic and French literature is also important.

KMPPL 69–73; AGEMPL 222–3. VK

dāstān, Turkish *destan*, Tajik *doston*, a type of lyric-epic form, usually a longer poem describing events, and characters and their actions. Lyrical passages are inserted to make the description more effective. The usual rhyme scheme is aa, bb, cc, etc. which is that of the *maṣnavī* (qv), and this has led to both the terms being used in the same sense. The *dāstān* has a long history, going back to oral folk narrative, perfected later in written form. In classical Persian and Tajik literature there is the love *dāstān*, *Vīs o Rāmīn* by Gorgānī (qv), while the heroic epic by Ferdousī (qv), the *Shāhnāme*, is also a *dāstān*, as well as the *Hasani arobakash* of M. Tursunzoda (qv) and the *Tāj va beyraq* of A. Lāhūtī (qv). JB

David (ruled c1000–961 BC), king of Israel, who is said to have been the author of the lament for Saul, and of the Psalms (qv), the majority of which are attributed to him by tradition. See also Bible. SZS

Davoyan, Razmik (b 1940 Mec Parni near Spitak), Soviet Armenian poet. The son of a peasant, he graduated from a health workers' school and a pedagogical institute. His two volumes of lyrics *Im ašxarh* (My World, 1963) and *Stverneri miǰov* (Through the Shadows, 1967), and the long poem *Ṙek'viem* (1969) are among the most original in contemporary Armenian poetry.

Trans.: L. Mardirossian, in OeO. LM

Deborah (late 12th century BC), Israelite prophetess, who praised in song the victory of Israel over Sisera, captain of the army of Canaan; the song was included in the *Book of Judges* (qv). See also Bible. SZS

Dede Korkut, a book of folk tales attributed to a wandering Turkish singer (*uzan*), Dede Korkut. In its present form the stories, which were previously handed down by word of mouth, date from the 15th century; the work itself is older, however, and probably formed part of the *Oghuznâme* (The Oghuz Legend) frequently mentioned in literature. It was brought from their Asian homeland by the eleventh century Turks of the Oghuz tribe who conquered Asia Minor under their Seljuq prince. The book contains twelve stories and tales, literally interwoven with proverbs and maxims. As far as the source of the motifs is concerned, the story of Tepegöz recalls that of Polyphemos the Cyclops in the Odyssey, while that of Deli Dumrul is reminiscent of Admete and Alcestis. E. Rossi has analyzed the contents and W. Ruben made a comparative study of the historical material; the latter gives the relevant bibliography.

Trans.: J. Hein, *Das Buch Dede Korkut* (Zürich 1958); W. Ruben, *Ozean der Märchenströme* (Helsinki 1944).

F. Babinger, *Die Geschichtsschreiber der Osmanen und ihre Werke* (Leipzig 1927); EI[1] II 1150; EI[2] II 200; P. N. Boratav, *L'étapée et la ḥikaye* in PHTF II 11–44, 405, 459; KLL II 688–81. OS

Dehkhodā, 'Alī Akbar (b c1879, d 1956), Iranian writer and philologist, one of the most enlightened men engaged in the campaign for national awakening. At the turn of the century in Vienna, continuing the studies he had begun at home, he met patriotic Iranian emigrés and became committed to their cause. Returning home he turned to journalism to serve his political ends, editing the literary page of the well-known progressive paper *Ṣūre Esrāfīl* (The Trumpet Call of the Angel of Death), and writing verse and prose. The former, reprinted in all the progressive papers of the day, was mainly rousing and didactic. Technically excellent, his verse reflected the incidents of the patriotic fight against a despotic regime. He was one of the pioneers of non-traditional forms in poetry, and the first writer to mould a modern Persian prose idiom in his short satirical sketches; these were very important for the development of modern literary Persian. These were collected and published in book form in 1958 under the original title *Charand-parand* (Pêle-mêle). Later however, Dehkhodā withdrew completely from political life, and devoted himself to philology and literature. Two outstanding works came from his pen during this period: *Amṣāl o ḥekam* (Proverbs and Aphorisms, 1929–32) and the first modern Persian dictionary, *Loghatnāme*, the work of a team chosen by him and working under his direction until his death.

KMPPL 37–40; AGEMPL 55–6. VK

Demirtchian, Derenik (b 1877 Akhalkalaki, d 1956 Yerevan), Armenian writer, poet and dramatist. He studied in Etchmiatsin, Tbilisi and Geneva, and from 1925 worked in Yerevan as teacher of language and literature; from 1953, Demirtchian was a member of the Armenian Academy of Sciences. After his first volume of intimate lyrics (1899), he turned more to stories, plays, novels, and criticism. Most popular is his historical novel *Vardanank'* (The Companions of Vardan, Vol. I, 1943, II 1946, also dramatized; French trans. Paris 1963), inspired by the 5th-century popular rising against Persia and aimed at encouraging nation-wide patriotism during World War II. Demirtchian helped to establish continuity of democratically informed Armenian literature with the mood of the post-revolution era.

Trans.: excerpts in SLM (trans. by A. Pyman), French in OeO (trans. by B. de Crest) and in ELA (trans. by P. Gamarra). LM

Dēnkart (Matters of Faith), 9th–10th century text, the longest in Middle Persian literature. The first author was the priest Āturfarnbag (c820), and it was completed by Āturpāt, son of Ēmēt, about a century later. The matters treated are dogma, moral questions, religious customs, traditions, and the history and literature of the Zoroastrian religion. The style is often obscure; nevertheless, besides valuable information, the *Dēnkart* often supplies the key to an understanding of social and economic conditions in ancient Iran.

RHIL 39ff; H. W. Bailey, *Zoroastrian Problems in the Ninth-Century Books* (Oxford 1943); P. J. de Menasce, *Une encyclopédie mazdéenne, le Denkart* (Paris 1958). OK

destan, epic poem, see **dāstān.**

Dib, Mohammed (b 1920 Tlemcen), Algerian writer and poet, writing in French. He studied in his native town, then in Oujda. His father was a carpenter; after his death Dib followed various trades. He married a Frenchwoman and has lived in France from 1959. Dib began writing at the age of fourteen. His novel on Algerian poverty, *La grande maison* (The Big House) was written in 1948, but published in 1952 after being rewritten as the first of a trilogy, *L'Algérie*; volume two described rural poverty and the necessity for revolt, volume three the unbearable life of textile workers in the town; the whole forming a social novel. Later novels describe the war of liberation as seen by widely different people in *Un été africain* (An African Summer, 1959); and as an apocalyptic vision in *Qui se souvient de la mer* (Who Remembers the Sea, 1962). His later work is dominated by symbolism, introspection and psychological problems, and the subjects are more general. From his youth onwards Dib has written many short stories, some linked to the novels and also poetry, but the first successful volume was published in 1961. He has been awarded literature prizes and his work has been translated. His latest novel, *Dieu en Barbarie* (God in Barbary, 1970) depicts the fusion of the nation after the liberation of Algeria and is thus once more on a patriotic theme. Other works: poetry, *Ombre gardienne* (Guardian Shadow, 1961); *Formulaires* (Forms, 1970); volumes of stories, *Au café* (In the Café, 1955); *Baba Fekrane* (1960); *Le Talisman* (1966); novels, *L'incendie* (The Fire, 1954); *Le métier à tisser* (The Weaver's Trade); *Cours sur le rive sauvage* (Run on the Wild Bank, 1964); *La danse du roi* (The King's Dance, 1968).

J. Déjeux, *La littérature maghrébine d'expression française* (Algiers 1970), 111–141. SP

diwān, *dīvān, divan,* a complete collection or selection of the work of a single poet, excluding *dāstāns* (qv); it thus contains his *qaṣīdas, ghazals, robā'īs, qit'as* (qqv), *ṃokhammases,* etc. The custom of collecting their poems in *dīwāns* became popular among poets in the 10th–11th centuries, but the earliest date from the 12th century (Anvarī, Mo'ezzī, qqv) and suggest that the fixed order which applied to the form later was not yet then established. It was not until the 13th century that the various forms (*ghazal,* etc) were arranged together, the poems in each group being then arranged alphabetically according to the last letter of the line (*bayt*). Each section of the *dīwān* (*baḥr*) was named after a letter, eg *baḥr alif.* JB

dobeyti, Persian poetic form, see **Bābā Ṭāher** 'Oryān and **robā'ī.**

Donish (Dānesh), Ahmad, also known as Makhdūm Kalla (b 1827 Bukhara, d 1897 Bukhara), Tajik writer and thinker, leader of the movement for enlightenment in Central Asia. Educated traditionally, he made his livelihood by copying manuscripts; although he was not popular at court, because of his great learning he was sent to St Petersburg three times on official missions. These journeys made him even more convinced that Central Asia was backward and that radical changes were needed. He wrote poetry in the traditional manner, but it is his prose, putting forward his ideas, which is outstanding in its significance. He wrote, for instance, that the ruler is the servant of his people, dreamed of a time when rulers would no longer make guns, proved that poverty was the outcome of tyranny and lack of education. He condemned the scholastic religious way of bringing up children, recommended that Russian should be taught, and showed the first signs of patriotism. In his *Tarjimai holi amironi Bukhoroi sharif* (Lives of the Amirs of Blessed Bukhara), also known as *Ta'rikhcha* (Little History) or *Risola* (Treatise), which was secretly copied and circulated, he condemned the ruling dynasty of Bukhara. Most of his ideas are to be found in the long treatise *Navodir ul-vaqoye'* (Precious events). His attitude to classical poetry is unusual; Bēdil (qv), who was very popular in Central Asia, is given a place of honour in the history of literature, but Donish warns writers against slavish imitation of the intricate style, which had become a fashion in nineteenth century Central Asia. Donish was the founder of a school, his chief disciples being Shohīn, Savdo, Asirī and others, including Sadriddin Aynī (qv). Towards the end of his life a pessimistic note entered his work. He wrote several theological and astronomical treatises: *Nāmūs ul-a'am* (The Supreme Book of Laws), *Mi'yor ul-tadayyun* (The Measure of Godliness), *Ar-risola fī a'mal ul-qurra* (On the Use of the Globe).

RHIL 529–32. JB

Doulatshāh b. Alā'oddoule Bakhtīshāh Samarqandī (d 1494/5), Persian literary historian of the Tīmūr period. In 1487 he wrote a popular general *tazkira* (qv), *Tazkerat osh-sho'arā* (Record of Poets), giving the lives of the Persian poets up to his own time, with examples of their work. Some biographies were translated into European languages.

SPL I No. 1093. FT

doston, poetic form, see **dāstān.**

Drakht(i) Āsūrik (The Assyrian Tree), date and author unknown, is a Middle Persian prose debate originally written in Parthian. It presents an argument between a Babylonian date palm and a he-goat, as to which of them is more useful to mankind. The goat wins. The social context of the text is not clear, and it is assumed to be religious in intent, but the allegory seems to portray a conflict of interests between the ruling Sassanian state represented by the goat (one of the incarnations of the god of victory) and the land of Babylon, which supplied Iran with food.

RHIL 49; TMSLZ. OK

ad-Dū'āji, 'Alī (1909–1949), Tunisian writer, poet, playwright and painter. Of an ancient patrician family, he was brought up to a life of leisure. His stories are like film-shots, often full of piquant comment and fine irony; he has adapted the 'camera eye' technique to the Arabic milieu. His novella *Jawla bayn ḥānāt al-baḥr al-mutawassiṭ* (Journey round the Bars of the Mediterranean, 1933), a non-traditional travel book out to amuse, was very popular. In 1942 he wrote his only (and as yet unpublished) novel *Shāri' al-aqdām al-mukhaḍḍaba* (The Street of Henna-dyed Feet), a penetrating and ironic observation of life in the working class quarter. His extensive works include plays in colloquial Arabic, and songs and poems which are still topical. Other work: short stories, *Sahirat minhu al-layālī* (She Couldn't Sleep because of Him, 1969). SP

dubaytī, dobeytī, Persian poetic form, see **robā'i** and **Bābā Ṭāher** 'Oryān.

Durayd b. aṣ-Ṣimma (b 530, d 630), Ancient Arabian poet, one of Muhammad's greatest Bedouin antagonists. As chief of the Banū Jusham tribe he took part in the stand of the Hawāzin against Muhammad, and fell in the battle of Ḥunayn. The Islamic tradition regards him as the representative of ancient Arabian paganism obdurately fighting against Islam. He is considered the greatest of the poet-knights (*fursān*). The most frequent themes in his verse, which is clearly Bedouin in character, are those of battle, love and friendship, elegy and panegyric. The verse which has been preserved is typical of the ancient Arabian pagan poets. His *dīwān* (qv) has not survived.

NLHA 83; BHLA 278; EI[1] I 1115. KP

Durian, Petros (1852–1872 Constantinople), Armenian poet, dramatist and actor. His slight volume of work sprang from despair at the tragic plight of the individual and the whole people to whom he belonged, the Armenian minority in Turkey. His vivid, musical verse, with the recurring motifs of death and ruin, unrequited love, helpless anger at social injustice and hatred of all despotism, was a break with the classicist poetry of his forerunners and opened the Romantic period in modern West Armenian literature (to, for example, Metsarents, D. Varuzhan, qqv). Through him, French and English literary influences reached the East Armenian poets of the 90s. Durian also wrote historical tragedies and attempted the social drama on contemporary themes, but this side of his work has fallen into oblivion. Works: *Taǵk' ev tatrergut'yunk'* (Songs and Dramas, 1872).

Trans.: A. S. Blackwell, *Armenian Poems* (Boston 1896); TPAAM.
THLA. LM

Dzanayty, Ivan, Ossetian poet, see **Niger.**

E

Ebne Sinā, Central Asian philosopher, see **Ibn Sinā.**

Ebne Yamin, real name Amīr Maḥmūd b. Amīr Yamīnoddīn Toghrā'ī Faryūmadī (b 1286 Faryūmad, Khorasan, d 1368 Faryūmad), Persian poet of the period of the Mongol domination of Iran. He was a master of the *qet'e* (see *qit'a*) poetic form, with the content of practical philosophy and didactics presenting also humanistic ideas on the brotherhood of man, resistance to oppression, etc.

RHIL 261. JB

Ecclesiastes (Qohelet), a Hebrew book of the Bible (qv) containing meditations (often pessimistic in tone) on the life of man and human society. Wrongly attributed to King Solomon, it dates from the 4th or 3rd century BC. See also Five Festal Scrolls.

R. B. Y. Scott, *Proverbs, Ecclesiastes* (Garden City 1965). SZS

Ecclesiasticus (Sirach), in full the '*Wisdom of Jesus the Son of Sirach*', a Hebrew collection of proverbs and good advice in the tradition of the Old Testament (qv) wisdom literature. The Hebrew original was lost for a long time; it was discovered partly at the end of the 19th century in Old Cairo, and in the Qumran caves in our century; Greek, Latin and Syriac translations were preserved. SZS

Edib, Adıvar Halide, see **Adıvar,** Halide Edib.

Ekrem Recaizade (1847–1914 Istanbul), Turkish literary historian and critic. A teacher of literature, he worked in various educational institutions and exercised a great influence on young people who were interested in literature. His textbook of literature *Talimi Edebiyat* established the standards of literary writing. In his three-volume poetical works *Zemzeme* (Murmuring, 1883–5) he introduced new poetic forms and ideas. His verse is sentimental, often elegiac and pessimistic, especially the poems written to mourn the death of his young son, Nejad Ekrem, in 1911, which he never got over. With Tevfik Fikret and Halit Ziya (qqv) he was a leading figure in the *Sereti Fünun* period.

HGTM 12–30; M. Hartmann, *Dichter der neuen Türkei* (Berlin 1919) 52–4; PHTF II 479–82. OS

Elley (Ellyay; real name Seraphim Aramaanabys Kulaas'kar, Russian: Seraphim Romanovich Kulachikov; b 1904 Nizhniy Amginsk, Alexeyev district), the greatest living Yakut poet. Of a peasant family, he became a teacher; his first publication appeared in 1924 and he graduated from the Moscow Journalists' Institute in 1928. In 1929 he edited the paper *Young Bolshevik*. From 1941–4 he served in the Red Army. He has published about 20 volumes of verse including *Küögeyer künnerler* (The Best Years of His Life, 1929), *Sanga yryalar* (New Songs, 1932), *Jollokh olokh* (A Happy Life, 1938), *Syrdyk uobrastar* (Clear Pictures, 1939), *Ayan uottara* (Will o'the Wisps, 1941), *Kyayiy yryalara* (Songs of Victory, 1950), *Ayar üleghe aykhal* (Glory to Creative Work, 1959), *Kihi-kihiekhe* (Man to Man, 1962). His favourite heroes in this poetry are the builders of new towns, fur hunters, and kolkhoz farmers wringing their crops from the everlastingly frozen earth of the Yakut north. Most of his poems have become the songs of the Yakut people; he used rhythmical syllabic verse, for the most part. In the 30's he turned to folk poetry, eg in the fairytale *Churum-Churumchuku* (1938). The ballad *Prometheus* (1949) is devoted to the time spent by N. G. Chernyshevskiy in Yakutsk. The ballad *Sakha sargyta* (The Fortune of Yakut, 1947) is a poetic version of the history of his land and people. His theoretical writings stress the need for serious literary education for Yakut writers.

MM

Emin, Gevorg (real name Karlen Karapetian, b 1919 Ashtarak), Soviet Armenian poet. The son of a provincial teacher, he went to the Polytechnic in Yerevan. His early poems were popular among young soldiers in the last war (also in Russian). Later, especially in his last two volumes *Ays tarik'um* (In this Age, 1968) and *XX-rd dar* (Twentieth Century, 1970), motifs of rationalism and intellectual meditations on the conflict between man and technical civilization are stressed. His work also includes literary criticism and poetical essays like the cycle *Yot'n erg Hayastani masin* (Seven Songs of Armenia, 1965) devoted to the rebirth of his country.

Trans.: A Grigoryan, in OeO; D. Rottenberg, in SLM; J. Gaucheron, in ELA and OeO. LM

Enmerkar, mythical Sumerian ruler of the early dynastic period (28th century BC), hero of the epic *Enmerkar and the Lord of Aratta,* loosely connected with the Lugalbanda (qv) epics. The historical background to all these poems is provided by the rivalries between the Sumerian city-states and the mountainous regions of Iran. Enmerkar, son of Utu, the sun-god, and ruler of Uruk-Kulaba, wants to force the mountain state of Aratta to supply gold and precious stones for the building of a temple to Enki in the Sumerian city of Eridu. Following the counsel of his protectress, Inanna, he sends a messenger up to the mountains, but the Lord of Aratta refuses to send the gifts because he considers Inanna to be on his side. She bestows her favours on Enmerkar, which confuses the Lord of Aratta; at first intending to fight, he nevertheless then asks Enmerkar for grain in return for the gifts asked of him. In Uruk a messenger is sent out with the grain, but the Lord of Aratta refuses to accept it, and hostilities break out. For the third time Enmerkar asks for the gold and jewels, in a message so complicated that the messenger is incapable of repeating it; the words are therefore inscribed on a clay tablet, for the first time in the history of man. (This is the Sumerian tradition of the origin of writing.) The messenger is accompanied to Aratta by the storm-god Ishkur, carrying the promised grain in his clouds. The end of the poem is not clear; the Lord of Aratta fears an enemy who is helped by the storm-god, and gives in. Other Sumerian poems, however, suggest that Uruk and Aratta did in fact go to war. The struggle between Uruk and Aratta for the favour of Inanna, and thus for the right to a 'sacred marriage' with the god of love, is described in the poem *Enmerkar and Sukheshda'anna*, written in the style of a 'dispute' (see Sumerian Disputes). The two cities fight, the outcome being decided by the goddess Nisaba and an old woman, Sagburru, who murder the Aratta general. After this the ruler of Aratta, Sukheshda'anna, acknowledges the dominion of Uruk.

S. N. Kramer, *Enmerkar and the Lord of Aratta* (Philadelphia 1952). BH

Enuma elish, opening words of the Babylonian myth of the creation, 'when the sky above had not yet been given a name'. Covering seven tablets, it probably dates from the 11th century BC. It was recited every year at the New Year celebrations of the god Marduk, chief among the gods of Babylon. In the Assyrian version the role of Marduk was taken by the Assyrian national god, Assur. According to the poem, in the beginning there was a primaeval ocean in which the waters of the god Apsu and the goddess Tiamat mingled; they produced several generations, including the god Ea, god of wisdom and spells. Disturbed by the noise of the younger gods, Apsu decided to destroy them, but Ea's spells overcame him and he killed Apsu. Ea and his wife brought forth the god Marduk. Tiamat, aided by eleven monsters she had formed, and her new husband Kingu, set out to avenge the death of Apsu. The only god who can stand up to her is Marduk, armed with all the divine powers, an evil wind, and an arrow. When Tiamat opened her mouth to swallow him he blew his evil wind into her maw, so that she could not shut her mouth, and shot his sharp arrow into her belly. Half of her dead body became the sky, and the other half the earth, while Marduk set the stars in their courses and created man from the blood of Kingu; man was to work so that the gods could rest in peace. The poem ends with praise of Marduk.

H. Heidel, *The Babylonian Genesis* (Chicago 1963); E. A. Speiser, in ANET[3]; LRPOA 36–70; J. V. Kinnier Wilson, *Documents from Old Testament Times* (ed. D. Winton Thomas, London 1958), 3–16. LMa

'Erāqī, Fakhroddīn Ebrāhīm b. Shahryār (b ?Hamadan, d 1289 Damascus), Persian mystic poet, one of the most remarkable personalities of his time. Most of his life was spent travelling in India, to Mecca and Konya, in Egypt and Syria. He was a disciple of several famous saints. There are many legends about his life, particularly about his birth and his travel to India; his work reveals an ecstatic and ardent soul. His mystic beliefs are comparable to those of Jalāloddīn Rūmī (qv). The most striking of his works is the *Lama'āt* (Flashes), a prose treatise with verses, influenced by the theosophical ideas of Ibn al-'Arabī (d 1240). 'Erāqī's conception of mystical love had a wide response and many commentaries on the *Lama'āt* were written; the best known is by the poet Jāmī (qv). 'Erāqī's charming *maṣnavī* (qv) on love interspersed with *ghazals* (qv) is more easily understood; it is called *'Oshshāqnāme* (The Song of Lovers) or *Dah faṣl* (Ten chapters).

Trans.: A. J. Arberry, *The Song of Lovers* (London 1939). HS

'Eshqi, Moḥammad Reẓā (b 1893/4 Hamadan, d 1924 Teheran), Iranian poet. His whole life seems somehow marked by its short duration. After attending (presumably only for a short time) Islamic and French mission schools, he worked as a French interpreter in Hamadan from the age of seventeen. He soon began to take an active part in politics, and established his own paper in Hamadan, *Nāmeye 'Eshqī* (1914–15). Like many of his fellow patriots, he came out against the intervention of England and Russia in Iran. Soon after, the Allies occupied Iran, and the patriots had to go into exile. 'Eshqī went to Istanbul, already a haunt for Persian emigrés. Here he wrote two long poems full of romantic patriotism: *Rāstākhīze salāṭīne Īrān* (Ressurrection of the Kings of Iran), really an opera libretto in which, besides the legendary kings, the prophet Zoroaster appears; and *Nourūznāme* in which patriotism finds its expression in the evocation of the ancient rites and the beauties of the Persian spring. After his return home 'Eshqī lived mostly in Teheran, as poet, actor and journalist, with unflagging zest proclaiming his decided (though changing) opinions and provoking official circles to retaliate. He published his own paper (*Shafaqe sorkh*), contributing political articles and satirical poems both to it and to other papers. These writings made him enemies among those in power, and he died the victim of political assassination. Apart from the two poems already mentioned, he wrote two other works of literary importance: *Kafane siyāh* (Black Shroud), a romantic vision of pre-Islamic Iran, combined with a progressive attitude condemning the veiling of women introduced by Islam; and the three-part *Ideāle 'Eshqī* ('Eshqī's Ideal), a lyrical epic composition about a girl betrayed and forced to suicide. There is a strong note of social criticism which is an innovation, as is the use of a strophe form instead of the traditional *maṣnavī* (qv).

AGEMPL 89–97; MLIC II, 132–52; MRPRPV 66–73. VK

Esther, Hebrew book of the Bible (qv), named after the young Jewish heroine who became Queen of Persia and helped to save her fellow Jews from destruction. The dramatic story written in the 3rd or 2nd century BC is the legend of the feast of Purim. See also Five Festal Scrolls. SZS

Etana, legendary king of the Sumerian city of Kish. He is the hero of an Akkadian myth preserved in three versions of different dates: from the first half of the second millennium BC, in Babylonia and Elam; from Assur c1200 BC; and a third version in several copies made for Assurbanipal's library in Nineveh (1st half of the 7th century BC). There are some differences in the treatment, but all three versions keep to the same main lines of the story. After a short introduction the hero Etana was probably brought into the story, but the opening (except for the introduction) has not survived. Etana worships the souls of the dead, worships and sacrifices to the gods, but in vain; he is still childless. The Sun god tells him that the eagle would help him to find the plant of fertility.

The story of the eagle is then inserted into the myth like a 'fable'. The eagle and the snake had sworn friendship, but one day when the snake was out hunting the eagle broke his oath and devoured the snake's young. The unhappy snake complained to the Sun god, the Babylonian god of justice, and asked for the eagle to be punished. He was told to hide in the entrails of a dead wild bull and wait there until the eagle came to feast on them. Then he could catch the eagle, pluck him and throw him into a pit to die of hunger and thirst. The snake did so, seized the eagle and threw him into a pit. Now the eagle asked the Sun god for help, and it was granted. Etana was sent to free him from the pit, and in return the eagle promised to fly up into the sky with him to seek the fertility plant. The first flight was a failure, Etana and the eagle both fell to earth, but the second attempt was apparently successful (the text is very fragmentary here); Etana acquired the fertility plant and finally his wife bore him a son.

E. A. Speiser, in ANET[3] 114 ff, and A. K. Grayson, ibid., 517; LRPOA 294–305. JS

Evliyā Çelebi b. Derviş Muhammed Zillî (1611–1682 Istanbul), 17th century Turkish 'globetrotter'; his life is known only from his own words. His father was court jeweller, his mother sister of the famous soldier and Great Vizier, Melek Ahmed Paşa, who took him under his wing after a gay youth. He accompanied his uncle on innumerable military and diplomatic journeys described in a remarkable work written 1631–7, including campaigns in Europe, Asia and Africa. Entitled *Tarîhi seyyâh* (Chronicle of a Traveller), it is better known as *Seyâhatnâme* (Description of Journeys). It is more than a description of travels and campaigns, however, giving a vast quantity of observations on culture, folklore, geography, with examples of many languages (particularly phrases needed by soldiers) and descriptions of remarkable buildings, beautiful cities and landscapes. He rarely quotes his sources; some facts can be confirmed in chronicles, but elsewhere his fantasy runs wild. *Seyâhatnâme* is in ten volumes; six were published in the 1890's, the next two appeared in a critical edition, 1929, and the last two volumes, in Latin script, were published in 1936. Vol. 1 describes Istanbul and its environs; the following deal with journeys to Bursa, Batumi, Abkhazia, the Cretan campaign; pilgrimages to the holy places; journeys to Baghdad, Bulgaria, Dobrudje; the campaign against Rakoczy, those in Transylvania, Hungary, today's Slovakia, and elsewhere. The last two volumes describe journeys through Anatolia, to Syria, Hijaz, the Sudan and Abyssinia.

PHTF II, 456; EI. HT

Exodus, second book of the **Pentateuch.**

Ezekiel (early 6th century BC), Israelite prophet, taken to Babylon as a captive; here he reproved his fellow Jews who had learned nothing from the destruction of their country, and turned their thoughts towards a new life in a theocratic state. The prophetic *Book of Ezekiel* follows the chronological order, and ends with his vision of the New Jerusalem. The style is systematic rather than poetic, but the detailed account of the visions makes it very effective. See also Prophets and Bible. SZS

Ezra (458 or ?398 BC), Jewish priest. He reformed worship in the Temple of Jerusalem, and introduced the reading of the Law (see Pentateuch). His name is given to 1) the Old Testament *Book of Ezra*, describing the return of the Jewish captives from Babylon and the rebuilding of the Temple; it originally included the *Book of Nehemiah* (qv); 2) the Apocryphal *Third* (or *First*) *Book of Ezra*, preserved in Greek which, besides a chronicle, includes a fictitious account of a contest between pages at the Persian court (with what appear to be Indian motifs) from which the Hussites took their motto: *Veritas vincit* (Truth will prevail); 3) the *Fourth Book of Ezra*, a pseudo-epigraph preserved in Latin, Syrian and other versions, written down in 70 AD after the destruction of Jerusalem and the Temple by the Romans. In the form of a vision, at times in poetic style, it considers the present and future fate of Israel and of individual people. See also Bible.

J. M. Myers, *Ezra, Nehemiah* (Garden City 1965). SZS

F

fahrîye (Turkish), **fakhriyye** (Arabic), panegyric poem, see **Nef'ī.**

Fākhūri, 'Umar (1895–1946), Lebanese journalist, essayist and critic, well known in public and cultural affairs. Born in a modest merchant family, he was active in the cause of Arab liberation from Turkish rule while still a student at the Osman University, Beirut. He graduated in law in 1913, and then at the American University, Beirut, and the Osman Medical Faculty. After the war he studied law and literature in Paris. He gave many public lectures and was active as a journalist in Lebanon. In 1927 he became a member of the Arabic Academy in Damascus. During World War II he worked with leading communists and in 1941 joined the League against Fascism and Nazism, founded the League paper *aṭ-Ṭarīq* (The Way) and was selected chairman of the Lebanese Society for Friendship with the Soviet Union. Fākhūrī was one of the most active propagators of progressive anti-fascist matter in Arab culture and literature. The breadth of his interests and the variety of his work arose from his profound all-round learning. His books and essays contain many stimulating ideas on the theoretical aspects of literature and its relation to reality, on literary techniques, on aesthetics, the place of the writer in society, the place of art in modern life, etc. His chief work, *Adīb fī's-sūq* (The Writer in the Market, 1944), reveals the development of his ideas from 'pure' literature to the literature of life and literature for the people. He has written many political essays sharply attacking fascism, Hitlerite propaganda, and racial and religious intolerance. His translations from the French are important, as are his studies in Arabic literary history. Further works: studies in literature, *al-Fuṣūl al-arba'a* (Four Seasons, 1941), *al-Bāb al-marṣūd* (The Spellbound Door, 1938); in politics, *al-Ḥaqīqa al-lubnānīya* (Lebanese Reality, 1944), *Lā hawāda* (No False Consideration, 1942). JO

Fānī, literary pseudonym of **Navā'ī,** Alīshēr.

al-Fārābī, Abū Naṣr Muḥammad b. Muḥammad b. Tarkhān Uzlagh (b c870 Farab, d 950 Damascus), important mediaeval Islamic philosopher, active particularly in Baghdad and Aleppo. His extensive works have survived in the Arabic original and in part in Hebrew and Latin translation. They are marked by polyhistoric breadth; they include philosophical, scientific and political works. Fārābī's philosophy draws much from Aristotle, and he was styled the Second Teacher, ie second to Aristotle, the First Teacher. Another basic element was neo-Platonism, drawing on neo-Platonist writings wrongly attributed to Aristotle, particularly the Theology of Aristotle. Fārābī attempted to bring the teachings of Aristotle and of Plato into harmony. Later thinkers took over the neo-Platonist principles of his cosmology, in particular the belief in emanations, and the transition from divine spiritual unity to the differentiated material world. In his view of man Fārābī also drew on Greek philosophy, elaborating the Greek idea of various categories of reason. His discussion of the ideal state is of great interest, analyzing as it does the imperfect and undesirable forms and depicting the ideal form of government to ensure the maximum development of individuality. Here Fārābī unites the Greek conception of a government of philosophers with Islamic theories of the state. Of Fārābī's other works his treatises on logic and on the classification of knowledge are worthy of note.

F. Dieterici, *Alfārābī's philosophische Abhandlungen* (Leiden 1892) and *Der Musterstaat von Alfārābī* (Leiden 1900); P. Brönnle, *Die Staatsleitung von Alfārābī* (Leiden 1904); E. J. J. Rosenthal, *The Place of Politics in the Philosophy of al-Farabi* (Cambridge 1971); CHPI 222–33, 356–7. vsd

Farrokhī, Abu'l-Ḥasan 'Alī (d 1037), Persian poet. He grew up in Sistan, but went to the court of Chaghanian in Transoxiana. From there he left for Ghazna where he wrote panegyrics to Maḥmūd of Ghazna, and to the members of his family and his court. Farrokhī's main fame rests upon his *qaṣīda* (qv) on 'the branding-ground of the horses', a poem that was composed before he had seen the actual scene and is, thus, a fine proof of his imaginative power. The descriptive poems about the ruler's adventures on the battlefield and at hunting-parties are very lively and full of interesting facts. His elegy on

Maḥmūd's death is certainly one of the most touching *qaṣīdas* in the Persian language. Farrokhī is fond of inserting at the end of his *qaṣīdas* long chains of blessings and wishes, with the phrase 'as long as . . .', a form in which he displays great variety. His style is charming and musical (he was an accomplished musician) and has a direct appeal to the reader even today, quite contrary to most of the *qaṣīdas* composed by Farrokhī's poetical master, 'Onṣorī (qv), or by later Persian court-poets. The fact that more than 9,000 verses of his have been preserved shows the popularity he enjoyed in his lifetime. None the less, a full evaluation of his poetical achievements is still lacking.

Ch.-H. de Fouchécour, *La description de la nature dans la poésie lyrique persane du XIe siècle* (Paris 1969). AS

Farrokhzād, Forūgh (1934–1967), Iranian poetess. One of the striking personalities of the younger generation of poets, she followed the innovator Nīmā Yūshīj (qv). She dared to reveal the most intimate feelings of a woman's heart, touching on subjects which had never appeared in Persian poetry; she used words previously unthinkable in a Persian poem. Poetry was life for her, and life was poetry; every word was a poetic word. Her early verse (1954–5) in simple four-line stanzas, not yet of constant literary value, was published in the volume *Asīr* (The Prisoner). In *Dīvār* (The Wall) and '*Osyān* (Rebellion), written between 1955 and 1958, she was still seeking to perfect both form and expression. The most marked notes heard throughout her work, ardent passion, fear, melancholy, despair, tormented longing, resounded with renewed force in the volume *Tavallodīye dīgar* (Rebirth), in which she attained to the form most suited to her temperament, in free verse. Her literary creed, introducing a selection of her poetry *Bargozīdeye asch'ār* (1964), in fact became her epitaph. She met a tragic end in a motor-car accident early in 1967. HS

Farzād, Mas'ūd (b 1906 Sanandaj), Iranian poet and translator. Like most of the writers of his generation he enjoyed a European as well as a traditional education. He studied in England and later settled there for many years. He has translated from the English, particularly Shakespeare (*Hamlet*, *A Midsummer Night's Dream*). At the beginning of his literary career he belonged to the well-known Teheran group *Rab'e* (The Four). The book of pamphlets and satirical pieces on which he collaborated with S. Hedāyat (qv) dates from this period. As a poet he is a lyricist of pessimist cast, aiming primarily at expression of the 'melancholy of his own soul'. In this he is essentially a modern poet, but in his forms and expression he follows the Persian classics, especially Ḥāfeẓ (qv); he is the author of a critical edition of the latter's works. He has published only three slim volumes to date, their titles appropriate to the atmosphere of the poet's world: *Gole gham* (Blossoms of Melancholy), *Bazme dard* (Feast of Pain, 1953/4), and *Kūhe tanhā'ī* (Mountains of Solitude).

AGEMPL 205; MLIC II 125–7. VK

al-Fāsī, 'Allāl (b 1910 Fez), Moroccan political leader and writer, journalist, historian and poet. A patriot and activist from a very early age, he was exiled to Gabon 1937–46. Yet he helped to create the *Istiqlāl* (Independence) party in December 1943, and soon became its President. He directed nationalist agitation from Cairo 1948–53. Since independence, however, his movement has not held the wide loyalties it once commanded. His creed is grounded in Islamic reformism and has pan-Arab sympathies, but insists on the territorial basis of the nation-state. Forceful as his prose can be, it is no mere propagandist tool; there is substance and thoughtful exposition in his historical presentations. His poetry is classical in form and loftily patriotic in tone. Main works: *al-Ḥarakāt al-istiqlāliyyah fī'l-Maghrib al-'Arabī* (1948; trans. H. Z. Museibeh, *Independence Movements in Arab North Africa*, Washington 1954); *an-Naqd adh-dhātī* (Self Criticism, 1952). PC

Faustus of Byzantium, Armenian historian, see **Phavstos** Buzand.

al-Faytūri, Muḥammad Miftāḥ (b 1930 Alexandria), Egyptian poet and playwright. His father, a pious man of South Sudanese extraction, sent him to the al-Azhar where the only study he found congenial was that of the classical poets, especially late Abbasid ones. Later he was attracted to the Romantic poets of the Syro–American and Apollo schools, and to Baudelaire. Conscious of being 'short and black and ugly', he found release at 18 by assuming an African identity. This concern with Africanism and therefore with racial discrimination and exploitation has dominated his work. His poetry is full of strong themes, extremely vivid images, and intense emotion. His one play, in verse, *Aḥzān Ifrīqiyā* (Sorrows of Africa), has a succession of powerful scenes on basic aspects of the slave trade, but is not perfectly integrated. Poetry: *Aghānī Ifrīqiyā* (Songs of Africa, 1955); *'Āshiq min Ifrīqiyā* (A Lover from Africa, 1964); *Udhkurīnī yā Ifrīqiyā* (Remember me, Africa, 1966).

Trans.: ALAC-P. PC

Ferdousī, Abolqāsem Manṣūr b. Ḥasan (b c940 Bāzh, near Tus in Khorāsān, d c1020), Persian poet. Of a not very wealthy land-owning family (*dehqān*), he was interested in history and mythology from his youth, as well as in literature and poetry. Although there are many legends about his life, few indubitable facts are known. Almost the only source of information are the autobiographical allusions in his epic *Shāhnāme* (Book of Kings), but the authenticity of this information is reduced by the frequent interpolations by generations of copyists and interpreters. In spite of the dearth of information, however, it seems clear that he did not begin work on his epic until he reached manhood. There is no evidence to show that he was moved to do so by the news of the death of the poet Daqīqī (qv) who was the first Persian poet to write an epic on a subject from the history of Iran. Daqīqī's work, however, did give Ferdousī a formally perfect literary model (in epic style, rhyme and metre). It is remarkable that only a fragment of Daqīqī's work survives (c1,000 lines on the introduction of Zoroastrianism under King Gushtāsp) embodied in Ferdousī's *Shāhnāme*.

In the tenth century the pre-Islamic custom of collecting the national history and traditions, in the spirit of the Pahlavi *Khvatāy Nāmak* (Book of Rulers) written at the end of the Sassanian dynasty, was revived. Besides Daqīqī's epic Ferdousī could draw on Abolmo'ayyad Balkhī's prose, *Shāhnāme*, the verse of the Marv poet Āzād Sarv, and especially on the prose elaboration of a *Shāhnāme* with which the ruler of Tus, Abū Manṣūr Moḥammad b. 'Abdarrazzāq, had charged four East-Iranian Zoroastrians (compiled in 957). Oral tradition and folklore was an equally important source. Ferdousī began work on his *Shāhnāme* about the year 980, when Iran was still ruled by the Samanid dynasty (overthrown in 1005). Although he had the support of his patrons the lords of Khorasan, Ferdousī dedicated the final version of the epic, completed about the year 1010, to Sultan Maḥmūd of Ghazna (999–1030). By his skilful cultural policy Maḥmūd had acquired the reputation of being the greatest patron of Persian literature, and the poet saw in him the ideal leader for the revival and unification of Iran. The dedication did not have the desired effect, however, and the Sultan did not pay due attention to the poet's lifework. Legend has it that instead of the promised *dinar* for every line, he sent the poet only a *dirham*.

It is probable that religious controversy helped to make the poem's reception so cool. Tradition described Ferdousī as a Shiite, even one of the Ismā'ilite sect, while Sultan Maḥmūd was a fanatical Sunnite and enemy of Shī'a. This brings up the question of whether Daqīqī's fragment on the victory of Zoroastrianism was included in the work as a tribute to the poet's precursor or from motives of cautious religious policy. There are traditions of open discord between the poet and the Sultan, the well-known 'satire on Maḥmūd' being quoted as evidence. The suggestion that conflict actually occurred must be dropped along with the 'satire's' claim to authenticity. Contemporary scholars regard it as a compilation of lines taken from different contexts in the *Shāhnāme* (c100 distichs). Equally imaginary is the tragically sentimental legend of the Sultan's tardy decision to reward the poet, his emissaries bearing gifts entering Tus through one gate as the funeral cortège bore the poet's body out through another. The legend reflects the last phase of Maḥmūd's political thinking when he was trying to establish a Greater Islamic Empire by invasions of Central Asia and western Iran.

In most manuscripts the *Shāhnāme* consists of about 50,000 distichs divided into 50 chapters (*dāstān*) of varying length, according to the individual reigns. It gives the history of the world, which is for the poet the story of Iran. Beginning with the creation of the world and of the first man, Gayomart, it continues with the mystical Indo–Iranian Pishād dynasty and the first Iranian dynasty of the Kayānids, up to the last ruler of the last Iranian dynasty, the Sassanians (Yazdagerd III, d 651), before the Arab invasion of Iran. In a purely Zoroastrian view of the world the mystical rulers represent the forces of Good fighting forces of Evil; in Ferdousī's presentation they are the settled Iranians fighting the nomadic Turanians of Central Asia. Legend, myth and historical fact are intertwined in the epic, welded into a poetic whole. The division of the work according to individual reigns is a purely chronological device which does not seriously affect the inner structure of the poem. The royal courts are often only incidental to the story, which centres round individual heroes, particularly the heroes of the Sistan and Zavolestan saga, the knights Sām, Zāl, and above all the Iranian national hero, Rostam. The plot proceeds through incidents which are not interpreted with a view to synthesis; it is more a chain of episodes in chronological order. Formal unity is secured by the use of a uniform metre throughout (the four-foot catalectic *motaqāreb* (see 'arūḍ), distichs rhyming in pairs (*maṣnavī*, qv), archaic diction (compared to the language of contemporary lyrics) and a consistent adherence to a style characterized by a number of constant phrases and rhetorical devices.

As far as the philosophical conception of the poem is concerned, we find the constant struggle between good and evil, with faith in the victory of the former, and the autochthonic Iranian principle of legitimate descent of the ruler, giving the ruler a halo of divinity. Ferdousī's conception of Fate is remarkable; blind, cruel and inexorable, yet it is necessary to fight against it. The theme of revenge is equally ineluctable. The story is often interrupted for lyrical and romantic scenes and descriptions of nature, which Iranian literary criticism regards very highly. Ferdousī is therefore considered not only the founder of the heroic epic (in which genre he is still the model right up to the twentieth century), but also of the romantic epic. The *Shāhnāme* has always been considered in Iran as the national historical chronicle; modern research suggests that the historical value of the work lies not so much in chronological facts as in the information supplied on the way of life and the social conditions of ancient Iran. The lyrics attributed to Ferdousī (*qiṭ'a*, *ghazal*, qqv) are of very doubtful authenticity. Modern research (by Ẓabīḥollāh Ṣafā, S. Nafīsī, qv, and others) on the basis of an analysis of the style and lexicon of the work, has also shown that the epic *Yūsof o Zalīkhā* was wrongly attributed to Ferdousī. For the literary and aesthetic value of his work, and even for his profound love of Iran, which fills the epic, Ferdousī is regarded as the immortal national poet of Iran, in spite of any changes literary taste has undergone in his homeland. In the context of world literature the *Shāhnāme* has a worthy place alongside the ancient Indian and the Homeric epics.

Trans.: J. Champion, *The Poems of Ferdosi*, I. (Calcutta 1785); A. G. Warner and E. Warner, *The Sháhnáma of Firdausí* (London 1905–1912); J. Mohl, *Livre des Rois*, I. (Paris 1883).
RHIL 154–62; BLHP II 129–47; T. Nöldeke, *Das iranische Nationalepos*, GIPh II (Stuttgart 1896–1904); H. Massé, *Les Epopées Persanes. Ferdausi et l'épopée nationale* (Paris 1935). JOS

al-Fersi, Muṣṭafā (b 1931), Tunisian writer, poet and dramatist. He studied at the Sorbonne, worked in the Ministry of Information, in radio and served as general director of the Tunisian film company; in 1970 he became General Secretary of the Tunisian Writers' Union. Al-Fersī began writing plays while still at school. His novel *al-Munʿarij* (The Bend), drawn from the life of Tunisian intellectuals in the early years of independence, weighs the alternatives; personal life or public commitments in pursuit of political ideals. His short stories have been published in *al-Qanṭara hiya al-ḥayāt* (The Bridge is Life, 1970); two have been filmed. His style is original, at times symbolic, at times using philosophical digression. He writes Arabic prose, French verse, and is also a critic. Other works: collected plays, *Qaṣr ar-rīḥ* (Castles in the Air, 1961). SP

Fikret, Tevfik (b 1867 Istanbul, d 1915 ibid.), Turkish poet and painter. Of a Circassian officer's family, he was an official and later French master at the Istanbul lycée. As a consequence of trouble with the Sultan's court he was forced to change his employment several times, and often endured poverty. He began to write and paint at an early age, and became a painter of note. In 1893 he became editor of the periodical *Serveti Fünun* (Wealth of Art), around which Turkish literary life centred at that time. The contributors were trying to carry on the renaissance of Turkish literature which had flourished at the time of the first constitution (1876); however, the patriotic enthusiasm and optimism of that time had been replaced in their work by hopeless escapism, and their writing was no longer closely bound up with journalism (see Ibrahim Şinasî). On the other hand, the *Serveti Fünun* writers paid more attention to form, and their works were not simply imitations of French models, but independent development of the native tradition, based on those models. Nor did they endeavour to keep the language pure, but regarded the assimilation of foreign words as a way of enriching it. Fikret's first volume of verse was written in this spirit: *Rebabi Şikeste* (The Broken Lute, 1899).

During the attempts to suppress the Young Turk movement *Serveti Fünun* was forbidden, and literary life came to a stop, most writers retiring from the field in disgust. It was at this time that Fikret took up a clear stand against the Sultan's rule and began to write poems attacking it; they were copied for distribution. He greeted the Russian revolution of 1905 with the poem *Sabah Olursa* (If Morning Comes), and was equally enthusiastic in his welcome for the Young Turk revolution in 1908. He established the daily *Tanin* (Echo), which was to educate the people for freedom. He was disillusioned by the results of the revolution, however, and as the newly emerging poets were singing the praises of the new freedom and the strength of the Turkish people, Fikret began to voice his criticism. Accused of insufficient patriotic feeling, he answered in poems for children, *Halukun Defteri* (Haluk's Notebook, 1912); it is they who will bring to fulfilment their forefathers' dreams of humanism. Fikret died while the Sultan was still on the throne, but as late as 1945–6 his work was capable of touching off conflict between progressive and reactionary forces; those who praised it were victimized as politically unreliable. He is also one of the best Turkish poets for his important formal contribution, using the classical Persian–Arabic metres in a freer form. He rejected grammatical rhymes and introduced rhymes based on the sound of the words. He was the first poet to choose his words to suit the melody of his lines and the mood of the whole poem. He also enriched the language with many images. Further works: *Hasta Çocuk* (The Sick Child); *Şekvayi Firak* (Pain of Parting); *Hani yağma* (The Rule of Rapace); *Şermin.*

PHTF II, 514 ff. ZV

Fitrat, Abdurrauf, (b 1884 Bukhara, d ?1947), Tajik and Uzbek writer, dramatist and ideologist of the reform movement in Central Asia. In 1923 he was Minister for Foreign Affairs in the Bukhara Republic, and later Professor in Tashkent and Samarkand. A victim of the Stalin cult period, he was later rehabilitated. He wrote both in Tajik and Uzbek, and his importance lies in his pioneer work in the use of simple comprehensible language in prose. He was well-known for his writings pillorying backwardness in Central Asia, particularly in education, such as *Munozira* (Discussion, 1910–11) and *Bayonoti sayyohi hindī* (The Story of an Indian Pilgrim, 1912–13). He also published a volume of verse, *Sayha* (The Cry, 1911), a satirical novella, *Qiyomat* (Day of Judgment, published 1964), and many plays in Uzbek and Tajik, eg *Chingizkhān*, *Ulughbek*, *Begijon*, *Hind ikhtilalchilari* (The Indian Rebels, 1923), *Shūrishi Vose'* (The Vose' Rebellion, 1927). Most of the plays were criticized for overemphasizing the nationalist point of view. JB

Five Festal Scrolls, five shorter Hebrew literary works included in the Old Testament (qv) canon: *Song of Songs* (qv), love poems; the folk-tale narrative of *Ruth* (qv); *Lamentations,* wrongly attributed to Jeremiah (qv), bewailing the destruction of Jerusalem and the Temple; *Ecclesiastes* (*Qohelet*) (qv), reflections wrongly attributed to King Solomon; and *Esther* (qv), the dramatic story of the defeat of conspiracies against the Jews in the Persian empire. See also Bible. SZS

Frik (b 1230–40, d 1310), Armenian poet. He lost his family and property in the Mongol invasions and could not become a priest as he wished. A wandering singer, he entered a monastery as a lay brother in old age. Author of about fifty lyrics full of bitter discontent, rebellion and irony at the failure of fate and God, and written in the vernacular Armenian of the time, mostly in the '*hay*[*e*]*ren*' form (15-syllable lines usually in quatrains, frequent in folk poetry). Compared to other *ashughs* (minstrels, from Perso-Arabic '*āshiq*, see *âşvk*), his verse is less polished, but livelier with more social motifs.

Trans.: TRA II. LM

Fuat, Memet (b 1926 Istanbul), Turkish critic. He graduated in 1961 in English language and literature at the University of Istanbul. In October 1960 he founded the publishing house *De*, which specializes in modern literary theory and criticism, poetry and drama, publishing both Turkish works and foreign translations. Amongst the most outstanding works he has published are the writings of Nazım Hikmet Ran (qv). Since October 1964 he has brought out a monthly review of literature and art, called *Yeni Dergi* (The New Review), at present the best review of its kind. In 1963 he initiated a series of anthologies with the title *Türk Edebiyati* (Turkish Literature) comprising his own choice of the most promising works of the preceding year, his commentaries on them, and a survey of the general literary production of that year. Since January 1967 he has written literary criticism and essays for the weekly paper *Ant*. Besides criticism, Memet Fuat also writes short stories and translates modern American short stories and plays. Works: short story, *Yaşadığımız* (What We Have Lived Through, 1951); criticism and reviews, *Düşünceye Saygı* (Respect for Ideas, 1960). DL

Furqat, Zokirjon (b 1858 Kokand, d 1909 Yarkent), Uzbek poet and propagator of Russian culture. Furqat began writing in Kokand, where he met progressive writers like Muqimī (qv). He went to Tashkent in 1889 and began to study Russian language, literature and history. He travelled in India, Arabia, Turkey, Greece and Bulgaria, and settled in Yarkent, Sintiang province, from where he kept up a flow of contributions to the Uzbek paper of Turkestan, *Turkiston viloyatining gazeti* (Newspaper of the Turkestan Vilayet). Lyric poetry occupies the most important place in his work; he also wrote many volumes of verse in admiration of Russian culture and history (Gymnasium, Suvorov, and others).

PHTF II, 400. NZ

Füzuli, Mehmed bin Süleyman (b end 15th century, d 1556 or after 1580), great Turkish poet who laid the foundations of Azerbaijan poetry. There is little (and that contradictory) known about his life, spent in Iraq (Baghdad, Najaf and Kerbela) during the Persian–Turkish fighting. Unlike many literary men of his time he never attained high office; he seems to have died in a plague epidemic. Coming from a multilingual background, he wrote both in Persian and Arabic, but his best work is in

Turkish, in the Azerbaijan dialect which he raised to literary level. His best and best known lyrical work is the romantic epic, *Leylâ ve Mecnûn*, an Arab theme set in his native parts, with vivid descriptions of the people and landscape. His three *dīwāns* (qv) of lyrics (Turkish, Persian and Arabic) resemble folk songs in their light touch and apparent simplicity, and became widely popular in the Turkish literary sphere. His philosophical works are no less significant, written in support of the *batin* movement, whose followers believed the aim of life to be the immersion of the soul in the totality of things (*Matlâ' ul-i'tikâd*, The Foundation of Faith). A Shiite, he also devoted one work to the Shiite saint, Ḥusayn, conceiving him not as a martyr in the cause of the Prophet's family, but as the symbolic hero deliberately fighting the violence whose victim he ultimately became. Füzuli's influence on Turkish literature came later, when realism and folk poetry began to affect poetry; he is considered the founder of Azerbaijan poetry, for helping Azerbaijan Turkish to oust Persian as the literary language of the region. Further works: *Bang o bâde* (Hashish and Wine, in Persian); *Ḥosn o' eshq* (Beauty and Love, in Persian); *Şah-ü-gedâ* (The Shah and the Beggar, in Turkish); *Şikâyetnâme* (Book of Complaints, in Turkish).

GHOP III 70 ff; PHTF II 254 ff. ZV

G

Gachechiladze, D., Georgian poet, see **Chikovani,** Svimon.

Gadiyev, Taomek, Ossetian poet, see **Gaediaty,** Taomaq.

Gaediaty, Taomaq (Russian Gadiyev, Taomak Sekayevich, b 1882 Ganis, northern Ossetia, d 1931 Ordshonikidze), Ossetian poet, journalist and literary critic. Gaediaty studied at the Arts Faculty of the University of Dorpat (now Tartu), participated actively in the revolution of 1905–7, was arrested in 1908 and banished to Siberia in 1910. After the February revolution of 1917 he was set free. In 1919 he became a member of the Communist Party. During the following years he worked as an editor, as the head of the North Caucasian Mountains Pedagogical Institute and as the director of the North Ossetian Scientific Research Institute. His first poems were a call for social liberation. The cycle *Akchaestony fystytae* (Writings of a Prisoner, 1908–9) describes the hardships and longings of an imprisoned revolutionary. His prose tales (eg *In Banishment*, *Vanya's Disease*, *Music*) were written in Russian during Siberian exile. He also composed two Ossetian dramas (*Os-Paeghatyr*, 1929, and *Amondmae tsaeujytae*, Going to the Fortune, 1928), many journalistic and critical articles and translations from Russian. Gaediaty has an excellent style; he occasionally wrote 'free verse' and 'poems in prose'.

ML

Gâlib Dede, Şeyh (real name Mehmed Esad; b 1757/8 Istanbul, d 1799 ibid.), Turkish mystic poet. Of an official's family, he first became an official himself, and then joined the order of whirling dervishes (*mevleviye*); he later became head of the largest monastery in Istanbul. His first great mystic-romantic *mesnevi* (see *maṣnavī*), *Hüsn-ü-aşk* (Beauty and Love), made him the last great classical poet of Turkey; it is considered the climax of the classical tradition. He drew some of his philosophical ideas from Jalāloddīn Rūmī (qv), but is entirely original in his magnificent poetic imagery and the account of the love affair between Beauty and Love, which crystallizes in theopanistic ideas (the opposite of pantheistic, ie the belief that god is to be found in all things). The verse he wrote after this, collected in his long *dīwān* (qv), does not rise above the average. His prose writing is didactic in character (for example, on the *mevleviye* poets). Although his *mesnevi* was popular reading right up to recent times, he had no important imitators because the trend towards modern forms prevailed soon afterwards.

PHTF II 450 ff; GHOP IV 175 ff. ZV

Gamsakhurdia, Konstantine (b 1891 Abasha), Georgian writer. He studied in Königsberg, Munich and Berlin (PhD 1919). The same year he returned to Georgia and founded the literary society Academic Association, which he headed until it was disbanded in 1931. During the twenties he edited in turn *Ilioni*, *Sakartvelos samreklo*, *Kartuli sitqva* and *Lomisi*. He translated Dante's *Divine Comedy* and Goethe's *Young Werther* into Georgian, and also engaged in research, particularly into the history of Georgia. Since 1944 he has been a member of the Georgian Academy. Gamsakhurdia's novels and short stories are marked by intricate well-developed themes, many literary and historical allusions, and archaic language. He pays great attention to formal structure and the development of ideas, and often intersperses the plot with philosophical meditations. The novel *Dionisos ghimili* (The Smile of Dionysos, 1925) describes a young Georgian intellectual going from one disillusionment to another, until the vicious circle is broken when he returns to his homeland. *Mtvaris motatseba* (Eloping with the Moon, 1938) deals with the social changes seen in Abkhazia since Soviet rule was installed. *Didostatis marjvena* (The Great Master's Right Hand, 1938) could be called a historical novel of adventure; its involved plot made it possible for a French publisher to present it in the form of 'comics'. The novel *Vazis qvaviloba* (Vine Blossom, 1959) deals with a Georgian village shortly before World War II.

Trans.: V. Eristavi, *The Hand of a Great Master* (Moscow 1959); R. Domec, *La dextre du Grand Maître* (Paris 1957); *Mindia, the Son of Hogay and Other Stories by Georgian Writers* (Moscow 1961); *Mindia, des Chogais Sohn*, in: Ruth Neukomm, *Georgische Erzähler der neueren Zeit* (Zurich 1970). VAČ

Garip, A., Turkish folk singer, see **âşık.**

gathas, verse parts of **Avesta** (qv); see also **Zarathushtra.**

Genesis, first book of the **Pentateuch.**

Ghani, S., Tajik writer, see **Niyozi,** Foteh.

ghazal, a very common poetic form in the Persian, Arabic and Turkish poetry, domesticated during the past centuries in other literatures of the Near East, such as Pashto, Kurd, Azerbaijan, Uzbek, Urdu, etc. The origin of the *ghazal* is in lyrical folk songs, of the 7th century AD; it was originally sung to the accompaniment of a string instrument. Its definitive form was acquired around the 13th–14th century. The original *ghazal* consisted of 4 to 12 (later even more) couplets (*bayt*), having the same metre and the same rhyme. The first *bayt*, called *maṭla'*, has an extra rhyme in the first half-line (*misrā'*), so that the half-lines in a *ghazal* have the following rhyme pattern: *aa ba ca da* etc. The last couplet, called *maqṭa'*, usually contains the poet's literary name, the pseudonym called *takhallus*. Each *bayt* is sometimes concluded by a refrain called *radīf* consisting of one or more words. Each *bayt* usually has a complete idea and a self-dependent meaning. The theme was originally confined to love, but later on, philosophical, mystical, didactic and social subjects were added. Among the poets of the *ghazal* in the classical period should be mentioned the Persians Ḥāfeẓ, Neẓāmī, Rūdakī and Sa'dī and the Turkish poet Bāqī (qqv); in the modern period the form was cultivated by the Persian–Tajik poet Lāhūtī, the Uzbek Ghafur Ghulom (qqv) and many others.

RHIL. JB

al-Ghazzālī, Abū Ḥāmid Muḥammad (b 1059 Khorasan near Tus, d 1111 Tus), foremost mystic, theologian and religious philosopher of mediaeval Islam. He was a teacher (1091–95) at the Islamic university (*madrasa*) in Baghdad, the Nizāmīya, founded by Neẓāmolmolk (qv). He carefully studied the main philosophical trends of his time (the polemic with heretical Shiite thinkers, etc), although he did not agree with Arab philosophy. The involved progress of his own views, described in his autobiography *al-Munqidh min ad-dalāl* (The Liberator from Error) finally brought al-Ghazzālī to a profound intellectual scepticism, from which he found refuge in Islamic mysticism, Sufism (qv). At the age of thirty-six he suddenly left his teaching post and travelled about the Islamic world as a Sufi; at the end of his life he taught for a while in Nishapur, and then died in his native Tus.

His fundamental work is the long *Iḥyā'ʻulūm ad-dīn* (The Revival of Religious Knowledge) written in his final Sufi period and expounding his system of ideas. He aimed to fill the traditions of Islam with more profound religious feeling, based on the mystical ideas he worked into the structure of Sunnite thought. He also played an important role in the development of Islamic theology (*kalām*, qv), incorporating elements of Greek dialectics and giving it its culminating form. He regarded the Aristotelian trends in Arabic philosophy with profound scepticism, however, having no faith in the possibility of reaching intellectual philosophical knowledge. In his *Tahāfut al-falāsifa* (traditionally translated as The Destruction of the Philosophers) he sharply criticized this philosophy for being based on the contradictions and inconsistencies of philosophical systems. In particular he rejected philosophical metaphysics (theories of the eternity of the world, etc) and the philosophical principle of causality; like al-Ashʻarī (qv) before him, he thought it a mere fiction, for the only force and the cause of all phenomena is God alone and the criterion of faith is not philosophical speculation but the evidence of religious (ecstatic) experience. In spite of his condemnation of Arabic philosophy (to which Ibn Rushd, qv, replied in his *Tahāfut at-tahāfut*, The Destruction of the Destruction), al-Ghazzālī was nevertheless often suspected of secret sympathy for philosophical studies. The mediaeval Christian scholastics often erroneously classified al-Ghazzālī as a philosopher of the Aristotelian school, along with al-Fārābī and Ibn Sīnā (qqv); they were acquainted for the most part with his *Maqāṣid al-falāsifa* (The Intentions of the Philosophers) which summarizes the philosophical trends, although it was written to introduce an anti-philosophical polemic.

Trans.: M. Montgomery Watt, *The Faith and Practice of al-Ghazzālī* (London, New York 1953 and repr.); Toufic Sabbagh, *Lettre au disciple* (Beyrouth 1959); Mac-Kane, *Al-Ghazáli's Book of Fear and Hope* (London 1962). Carra de Vaux, *Gazali* (Paris 1902); I. Goldziher, *Streitschrift des Gazáli gegen die Bátinija-Sekte* (Leiden 1916); J. Obermann, *Der philosophische und religiöse Subjektivismus Ghazalis* (Wien, Leipzig 1921). vsd

Ghazaros Pharpetsi (Lazar of Pharpi, second half of 5th century), Armenian historian. He studied theology in Constantinople and became abbot and famous preacher. Unjustly accused, he fled to Amid whence he sent to a patron of the Mamikonian family the letter known as 'The Accusation of Lying Monks', in his defence. Recalled, he was charged with the task of writing a history to celebrate the Mamikonian family. Following Yeghishe (qv) and Pavstos Buzand[atsi], he recorded an eyewitness account of the second rising against Persia, which was victorious, in his *Patmut'-yun Hayoc* (History of Armenia, French trans. by S. Ghésarian, in LCHAMA II, and by P. G. Kabaragy, Paris 1843). His style is simple, keeping to historical chronology and verisimilitude, describing nature and human ways in a readable manner, and covering enemies and traitors with ridicule and scorn.

HO 7; THLA; AAPJ. LM

Ghulom, Ghafur (b 1903 Tashkent, d 1966), Uzbek poet, writer and critic; he began work as a compositor, later trained as a teacher, and started writing poetry about 1919, publishing verse, humorous stories and feuilletons in the satirical periodical *Mushtum.* His early verse showed him turning away from traditional Uzbek prosody; later he translated Mayakovskiy, whose work influenced him profoundly. Ghulom also wrote prose, and was an outstanding critic (studies of Navā'ī, Muqimī and Furqat, qqv). He translated Pushkin, Lermontov, Shakespeare (*Othello, King Lear*), Pablo Neruda and others, and was himself translated into Russian and other languages. Chief works: volumes of verse, *Dynamo* (1931); *Biz seni olkishlaimiz* (We Celebrate Thee, 1958); *Ballada ve poemalar* (1934); *Khitoidan lavhalar* (Pictures from China, 1932); *Ikki vasika* (Two

Letters, 1934); stories, *Yodgor* (Memorial, 1936).

PHTF II 712. NZ

Gilgamesh, hero of the Akkadian epic known principally from Assurbanipal's library (1st millennium BC) under the title 'He Who Saw the Depths', in twelve tablets or books. It combines several early Sumerian epics about King Gilgamesh of Uruk which are not a connected narrative. Parts of the Akkadian work are also represented by 2nd millennium tablets. The prologue and epilogue tell how Gilgamesh had walls built round his city. The walls of Uruk were of special plano-convex shaped bricks, as excavation has shown, used only in one period. It is thus possible to place the reign of Gilgamesh at c2700 BC. According to the historiographic text known as the 'Tummal inscription', he was a contemporary of the founder of the First Dynasty of Ur (c2550 BC).

Gilgamesh was a historical person, as is also testified by the Sumerian epic *Gilgamesh and Agga*, describing the war between King Agga of Kish and Gilgamesh of Uruk. A whole series of loosely connected poems about the heroic deeds of Gilgamesh, and events in his life, soon grew up round his name, as they did round his predecessors, Enmerkar (qv) and Lugalbanda (qv). Besides *Gilgamesh and Agga*, four other Sumerian epics of Gilgamesh are known, primarily mythological in content. The most important are *Gilgamesh and Huwawa*, dealing with his campaign against the giant Huwawa in the mountains of Lebanon, and *Gilgamesh and the Oak*, which gives the Sumerian idea of life after death; both were woven into the Akkadian Gilgamesh epic, as was the description of the Flood from the Ancient Babylonian poem *Atra-hasīs* (qv), dated in the 17th century BC.

The best preserved version of the Gilgamesh epic, in Assurbanipal's library, begins with a prologue summarizing the story. The story then follows. The great king Gilgamesh drives his subjects to labour on the walls round his city; to put an end to this unbearable tyranny the gods create a half-wild man, Enkidu, who grows up in the steppe among the wild animals. When the two met and fought, it is not clear who was the winner; however, they became friends for life. Gilgamesh's longing for fame and great deeds would not let him rest, and he decided to rid the world of all the evil incarnate in the giant Huwawa, guardian of the cedar forests of the mountains of Lebanon. The two friends, victorious, kill the giant and return to Uruk; the goddess Ishtar is so overcome by the beauty of Gilgamesh that she demands his love, but he rejects her with insults and she demands his punishment. One of the two friends must die, and the lot falls on Enkidu. The death of his beloved companion shocked Gilgamesh so deeply, and he became so obsessed with the thought of death, that he set out to visit the ancient king Utanapishtim, the only human being to have been granted immortality and allowed by the gods to live on the island of the blessed. After a long and dangerous voyage Gilgamesh lands on the island, passing over the waters of death. Utanapishtim tells him that he was the only human being saved by the gods from the great Flood. This can never be repeated, and so Gilgamesh can only gain immortality by remaining sleepless for six days and seven nights. Worn out by his journey, Gilgamesh fails, and only one more possibility lies before him: to go down to the bottom of the sea and pluck the plant of life. He does not hesitate, and bearing the magic plant he sets out for Uruk. Yet even now his bid for immortality fails; he lays the plant down by a spring, to refresh himself with the water, and a snake swallows it. Gilgamesh comes back to Uruk empty-handed, his only comfort the great battlements he has had built round the city. This is the end of Tablet XI, which concludes as the prologue began, with praise of the city wall. The twelfth tablet, added later, is an exact translation of another Sumerian poem and tells of Enkidu's descent to the underworld, and of the life led there by the dead. The Babylonian epic of Gilgamesh was known far beyond Mesopotamia; fragments written in Hittite, Babylonian and Hurrite have been found in the capital city of the Hittite kings, and a fragment of Tablet VII was discovered near Megiddo in Palestine.

P. Garelli, *Gilgamesh et sa légende* (Paris 1960); H. Heidel, *The Gilgamesh Epic and Old Testament Parallels* (Chicago 1966); E. A. Speiser, in ANET[3], 72–99; N. K. Sandars, *The Epic of Gilgamesh* (Penguin Books, 1965); S. N. Kramer, in ANET[3], 44–51; W. von Soden, *Das Gilgamesh-Epos* (Reclaim 1968); LRPOA 145–226. LM

Ginza (Treasure), a large collection of Mandean religious writings, the oldest (hymns) being probably from the 2nd century. The somewhat later prose treatises comprise religio-philosophical meditations, liturgical texts, ritual, etc. The texts were collected after the Islamization of Babylonia at the end of the first millennium. The *Ginza* (divided into the *Left* and *Right Ginza*), together with the later *Book of John* and the liturgical collection *Qolasta*, make up the literature of the Gnostic religious community of the Mandeans, who have preserved their religion and their special Aramaic dialect to this day, in southern Iraq and the adjacent parts of Iran.

Trans.: M. Lidzbarski, *Ginza, Der Schatz oder das grosse Buch der Mandäer* (Göttingen 1925).
E. S. Drower and R. Macuch, *A Mandaic Dictionary* (Oxford 1963), ix–xi. SZS

Gökalp, Mehmed Ziya (b 1875 Diyarbekir, d 1924 Istanbul), Turkish writer and poet, sociologist and philosopher, the ideologist of the pan-Turkish movement. Of an official's family, he was expelled from the Veterinary College for his part in the Young Turk movement. After the 1908 revolution he became a leading figure in the Young Turk party and, as a writer and professor of sociology at the University in Istanbul, was the leading propagator of Turkish patriotism. He was among Atatürk's closest collaborators. Influenced by the works of Ahmed Vefik Paşa, among others, he believed that the Turkish nations had a great role to play in history, and the idea of the cultural and economic unification of all the Turkish-speaking peoples became the principal theme of his prose and poetry alike. His ideas became popular in Turkey itself, which was to be the leading nation when the Turks were united; and in Tsarist Russia, in Kazan and Baku, journals were published which printed Gökalp's writings and propagated pan-Turkish ideas. These ideas provided the newly-awakening Turkish peoples there with a weapon against the national policy of the Tsarist government. In the social sphere Gökalp proclaimed the right of women to equality, and urged that the power of the religious leaders should be restricted. At the beginning of this century his ideas inspired new political and literary trends, and, especially after the revolution of 1908, his ideas became very influential; later, however, when they had been superseded in Turkey itself by more mature theories, the Turkish right wing and anti-Soviet elements took them up again, distorting and abusing some of Gökalp's sayings.

His literary works are not of particular importance in Turkish literature, but their significance lies in the vast influence of his pan-Turkish ideas, as they appeared in the work of many writers, both Turkish and foreign. Apart from poems, stories, and treatises on pan-Turkism (*Türkçülüğün esaslari*, The Fundamentals of pan-Turkism, 1923) he wrote on Turkish history and folk lore. Other important works: stories *Kızıl elma* (Red Apple); *Yeni hayat* (New Life); *Altın ışık* (Gold Ray).

Uriel Heyd, *Foundations of Turkish Nationalism* (London 1950); Niyazi Berkes, *Turkish Nationalism and Western Civilization* (New York 1959). ZV

Golshan, Persian poet, see **Qā'ānī,** Ḥabībollāh Fārsī.

Gōrān, 'Abdullāh (b 1904 Halabja, d 1962 Sulaimaniya), modern Kurdish poet. He went to school in Halabja and in 1921 went on to the teachers' training college in Kirkuk. For financial reasons he could not complete his studies. From 1925 to 1937 he worked as a teacher in Halabja, devoting himself to intense language study (Arabic, Persian, Turkish and English) and literature. English literature in particular (Shelley) influenced his poetry. From 1937 he worked as a civil servant. His first poetry was written at the age of thirteen; his early work was influenced by classical poetry in its form. His characteristic genre is the lyric, at first the romantic lyric, later coming under realist influence. He has been called a poet-painter. He travelled a great deal through Kurdistan, drawing inspiration for his work. Love poetry is also well represented in his lyrics. During World War II and the years which followed he turned more and more to the history of his own people and to social and political questions. He was often victimized, and spent several terms in prison, before the 1958 revolution in Iraq; then he became active in the political and cultural life of the country and in a wider context. Gōrān died in 1962, leaving behind a body of work in several genres from classical *ghazals* and *qaṣīdas* (qqv) to theatrical plays, novellas, and translations from French and English. Unfortunately his work has not yet been collected, classified and published. Besides their poetic value, his poems are distinguished by their excellent language. Four volumes of his poetry have so far appeared: *Bahasht a yādegār* (Paradise and a Memory, 1950), *Frmēsk u hunar* (Tears and Art, 1950), *Sirusht u darūn* (Nature and the Soul, 1968), *Lāvik u payām* (Ballad and Message, 1969). Many of his poems appeared separately, or not at all.

C. J. Edmonds, *Kurds, Turks and Arabs* (London 1957), 172–9 (excerpts from his poetry, with English translations: *Gasht la Hewramān*, A Tour in the Hewraman); FKH xlvii–xlviii.
AK

Gorgānī, Fakhroddīn As'ad (11th century), Persian poet. He held office at the court of Toghrol, the founder of the Seljuq dynasty. Not much of his poetry has survived, and most of that is in the anthology of 'Oufī (qv). Gorgānī's fame rests on his *maṣnavī* (qv) of 9,000 distichs, *Vīs o Rāmīn* (1057). The author claimed to have based it on an old Pahlavi work. Sometimes regarded as highly immoral, the poem is in fact a condemnation of the immorality that was ruining society. Some scholars believe that the poem served as the model for the mediaeval poem of chivalry, *Tristan et Iseult*; there are many similarities between the two.

Trans.: Henri Massé, *Le roman de Wis et Rāmin* (Paris 1959).
RHIL 177–9. JB

Grigor Narekatsi (Gregory of Narek, b 945–9, d 1003–1011), Armenian poet, son of a bishop. He studied theology and philosophy, and lived and died in the Narek monastery on Lake Van. He was famous as a poet, adored by the people, and beatified by the Church after death. He was the first to overcome the dogmatism, formal monotony and dependence of Armenian mediaeval poetry on church ritual. He wrote odes (*gandz*), songs (*tagh*) where the religious element is often superficial only, interpretations of the *Song of Songs* (qv), and prayers. He used monorhyme (perhaps derived from the folk poetry) and drew epithets, metaphors and neologisms from the colloquial language. His *Matean ogbergut'ean*, or *Agot'amatean* (Book of Lamentations, French trans. by I. Kechichian, Paris 1961, English excerpt in LACC), also called *Narek*, a cycle of 95 lyrics, 'talking with God in the depths of his heart', informed by the idea of the divine and the human merging through suffering and spiritual purification, contains many paraphrased passages from the *Psalms* (qv). It is the intimate confession of an involved spiritual life, disturbed by the contradictions of the human lot, and in its profundity and wealth of thought clothed in polished and infinitely varied expression, has the power to move the reader even today. This work, the keystone of reflective poetry in Armenian, is viewed as one of the first harbingers of the Armenian cultural renaissance.

Trans.: A. S. Blackwell, *Armenian Poems* (Boston 1917), *Armenian Legends and Poems* (London 1917).
NACL; THLA; HO 7; LACC. LM

Grigoris Aghthamartsi (Gregory of Aghtamar, b end 15th century, d after 1569 on Aghthamar Island, Lake Van), Armenian lyric poet and educated church dignitary. About thirty lyrics have been preserved, singing enthusiastic and even ecstatic praises to spring, the nightingale and the rose, the beauties of nature and of women. More influenced by Persian poetry than were his predecessors (Kostandin Yerznkatsi, qv, Yovhannes Thlkurantsi, etc).

Trans.: TTA; TRA III. LM

Grishashvili, Joseb (real name Mamulaishvili, b 1889 Tbilisi, d 1965), Georgian poet and literary historian. Son of a mason, he could not complete his secondary school studies on account of poverty. He earned a living as a prompter in the theatre, and later as an actor. After the Revolution he devoted himself to literature and the history of literature, particularly of the lesser genres. He wrote a book on the bohemian literary life of Tbilisi (Tiflis) (*Dzveli Tbilisis literaturuli bohema*, 1947) and in 1960 became a member of the Georgian Academy. Grishashvili wrote mainly reflective poetry and children's verses (*Sabavshvo leksebi*, 1948). His verse reflects his profound knowledge of the literary traditions of the city of Tbilisi and echoes the work of half-forgotten folk poets, yet at the same time it is intellectual and even, at times, learned poetry. In his work scholarship and art are well-balanced, making him a master of the essay form. VAČ

Gudea, Sumerian ruler (c2144–2124 BC) under whom literature reached a classical perfection in the south Mesopotamian city of Lagash. Of the numerous votive inscriptions on seals, steles and statues, dedicating new buildings to different gods, the most remarkable is a hymn about the building of the Eninnu temple, on two large clay cylinders now known as A and B. The Lagash god, Ningirsu, appeared to Gudea in a dream, telling him in symbols that the temple was to be built. Gudea did not understand, and the goddess Nanshe had to explain his dream to him; even then it is not clear, and during the ritual purification of the site for the temple Ningirsu again appears to Gudea in a dream, and promises his aid. The king is relieved and calls upon the people of Lagash to live in peace; the city is cleared of thieves and assassins, the foundation stone is laid, and material brought from abroad to build the temple. The description of the building of different parts of the temple is accompanied by prayers to the gods for a favourable augury (inscription on Cylinder A).

When the building is completed, Gudea asks the assembly of the gods that Ningirsu may enter Eninnu and that his wife, the goddess Baba, may enter her chamber. The land is calmed by the final preparations for the consecration of the temple buildings. The ceremonial entry of the divine couple and their train of many servants into the new shrines provide the occasion for ceremonies, sacrifices and feasts throughout the land, for seven days. In the last line the Sumerian gods Anu and Enlil thank Gudea and promise Lagash good fortune (inscription on Cylinder B). The drum song in honour of the goddess Baba probably originated also in the reign of Gudea, although the text as preserved is of a later date.

S. N. Kramer, *The Sacred Marriage Rite* (Bloomington 1969), 13 ff; A. Falkenstein, *Die Inschriften Gudeas von Lagash* (Analecta orientalia 30, Rome 1966); A. Falkenstein and W. von Soden, *Sumerische und akkadische Hymnen und Gebete* (ed. K. Hoenn, Stuttgart 1953). BH

Gulia, Drmit (b 1874 Warcha, Sukhumi, d 1960), Abkhazian writer. The son of a poor peasant, he studied at the Gori Teachers' Training College and then at Khoni in Georgia, but did not finish his course for health and family reasons. A village teacher up to the Revolution, after Soviet power was established in Abkhazia (1921) he attended the Sukhumi High School, and lectured at Tbilisi (Tiflis) University 1924–5. He founded Abkhazian literature in all its genres: poetry (he published the first volume of Abkhazian verse, 1912), prose, journalism (he founded the first newspaper in Abkhazia, *Apsne*, 1917), the drama and popular scientific writing. He also collected and published Abkhazian folk-lore, translated from the Russian (Pushkin, Lermontov, Krylov), from the Ukrainian (Shevchenko) and from the Georgian (Rustaveli, Tsereteli, Chavchavadze, qqv, and others). He published school textbooks and readers, including the first practical Abkhazian primer. In the literal and the wider sense, he was the teacher of most Abkhazian writers, and enjoyed vast authority right up to his death. His best work is to be found in his lyrics, especially in latter years, when the need for revivalist propaganda had disappeared. His work discusses the sense of life for the individual, and for a small nation (there are 80,000 Abkhazians); he comes to the conclusion that the latter finds its justification in the ability to develop an independent modern culture. His novel, *Kamachich*, the first in the language, is based on ethnographical material. His plays were strongly influenced by the Gorki tradition. VAČ

Güntekin, Reşat Nuri (b 1892 Istanbul, d 1956 London), Turkish writer. He made a career in government service, became Chief Inspector of the Ministry of Education, served on the staff of the Turkish Embassy in Paris, and retired in 1954. His work includes stories, novels, plays and translations. He achieved fame with his novel *Çalıkuşu* (Golden-crested Wren). The novel *Dudaktan Kalbe* (From Lips to Heart) is a masterly depiction of the decadent figures of the last century. He is skilled in realistic narration and the description of his characters' minds in simple, clear language. His novel has a touch of the sentimental, while the stories are spiced with a delicate humour. His principal theme is love and fidelity. He made many translations from the French. His works were published in 24 volumes, the last in 1961; several of his books have been translated into European languages.

Trans.: Max Schultz, *Zaunkönig. Der Roman eines türkischen Mädchens* (Leipzig 1942); Windham Deedes, *The Autobiography of a Turkish Girl* (London 1949).
O. Spies, *Die türkische Prosaliteratur der Gegenwart* (Leipzig 1943); PHTF II 586 ff. OS

Guramishvili, Davit (b 1705 Saguramo, d 1792 Myrhorod), Georgian poet. Of an ancient noble family, he was kidnapped in 1727 or 1728 by bandits and carried off to Dagestan. He managed to escape to Russia, and to reach Moscow on foot. Here he found his way to the exiled Georgian king, Vakgtang VI, who took him into his suite. On the king's death he adopted Russian citizenship and joined the Georgian Hussars. At the beginning of the Seven Years War he was taken prisoner by the Prussians and held in Magdeburg, from where he escaped. Reaching Russia again he settled in Myrhorod in the Ukraine and became a farmer and miller. In a sense he is a one-book writer, having collected all his verse in one volume entitled *Davitiani* (The Book of David). It includes a long historical autobiographical poem, personal and love lyrics written to Russian and Ukranian song tunes, and the intimate long poem *Vesela vesna* (Ukrainian) describing the troubles of a young shepherd and his bride, ending happily in marriage. *The Book of David* also includes philosophical laments about the ill-wind of the world, permeated by religious symbolism. The book influenced later Georgian poetry primarily by its patriotic spirit, while formally it enriched verse by the introduction of new metres and strophe forms.

Trans.: UAGP. VAČ

Gürpinar, Hüseyin Rahmi (b 1864 Istanbul, d 1944 ibid.), Turkish writer. He was immensely productive, and although his novels are not great works of literature, they are good reading in the best sense of the word, and some were very successful. To portray Turkish life and ideas he took figures from different classes of society, who were typical of the atmosphere at the turn of the century; his novels are thus valuable as historical evidence for the life of the period. His language is straightforward and easily understood, close to the normal speech of the people, and he used colloquial Turkish for the dialogue. Over 30 of his 70 works were novels.

Trans.: Imhoff Pascha, *Die Geschiedene* (Konstantinopol 1907); Muhsiné Hanim, *Der liebeskranke Bey* (Berlin 1907).
O. Spies, *Die türkische Prosaliteratur der Gegenwart* (Leipzig 1943); PHTF II, 555–8; HGTM 47–9. OS

H

Habakkuk (c600 BC), Hebrew book of prophecies which describes the threat of invasion by the Chaldaeans (ie Babylonians) and condemns their violence. An Essene commentary (2nd–1st century BC) thought the prophecies applied to the Greeks or the Romans. See also Bible, and Twelve Lesser Prophets. SZS

Habibi, 'Abdhulḥay (b 1910 Kandahar), Afghan writer and scholar, writing in Pashto and Dari. He studied in Kandahar, where he became a teacher (1931–1940) and editor of the Kandahar paper *Ṭulūi afghān*. In 1940 he entered the Pashto Academy (*Pashto tolena*) and held various posts in Kabul; in 1951 he worked with the progressive paper *Watan*. From 1953 to 1962 he lived in Pakistan. He is now a professor on the staff of the Faculty of Letters, Kabul University, and director of the Kabul Historical Society. He writes both prose and verse, in Pashto and Dari, but his verse has not yet been collected in a *dīwān* (qv). He is primarily a philologist of high standing, and a historian of literature whose editions of the Afghan classics are of great value. He has also published *Pashtāne shu'arā* (Pashto Poets), *Afghānistān ba'ad az islām* (Afghanistan after Islam), a history of Pashto literature, and a great number of monograph studies of Afghan writers in books and periodicals. JB

ḥabsiyāt, prison poems, see **Mas'ūd** b. Sa'de Salmān.

Haci Halifa, Katib Çelebi (real name Mustafa b. Abdüllah, 1609–1657 Istanbul), Turkish scholar and writer, who followed his father as an officer and spent many years in the field. Later he was in office in Istanbul, and, coming into an inheritance, he devoted himself to writing. His poetry is overshadowed by his scholarly works on history, astronomy, geography, theology and law. His great encyclopaedic *Keşf üzzunûn . . .* (Revelation of the Thoughts . . .), a history of Ottoman literature written in Arabic, has preserved long excerpts from works which have not otherwise survived. For his two editions of *Cihânnümâ* (Cosmography) he also drew on Latin writers, especially Mercator and Hondius, whose 1621 *Atlas of Asia Minor* he also translated into Turkish. His biographical encyclopaedia of famous men (*Sullam al-vusûl . . .*, Ladder of Attainment . . .) and his history of the Ottoman navy *Tuhfat ul-kibâr fi asfar ul-bihâr* (Gift to the Great Sea Voyages), are still used as reference books and sources.

Trans.: G. L. Lewis, *The Balance of Truth* (London 1957).
GHOP I and II passim. ZV

Haddad, Malek (b 1927 Constantine), Algerian writer and poet, writing in French. He studied at Constantine lycée, and in France, but turned from law to literature, writing for radio and many papers. From 1955 he was in exile in France, as a teacher, agricultural labourer and journalist. He returned to Algeria after the liberation, working as a journalist in Constantine until 1968, then in government office in Algiers. His first published poems were *Le malheur en danger* (Trouble in Danger, 1956), with revolutionary echoes. In the introduction to this and to the following volume *Écoute et je t'appelle* (Listen and I Will Call, 1961) he dealt with theoretical questions, including why Arab writers publish in French. Haddad takes the poetic style into his novels, the first (*La dernière impression*, Last Impression, 1958) showing a young Algerian intellectual awakening to political reality and taking part in the liberation movement. Later books depict the life of Algerian exiles in France, often concentrating on their personal problems. Stories and poems are to be found in various journals. Other works: novels, *Je t'offrirai une gazelle* (I Will Offer you a Gazelle, 1959); *L'élève et la leçon* (The Pupil and the Lesson, 1960); *Le quai aux fleurs ne répond plus* (The Flower Embankment is not Answering, 1961).

SP

ḥadith (lit. narrative, talk), in Arabic tradition, an account of what the Prophet said or did. His followers began collecting *ḥadīth* some thirty or forty years after Muhammad's death, and later they were used to support the legal or political ideas of various sects, and to confirm religious dogma. To this end many *ḥadīth* were invented. In the ninth century the various traditions were collected; of the many extensive collections those of al-Bukhārī (d 870) and Muslim (d 875) were the most widely read and accepted. Apart from the *Qur'ān* (qv) tradition, the *ḥadīth* forms the principal source of Islamic law.

A. Guillaume, *The Traditions of Islam* (Oxford 1924); J. Schacht, *The Origins of Muhammadan Jurisprudence* (Oxford 1950); M. Z. Siddiqi, *Hadith Literature* (Calcutta 1961). IH

Ḥāfeẓ, Moḥammad Shamsoddīn (b 1327 Shiraz, d 1390 Shiraz), Persian poet. Information about his life comes from two sources: tradition, passed on by Iranian biographers, and his own poetry. He is said to have been the son of a moderately well-to-do merchant from Isfahan, who settled in Shiraz. The children were still small when the father died. The mother then had to support the family, which in the Orient inevitably spelled poverty. According to the same (not altogether reliable) sources, Ḥāfeẓ became a baker's apprentice and took up poetry at an early age. At first, people ridiculed his poems. But then, according to legend, 'Alī appeared to him and let him partake of the food of Paradise. Ḥāfeẓ then became a seer and king of lyric poets, the 'Tongue of the Hidden', as the Persians call him. The fact that Ḥāfeẓ was from the start very poor is proved by a manuscript dated 1355, preserved in the National Museum in Tashkent; it is a work by an earlier classical writer, Amīr Khosrou (qv) copied by Ḥāfeẓ. In those days, copyists were poorly paid, and the manuscript bears witness to the poet's poverty. Still, by the late 1340's and the early 1350's, he had gained some sort of reputation as a poet, judging by a few poems dedicated to the viziers 'Emāndoddīn and Qavāmoddīn, of Abū Esḥāq, the sovereign of Shiraz in those decades.

The reign of the next sultan, Mobārezoddīn (1353–1358), brought severe hardship to Ḥāfeẓ. In his early poems under Abū Esḥāq, Ḥāfeẓ had revealed a tendency to glorify sensuous pleasures. Mobārezoddīn was a puritanical ascetic, who enforced to the letter the ban on the drinking of wine in the city from which he ruled his domain. Ḥāfeẓ's poetry becomes a veiled secret idiom; the leading character becomes '*moḥtaseb*', the secret police, a nickname which in Ḥāfeẓ's poems of that time refers to Mobārezoddīn. The next reign, that of Shāh Shojā', which lasted some 25 years, was the poet's most fruitful creative period. This Shāh was extremely liberal in his attitude towards wine, and poetry was cultivated by him. He even sometimes wrote important diplomatic correspondence in verse. Judging from his poems, Ḥāfeẓ had made enough of a name for himself by the time the new ruler ascended the throne to be invited to become the sovereign's drinking companion. The intimacy of their relationship is suggested by the very fact that at a certain stage the Shāh seems to have started to envy the superior gift of the poet. It remains unclear just what caused Ḥāfeẓ to be banished from Shiraz for six years (1368–1374), but the probability is that it was a combination of the Shāh's professional jealousy and intrigues by the priesthood against the nationally famous freethinker. Ḥāfeẓ, whose stanzas to this time had mixed the ingredients of love poems and princely panegyrics with the Shāh as the object of extravagant eulogy, now found a woman to write about, Dordāne, unknown in historical annals. During his period of banishment, she was his source of inspiration. When Shāh Shojā' died in 1384, Ḥāfeẓ withdrew into retirement, for his poems in honour of the next three rulers are few and far between. His star was then at its zenith. Even during his lifetime, he was celebrated as a great poet not only in his native land but also beyond its borders, in Mesopotamia and India. The Sultan of Bengal, Ghiyāṣoddīn, and the Sultan of Baghdad, Aḥmad, invited him by letter to their courts.

Eulogy of the sovereign, an abstract general experience of love, wine, a man of statuesque beauty, and a woman produced very much the same kind of inspiration in every poet. The poet's individual gift, of course, could make the expressive content differ. To those not well read in Persian poetry, the classical poets all seem very similar. Ḥāfeẓ used the same language of suggestion, mystical similes, surprisingly effective imagery and a cosmic sensibility imbued with an extremely tolerant outlook on life in his love poems, official panegyrics, drinking songs and Sufi verse. His poetry presents problems no less difficult than those involved in reconstructing his biography. He has been described as an Epicurean, a mystic, a panegyrist, and none of these labels alone fits him. The problems lie both in the sphere of poetics and that of the history of ideas. The poet's own patron complained that he wrote unclearly. The next significant comments on his poetry were made, oddly enough, by Goethe (see below). In his monograph (1892) Harald Rasmussen represented Ḥāfeẓ as a nature mystic who, when confronted with the beauty of natural scenes, lapsed into pantheistic visions. A new beginning in the study of Ḥāfeẓ and his art was made by Qāsem Ghanī (1943); thanks to a good knowledge of the history and vocabulary of Ḥāfeẓ's time, he succeeded in dating over a hundred poems on the basis of their dedications and references to contemporaries. Roger Lescot (1944) continued Ghanī's work. He divided Ḥāfeẓ's career into a youthful period up to the year 1358, a mature period up to 1366, the period when he was out of favour, lasting up to 1373, and the period of his old age from 1376 to his death.

One of the important works on Ḥāfeẓ is H. H. Schaeder's *Goethes Erlebnis des Ostens* (1938). Schaeder adds a new interpretation to the ambiguity of wine/mysticism in connection with Ḥāfeẓ: Goethe, as expounded by Schaeder, affixes to this ambiguity the concept of harmony between opposites, which he saw as the stylistic ideal of Ḥāfeẓ. We cannot distinguish between the wine and the mysticism (as well as the politics in Ḥāfeẓ's poetry). If we try to do so, we must perforce distort the poet's stylistic goal. Goethe was an ardent student of Ḥāfeẓ during the years 1814–1818. He analyzes Ḥāfeẓ by applying concepts that in content are equivalent to certain concepts of the New Criticism. In Russia, Ḥāfeẓ exercised some direct influence on the modern school of poetry. His style, therefore, seems to have a connection with the modern movement of poetry in Europe.

Among Iranian Ḥāfeẓ studies, the following are noteworthy. In his weighty work (1939), Mo'īn gives all the old biographical material based on tradition. Dashtī describes Ḥāfeẓ's style as a synthesis of the styles of Rūmī, Sa'dī and 'Omar Khayyām (qqv). Kasravī, displaying a poor instinct, condemns Ḥāfeẓ on political grounds. Bāmdād (1960) portrays Ḥāfeẓ as belonging to the Malāmatiyye Sufi sect, the members of which acted like drunkards and beggars. The extant editions of the text are detailed by R. M. Rehder in *New Material for the Text in Hafiz*, Iran (1965) III, p. 109. The best editions are those by Qazvīnī and Ghanī (1942) and Khānlarī (1959), along with the partial edition of Gulsum Galimova (Dushanbe 1971), which is based on a manuscript dating from 1403, or only 13 years after the poet's death.

Trans.: P. Avery and J. Heath-Stubbs, *Hafiz of Shiraz, Thirty Poems Translated* (London 1952); A. J. Arberry, *Ḥāfiz. Fifty Poems of Hafiz* (Cambridge, New York 1947, repr. 1968).

J. W. Goethe, *Noten und Abhandlungen zu besserem Verständnis des West-östlichen Divans* (various editions); H. H. Schaeder, *Goethes Erlebnis des Ostens* (Leipzig 1938); R. Lescot, *Essai d'une chronologie de l'oeuvre de Ḥāfiẓ* (Beyrouth 1944); H. R. Roemer, *Probleme der Hafizforschung und der Stand ihrer Lösung* (Mainz 1951); R. Levy, *An Introduction to Persian Literature* (New York, London 1969), 116–34; RHIL 262–74. HB

Ḥāfiẓ, Egyptian poet, see **Ibrāhīm,** Ḥāfiẓ.

haggadah (*aggadah*, Hebrew: story), mainly religious stories, especially the narrative parts of the *Talmud* (qv) and related works. They contain much folk material. The Easter *haggadah* (*haggadah shel pesach*) comprises the texts read at the Passover supper. The concluding passages, the numerical series and the story of the kid appear also in European folklore. See *Midrash.* SZS

Haggai, Hebrew book of prophecies inciting the people of Judah (520 BC) to rebuild the Temple after their return from exile in Babylon. In 1632 Commenius took it as his model for a book on the need to regenerate the Church, *Haggaeus redivivus* (Haggai Revived). See also Bible, *Twelve Lesser Prophets.* SZS

Haik, Vahe (real name Tinchian, b 1896 Kharberd), Armenian writer in the diaspora. He emigrated to Greece during the first world war, and later to the USA where he studied law, literature and history. Together with A. Tchobanian and B. Nurikian he helped to organize the Armenian press in the diaspora, especially in France and the USA. He published novels (eg *Gorgi vačarakan,* The Carpet Merchant), poems, stories, Armenian ethnographical and political studies (eg *Hayreni cxan,* Native Roof) in the journals *Anahit, Nor gir* and others. He translated classical Armenian works (eg Yeghishe, qv) into English for American Armenians. LM

Hājji Khalifa, Turkish polyhistor, see **Haci Halifa,** Katib Çelebi.

al-Ḥakim, Tawfīq (b 1898 or 1902, near Alexandria), Egyptian novelist and playwright. Raised in a village, he attended highschool in Cairo. He participated actively in the anti-British nationalist movement of Sa'd Zaghlūl and was arrested. He studied law in Egypt and literature in France. Upon his return from France, he spent four years working for the Egyptian Department of Justice and Social Affairs. His experiences in Egypt and France are reflected in his writings. He began writing his first play in Paris in 1926 and a year later wrote his first novel, *'Awdat ar-rūḥ* (The Return of the Spirit, 1933). This is mainly a description, with allegorical content, of the life and problems of the middle class and peasants in Egypt and their involvement in the 1919 nationalist struggle. His *Diaries of a Rural Prosecutor* record, in a humorous vein, but penetratingly, his own tribulations in Egypt's villages. Al-Ḥakīm is as famous for his plays as for his novels. The plays are mostly social or symbolic, with a few on historical subjects (including a long one on Muhammad). Despite the French influence on his plays, most of them are presented in a typically Egyptian environment. They are sprinkled with the playwright's ideas, mainly his defence of non-conformism in modern society. This is perhaps the main theme of his analysis of man, and his constant preoccupation. Al-Ḥakīm's style in his novels, stories and plays is polished and skilfully adapted to the personalities of his heroes.

Trans.: A. S. Eban, *Maze of Justice* (London 1947); R. B. Winder, *Bird of the East* (Beirut 1967). Many of his plays have been translated into French, Italian and Hebrew. Deny Johnson-Davies, *The Tree Climber* (London, Toronto, New York 1966); *L'oiseau d'Orient* (Paris 1960); *Souvenir d'un magistrat-poète* (Paris 1966).
BGAL S III 242–50; GLAIG 203–5; LSATC 138–47; WILA 290–1. JML

Hakobian, Hakob (b 1866 Gandza, now Kirovabad, d 1937 Yerevan), Armenian poet. Son of a cobbler, himself a factory worker, later accountant, and a revolutionary, he organized 'workers' press and publishing house and was active in public life after the Revolution. In 1899 his first volume of lyrics appeared. Before the Revolution his work depicted the hard life of the workers and the future possibilities for the working class (eg *Nor aravot,* New Morning, 1909). After the Revolution his lyrics and lyrical epics described the building of socialism, the industrialization of Armenia, and the new moral outlook. Hakobian translated Pushkin, Gorki and others, as well as workers' ballads and revolutionary songs. With S. Kurghinian and M. Arazi he is regarded as the founder of Armenian proletarian poetry.

Trans.: French excerpts in ELA (by J.-P. Rebec). LM

al-Hamadhāni, Arab poet, see **Ibn Fāris,** Abu'l-Ḥusayn, and **maqāma.**

Hamdî, Mehmed Hamdüllah (1449–1503 Göynük), Turkish poet; early well known as a poet, he did not achieve high office in Istanbul, and returned home disappointed.

His *dīwān* (qv) is full of Sufi thought (see Sufism) but also includes embittered verse complaining quite plainly of poverty and the lack of appreciation by those in power. His best *mesnevi* (see *maṣnavī*) is *Yusuf ve Züleyha*, the story of Joseph and the wife of Potiphar, in which Joseph is the incarnation of divine beauty and Züleyha of unlimited devotion. His *Leylâ ve Mecnûn*, the story of their love, is based on Jāmī (qv). Another versified love story, *Tuhfei Uşşâk* (The Lovers' Gift), non-traditional in conception and treatment, was only later appreciated and imitated. Other works: *Mevlidi Nebî* (Birth of the Prophet); *Muhammedîye* (Book about Muhammad); *Kiyâfetnâme* (Book on External Description).

GHOP II 138ff; PHTF II 436. ZV

Ḥamid Kashmiri (b end of the 18th century at Nawpag near Shāhābād, d 1844 ibid.), Kashmir poet writing in Persian. All that survives of his work is the verse chronicle *Akbarnāma* (The Book of Akbar, 1844) describing the heroism of the Afghans during the first Anglo-Afghan War of 1838–1842. The epic is written in the metre and form of Ferdousī's (qv) *Shāhnāme* and is popular in Afghanistan. JB

Hamit, Abdülhak Tarhan (b 1852 Istanbul, d 1937 Istanbul), Turkish poet and playwright. Son of the eminent historian and diplomat, Hayrullan, Hamit spent his early years in Paris, at Istanbul's Robert College, and in Teheran where his father was ambassador. After receiving a very thorough education both in the traditional Islamic and in the Western languages and cultures, he entered the foreign service and spent many years abroad. For some years he served as a member of the Senate. In Republican times he was granted a state pension, and at an advanced age was made a member of the National Assembly (1928–37). Of his four successive wives, two were Turkish, one English, one Belgian. Hamit's linguistic erudition enabled him to blend Ottoman style with Western refinements and to express his thoughts and poetic fantasies in lines of immaculate prosodic soundness and esoteric vocabulary.

His early poetry was revolutionary in form and content. In form, because he used the traditional form elements in new combinations, applying European rhyme schemes to them; sometimes he himself invented new metres. In content, Hamit's poetry was the first to draw heavily on the potential of a vividly painted landscape with symbolic and mood implications. His dramas, most of which are set among heroes of early Near Eastern history (early Islamic, Assyrian, Alexander the Great, some also from Turkish history), were intended for reading only. Also *Finten*, a blood-curdling romantic fantasy set in contemporary London high society, is no exception to this, although a stage version of it was produced in recent years. Because of their difficult language, Hamit's sonorous verses were relished by many a fascinated reader but understood by few only. Until the end of his life he remained a national symbol and was always referred to as 'the grand poet'. During his long lifespan Turkish letters went through many phases, new currents appeared and disappeared. He hardly took notice of these. Throughout he remained a romantic, an eclectic, and a highbrow. His poems appeared (since 1879) collected in many volumes. Most famous is the long ode he wrote on the death of his first wife (*Makber*, Cemetery, 1885). Of his many plays, best known are *Duhteri Hindu* (The Daughter of India, 1875) and *Tarik* (1879), both in prose; *Finten* (1918), but written c1897 (partly in verse); *Nesteren*, 1878, *Esber*, 1880, *Tezer*, 1881, *Hakan*, 1937 (all in verse).

PHTF II 479ff; GHOP V 77ff. AT

Hamzatov, Rasul (b 1923 Tsada, Khunzakh, Dagestan), Avar poet. A son of the poet Hamzat Tsadasa (qv), he studied at the Dagestan Pedagogical Institute and worked as a teacher, later studying in Moscow at the Gorki Institute. He held office in the Dagestan and Soviet writers' organizations and has received official recognition of his services. Besides his own writing he translates, mainly Russian, poetry. Formally he carries on the tradition of his father's poetry, deriving material from folk and classical Dagestan verse, and has revived some forgotten genres (records and laments). Laconic, subtle simplicity marks his verse. VAČ

handarz (advice, directions, also last will and testament), the Persian name for a certain type of Middle and Modern Persian literature, also known as *pand-nāmak* (Book of Advice). These texts contain moral instruction and practical advice for social life, and are usually no more than a collection of aphorisms and maxims written in a stereotyped style and rarely employing comparisons, proverbs or sayings. Some *handarz* present long treatises on religious duties and on points of dogma, so that they are something between a collection of instruction and a catechism. Some are formulated in question and answer style. The authors are not known, but *handarz* are ascribed to famous Sassanian writers (3rd–7th century) like Āturpāt Mahraspand, the editor of the *Avesta* (qv) (4th century), his son Zartusht, the vizier Vazurgmihr (qv), King Husrav I (531–579) and to others. The *handarz* reveal the nature and characteristics of the people of ancient times, and reflect folk wisdom.

RHIL 37ff; TMSLZ. OK

Hanno (probably 5th century BC), Carthaginian explorer. Sent on an official voyage of exploration round the coast of Africa, he seems to have reached Cameroon. A Greek account of the voyage (the *Periplus of Hanno*) is presented as a translation of a Phoenician inscription in a temple in Carthage, but is more probably a free version of a Punic original. This Greek text, and the fragments of Magon's work on agriculture, preserved in a Latin translation, are all that is left of the extensive African Phoenician (Punic) literature. The numerous votive and ritual inscriptions are composed of conventional formulas with no claim to literary style.

D. Harden, *The Phoenicians* (London 1962), 171–7; S. Moscati, *The World of the Phoenicians* (London 1968), 181–4; M. Snytzer, *Les passages puniques en transcription latine dans le 'Poenulus' de Plaute* (Paris 1967). SZS

Ḥaqqī, Yaḥyā (b Cairo 1905), Egyptian novelist, short-story writer, essayist and critic. He studied law and held various legal posts; he entered the Foreign Ministry in 1929, and served many years as a diplomat. Later he became Director of Arts and Advisor to the National Library; now he is editor of the Egyptian literary journal *al-Majalla.* His first and most important work was the novel *Qindīl Umm Hāshim* (The Lamp of Umm Hāshim, 1944), which had symbolist traits. The hero Ismā'īl is the type of Egyptian intellectual who has assimilated modern scientific ideas and culture in Europe and is faced at home with the problem of how to nurture what is valuable in the new, while living amidst the old traditions. He finally concluded that modern science and culture can only take root in Egypt if the native tradition, the cultural heritage and the mentality of the people are respected. From his legal experience Ḥaqqī has drawn material for humorous and serious sketches of the life and problems of ordinary people. His *Fajr al-qiṣṣa al-miṣrīya* (The Dawn of the Egyptian Short Story) describes the cultural atmosphere of Egypt in the 20's, when modern Egyptian literature and the new short story genre were coming into being. Other works: volumes of stories, *Dimā' wa ṭīn* (Blood and Clay, 1955); essays, *Dam'a . . . fa 'btisāma* (A Tear . . . and then a Smile, 1965). JO

al-Ḥarīrī, Abū Muḥammad al-Qāsim b. 'Alī (1054–1122), an important exponent of the Arabic *maqāma* (qv). Rich and well-educated, al-Ḥarīrī was a high-placed government official in Basra, serving the Seljuq sultan, Malikshāh, and his famous vizir, Niẓām al-Mulk. He wrote several books of philological interest, such as *Durrat al-ghawwās* (The Pearl of the River), in which he collected grammatical and dialectical errors of various authors. He is, however, mainly remembered for his *maqāmāt*.

Al-Ḥarīrī is not the inventor of this

genre, imitating the rhymed prose pattern set by al-Hamadhānī, but he made it a model of Arabic linguistic excellence. A virtuoso of the polished style, al-Ḥarīrī was certainly a consummate master of Arabic. His virtuosity is expressed in striking alliterations, ingenious use of the letters of the alphabet and philological devices, and these are evident in most of his fifty *maqāmāt*. Consequently, they are difficult to understand, and several commentaries were published on them during the 12th and 13th centuries. The hero of all fifty *maqāmāt*, and the real connecting link between them, is the fictional character, Abū Zayd of Sarūj (a town in northern Iraq). A prototype of the vagabond living by his own wits, this rogue invariably succeeds in duping the native townspeople and others and in fleecing them. A master at confidence tricks, Abū Zayd is an accomplished speaker and actor, with a truly marvellous mastery of Arabic. Amoral in his trickery, but magnanimous towards the poor, a clever combination of Robin Hood and Scapin, Abū Zayd undoubtedly has the sympathy of al-Ḥarīrī and probably of his readers.

Trans.: The *maqāmāt* of al-Ḥarīrī have been translated, partly, into English, French, Turkish, Persian and other languages, eg Venture de Paradis, *Les séances de Ḥarīrī* (Stockholm 1964).

FAL 164–70; BGAL 123–6; WILA 177–9.

JML

Haşim, Ahmet (b 1885 Baghdad, d 1933 Istanbul), Turkish poet. He went to Istanbul in 1896, remaining there for most of his life as a professor. He became the leading representative of the *Fecri Ati* group, embracing mainly the generation of poets who followed on after the *Serveti Fünun* group (see Tevfik Fikret). The group issued a manifesto, dated 24.2.1910, declaring their desire to carry on the cultural and literary work of their predecessors, to form closer contacts with Western literatures, especially with the poetry of the French symbolists, and to collaborate with Western literary organizations. The group disbanded in 1912, but Ahmet Haşim did not join any other group later. He introduced the criteria of contemporary western literatures into Turkish poetry, and his work brought the Turkish poetic idiom closer to that of world literatures. He was a sensitive dreamer, an aesthete alienated from his environment and achieving neither success in his career nor happiness in his family life.

His verse was the perfect example of 'la poésie pure', perfect in form and expression. He is more the poet of delicate feeling than of real emotion, a descriptive poet whose best verse is impressionist landscape painting suggestively purveying atmosphere. His love poetry is outstanding, addressed to an imaginary and unattainable love, and so are his melancholy verses remembering his childhood. It is poetry of a musical quality, specializing in soft half-tones, with a characteristic colour-scheme dominated by silver and blood-red; the sense of proportion and sincerity of inspiration are marked. In his endeavour to achieve musicality and colour in his verse, he created an original, personal idiom which is not really part of the colloquial language of his day. He used the Turkish poetic form '*arūḍ* (qv) in various verse forms, most frequently that of the free *müstezad* (a sort of distich with 4+2 foot) adapted to the free verse of the French symbolist poets. His most successful poems were short lyrics of four, three or even two distichs (*qit'a*, qv). Unlike his poetry, Haşim's prose (articles, essays, travel sketches) is clear, exact, masterly in style, and characterized by acute observation, subtlety and wit. His prose also reveals the poet and aesthete in his clear and direct reaction to the problems of contemporary society. Works: poetry, *Mehtapta Leylekler* (Storks in the Moonlight), *Göl Saatleri* (Hours by the Lake, 1921) and *Piyale* (Goblet, 1926, 1928). The two latter volumes appeared in successive editions, and also together in one volume; essays, *Bize Göre* (In Our View, 1928, 1960), *Gurabahane-i Laklakan* (1928); travel sketches, *Frankfurt seyahatnamesi* (Frankfurt Travel Notes, 1933, 1947).

PHTF II 563ff.

DL

Ḥassān b. Thābit (b c563, d c659/669/673), Arabian poet of the pre-Islamic period who was the first well-known poet to follow Muhammad, the founder of Islam. He later served Muhammad as a poet, defending his interests and those of the Muslim community. He gained his experience as a panegyrist at the pre-Islamic courts of the Ghassanids and Lakhmids. After the death of Muhammad he was given an official post under the Caliph Mu'āwiya. Many of the poems included in his *dīwān* (qv) are not authentic (eg some of the elegies, the eulogy on the Anṣār, etc), but they are good examples of the poetry which flourished in the entourage of the Prophet.

BGAL I 37–8, S I 67–8; NHLA 52–4; BHLA 313–6; GSLA 104–5; WILA 41; EI² 271–3. KP

Hātefi, 'Abdollāh (b 1440–50 Kharjerde Jām, d 1521 ibid.), Central Asian Persian poet in Herat. His mother was the sister of Jāmī (qv). Only fragments of his *dīwān* (qv) have been preserved, but five *maṣnavī* (qv) have survived, written on the model of Neẓāmī's (qv) *Khamse*. They are the love *dāstāns* (qv), *Leylī va Majnūn*, which some rate higher than the poems on the same theme by Neẓāmī and Amīr Khosrou Dehlavī (qv), *Shīrīn va Khosrou*; the didactic *maṣnavī Haft manzar* (Seven Panoramas), and the historical *dāstān Zafarnāme* (Book of Victories) or *Tīmūrnāme* (Book about Timur), on the conqueror's fourteenth-century campaigns, which pictures them as bloody and violent destruction. The last of the five, *Fotūḥāte shāhī* (Royal Victories) or *Shāhnāme* (Book of Kings), was not finished, for what reason we do not know. It was promised to the Shāh Esmā'īl, and was written in praise of the Safavi dynasty, but only about 1,000 distichs were completed. JB

Ḥāwi, Khalīl (b 1925), Lebanese poet. He studied philosophy at the American University in Beirut, where he now lectures; he took his doctorate at Cambridge. His field of study has influenced the subject matter of his poetry, which can be called philosophical or intellectual. His work reflects both Arabic and European poetry; the former is represented by the spiritual trend, in mediaeval poetry by al-Mutannabī (qv) and in modern Arabic poetry by 'Umar Abū Rīsha. Of the western philosophers and poets he draws mainly on Shelley, Ezra Pound, T. S. Eliot, Sartre and Nietzsche. External impulses find independent formulation in his poetry, with great pains spent on the choice of figures of speech, the expressiveness of the words, and the close association of the phonic and semantic aspects of verse. His favourite symbol is that of the sailor seeking for the truth, Sindibād; it appears in his first book *Nahr ar-rimād* (River of Ashes, 1961), and later in *an-Nāy wa'r-riḥ* (The Flute and the Wind, 1961). Sindibād from the East weighs both Eastern and Western values, the spiritual values of the East and the philosophy and science of the West, and finds neither capable of giving man the final solution to his problems. That solution lies in rejection of the present for a future renewal of life.

ALAC-P 182; CAR 275. KP

Haykal, Muḥammad Ḥusayn (b 1888, d 1956), eminent Egyptian journalist, novelist and critic. He was one of the younger supporters of Aḥmad Luṭfī as-Sayyid, and gained his early journalistic experience writing for *al-Jarīda* (The Review), Luṭfī as-Sayyid's periodical. After World War I, he was closely associated with the Wafd political party, and became editor of *as-Siyāsa* (Politics), the party's newspaper, which soon achieved a status comparable with the older Lebanese-owned *al-Ahrām* and *al-Muqaṭṭam*. Haykal is perhaps best remembered for his authorship of one of the earliest novels to appear in modern Arabic: this was *Zaynab*, published in 1914, and written while the author was a law student in Paris. Although the book presents a highly idealized and romantic view of life in the Egyptian countryside, it is by no means devoid of serious sociological comment and criticism. RCO

hay[e]ren, poetic form, see **Frik.**

Hazār afsāna (Thousand Tales), Pahlavī book, see **Alf layla wa layla.**

Hazhār (Poor Man), real name Sharafkandī, 'Abdurraḥmān (b 1920 Mehābād), progressive Kurdish poet, writing in the Mukrī dialect. Of a well-educated and profoundly patriotic family, he received his earliest education from his father, in church school in Mehabad and later elsewhere. Since 1941 he has been a member of the *Komeley Zhiāni Kurd* (from 1945 the Democratic Party Kurdistan). He became active in political life after the Kurdish Republic was proclaimed in Mehabad (23.1.1946), emigrating to Iraq after it was defeated. Later he lived in Syria, returning to Iraq in 1957. He has written verse since his youth, publishing in periodicals (*Nishtimān, Kurdistan*). His first volume of verse *Ālakōk* (A Mountain Flower) appeared in Tabriz in 1945 and another volume appeared illegally in Syria in 1958: *Baytī saramar u lassayi sag u mānga shaw* (The Story of the Rams' Heads, and a one-act-play, The Dog and the Moon). Patriotic themes predominate in his verse, in which he uses not only the traditional genres but also the fable, which has no tradition in Kurdish literature. AK

Hedāyat, Ṣādeq (b 1903 Teheran, d 1951 Paris), Iranian writer whose stories have won him a world-wide reputation. Of an old aristocratic family whose members held high office at court, he was meant for a similar career. After leaving the French school in Teheran he went to Europe, spending some time in Belgium and longer in France, to complete his education. These years, 1926–30, determined his future; after some hesitation he concentrated on literature. He enthusiastically read the literature of the West, and made his first attempts at writing in Persian. Returning home, he refused the traditional career, taking only insignificant posts in order to be free to write and to devote his time to cultural pursuits. With three of his friends, M. Farzād, B. 'Alavī (qv) and M. Mīnovī, he organized the famous *Rab'e* group, and their militant efforts to modernize Persian literature and other fields of culture attracted other young people who shared their ideas. The *Rab'e* group won people over not only by their skill in polemic against conservative opinion, but by their own writing. The first half of the 30's up to the visit he paid to Bombay in 1936 represents the most fertile period in Hedāyat's life. Besides his own writing, he devoted himself at that time to to the study of the pre-Islamic period of Iranian culture, especially Pahlavī; he later translated several tracts from Pahlavī into modern Persian. In 1941, after the abdication of Reẓā Shāh, there was a period of relative liberalization; Hedāyat took part in the cultural ferment which ensued, mainly with his contributions to the new literary magazines (especially *Sokhan*, The Word) publishing theoretical studies on literature, translations (Kafka, Sartre, etc) and, more rarely, his own work. In 1950 he went to Paris, where his third attempt at suicide (he had made two while a student) was successful. He is buried in the Père Lachaise cemetery.

His work, the core of which consists of his short stories, reflects his broad interests and his extensive knowledge of European literature as well as the cultural traditions of his own country. He developed further the trend which arose with Jamālzāde's (qv) work. On the one hand he drew his themes from very different social strata, and even from the lowest of the low, the outcasts of society (an environment hitherto untouched by literature) but where he found many unforgettable figures. On the other hand he enriched the language not only by using colloquial speech, but by adapting folk sayings and proverbs, and thus opening up a wealth of new creative possibilities for modern Persian prose. No writer before him had shown such a broad and profound interest as an artist in his fellow-men. He searched out fundamental human relations and emotions, and was unusually bold in his exposure of the hidden motives swaying man, deforming his character, and leaving him no chance to resist or to change the course and significance of events, determined by fate. Many of his heroes, the intellectuals and the simple alike, figures from the far past, the present or the future, all resolve their situation by suicide. This puts Hedāyat's work in the context of a specific attitude to life which draws on certain existentialist aspects of Sufi mysticism (see Sufism), and at the same time in the context of modern world literature. Translation into the major languages of the world has ensured Hedāyat his rightful place in world literature. Main works: volumes of stories, *Zende ba gūr* (Buried Alive, 1930), *Se qatre khūn* (Three Drops of Blood, 1932), *Sāyeroushan* (Twilight, 1933), *Būfe kūr* (Blind Owl, 1937), *Sage velgard* (Stray Dog, 1942), *Hājjī Āqā* (1945); plays, *Parvīn dokhtare sāsān* (Parvin the Sassanian, 1930), *Afsāneye āferīnesh* (Legend of the Creation, 1946).

Trans.: D. P. Costello, *The Blind Owl* (London, New York 1957, New York 1969); Roger Lescot, *La chouette aveugle* (Paris 1953).
V. Monteil, *Sâdeq Hedâyat* (Teheran 1952); KMPPL 137–208; RHIL 410–3. VK

Ḥejāzī, Moḥammad (b 1900 Teheran), Iranian writer. Early private tutoring was succeeded by study in France, and on his return home he became a government official. In recent years he has devoted himself entirely to writing and to public life. He is primarily a novelist, although he also uses shorter literary forms. The story of most of his novels is erotic, and his heroines women of the middle classes. The general tone of the novels is a moralizing one, with a good deal of sentimentality and a tendency to didactic sententiousness, all of which detracts from the balanced literary structure. His language preserves the traditional poetic style, which has helped to make his novels and essays very popular with the broad reading public. His work: novels, *Homā* (1927), *Parīchehr* (1929), *Zībā* (1931), *Parvāne* (1953), *Sereshk* (Tears, 1954); essays, *Ā'īne* (Mirror, 1933), *Andīshe* (Reflections, 1940).

RHIL 409–10; KMPPL 75–8; AGEMPL 149–52. KB

Helālī (also Hilālī or Hilolī), Badroddīn (b c1470 Astarābād, d 1529 Herat) a Central Asian Persian poet of Chaghatay (Turkish) origin, originally of the Herat poets of Jāmī and Navā'ī (qqv) school. He was not recognized at court, and was executed in 1529 on the grounds that he was a Shiite. He wrote three long *maṣnavī* (qv) of which the best known is *Laylī va Majnūn*; unlike the version of the same name by Neẓāmī (qv), Helālī's has a happy ending. *Sefāt ol-'āsheqīn* (Description of Lovers) is a didactic work and the well-known *Shāh o gadā* (The King and the Beggar) is didacto-mystical. His *ghazals* (qv) have made him well-known to the people of Central Asia, and most of them are still sung to folk tunes, as *shashmaqom* (a long instrumental-vocal Tajik form).

RHIL 434–5. JB

Hikmet Ran, Nazım (b 1902 Salonica, d 1963 Moscow), Turkish poet and dramatist. Of an aristocratic Turkish family, he was the grandson of the poet and critic Nazim Pasha, who introduced him to classical Turkish literature at an early age. He was more profoundly influenced, however, by his mother, who had enjoyed a European education and helped him to understand French literature and European music. He was expelled from the Naval College of Istanbul for taking part in a student strike, and after going into hiding moved to Ankara, where he became a communist and took part in the liberation movement. In 1921 he went to Moscow, where his friendship with Mayakovskiy and other Soviet poets left a lasting mark on his work and his personal life. Returning to Turkey in 1924, he wrote for progressive papers, and in 1925 was sentenced to 15 years in prison for his poetry. He went back to the Soviet Union, where he studied and in 1928 published his first volume of verse in Baku. That year he went back to Turkey and collaborated with the journalist family, Sertel, on the periodical *Resimli Ay* (Pictorial Monthly), around which the progressive intellectuals had gathered. In 1937 he was sentenced to 28 years' imprisonment for his anti-fascist *Madrit Kapilarinda* (At the Gates of Madrid). His long term in prison ruined his health, but did not prevent him writing. In 1951 he was finally released, under pressure of world public opinion, and emigrated in the same year. Living in the Soviet Union and other socialist countries, he was active in public life and in the international peace movement. He often visited Czechoslovakia, which inspired some of his works; he took Polish citizenship, and adopted the the name of the family from which one of his ancestors had come: Borzęcki.

In Hikmet's first poems, 1919, the romantic melancholy and artificial language were those of the average verse of the time. After the occupation of Constantinople, however, the poet began rousing the people to fight for their national freedom. Living in Anatolia, he introduced the villagers as the heroes of his verse. The poverty in which they lived convinced him that the call for national independence must go hand in hand with the struggle for social justice. This conviction became clearer in the next period of his work, culminating in the volume of verse *Güneşi İçenlerin Türküsü* (The Song of Those Who Drink the Sun, 1928), joyful praise of work in a free community. Although the struggle for a better world never ceased to be Hikmet's main theme and inspiration, his heroes gradually lost their superhuman perfection and became human beings, ordinary people. He came to the conclusion that the fight for the ideals of all humanity is won only when the least important individual has been freed. He therefore wrote intimate lyric poetry, alongside his verse reacting to political events; however, in conception and impact the lyric poetry is as thoroughly political as all his work, as thoroughly humanist and optimistic. Not until the last period, when he was writing in exile, cut off from Turkish readers and profoundly homesick, did a deep melancholy mark some of his best poems.

Hikmet wanted his verse to speak to the broadest public, and therefore wrote in forms and language that were comprehensible. His prime models were folk songs, the work of the *âşiks* (qv) and the vernacular he got to know during the fighting in Anatolia. He developed a poetic form which is an integral whole. In it the gradation is evident, but the rhythm varies in order to underline the idea in different parts of the poem. He did not avoid colloquial phrasing, and made good use of the Turkish vocal harmony for poetic ends. The collection, *835 Satir* (835 Lines, 1929), written in this manner, aroused great interest. Later, and particularly while in prison, he developed the style still further. His poetry was passed round by word of mouth only, and that meant it had to be simple and effective. At this time he wrote poetry to be heard. This stress on the melodic aspect of the poetic idiom gives a remarkably poetic character to his otherwise realist and civilian themes.

His first play was published in 1932; he also wrote plays in prison, and later returned to the genre more and more frequently in an attempt to communicate with ordinary people all over the world. His first novel dealt with the fight against colonialism, and shortly before his death he again turned to prose, with the novel, *The Romantic*, a portrait of Turkey in the 20's. Although the works he wrote in exile and in prison were outside the current of Turkish literary life, his earlier work left its permanent mark on the next generation of writers and was an important stage in the development of Turkish poetry. Hikmet also exercised a strong influence on the literature of Azerbaijan and other Turkish peoples in the Soviet Union. Principal works: verse, *Anadolu* (Anatolia, 1922); *Seçilmiş Yazilar* (Selected Works, 1937); *Varan 3* (The Existing 3, 1930); *1+1=1* (1930); *Taranta Babu'ya Mektuplar* (Letters to T. Babu, 1935); *Simavne Kadısı Oglu Şeyh Bedreddin Destani* (Destan on Bedreddin, Son of a Judge from Simavne, 1936); novels, *Benerci Kendini Niçin Öldürdü*? (Why Did Bernerci Die? 1932); plays: *Kafatası* (The Skull, 1932).

Trans.: Taner Baybars, *Selected Poems* (London 1967); *Anthologie poétique* (Paris 1964). ZV

Hilāli (Tajik **Hiloli**), Badruddīn, Persian poet, see **Helāli,** Badroddīn.

Hodayot (Songs of Praise), the most extensive of the Essene Hebrew writings found in the Qumran caves by the Dead Sea. Of the greatest literary value, they comprise exhortations and meditations in the manner of the Old Testament *Psalms* (qv). The author of the collection, or at least of part of it, is believed to have been the Teacher of Righteousness, founder of the Essene community, active in the second half of the 2nd century AD. The *Hodayot* are not as original as the *Psalms*, but use traditional imagery to give effective expression to the conflicts, difficulties and hopes of the disturbed time after the Wars of the Maccabees. There are similar poems in other longer writings found in Qumran, in the *Manual of Discipline* and the *Manual of War*. Of literary interest, too, is the fragment of the *Book of Mysteries*; but most of the prose writings, topical interpretations of the *Old Testament* (qv), orders, rules and liturgical texts do not appear to have been written with literary ambitions.

Trans.: Th. H. Gaster, *The Dead Sea Scriptures* (Garden City 1956); A. Vermes, *The Dead Sea Scrolls in English* (Harmondsworth-Baltimore 1962). SZS

Hosea (late 8th century BC), Israelite prophet. The description of unhappy marriage with which his prophecies open may have been drawn from his own life; adultery is for him the symbol of Israel's faithlessness towards God. Hosea inveighed against heathen ways and the kingship that imitated them. The prophesies of Hosea are remarkable for the feeling and even passion with which they are imbued. His northern Israelite origin can be seen in his language, which differs from the Jerusalem Hebrew of the other books in the Bible (qv). See also Twelve Lesser Prophets.

G. A. F. Knight, *Hosea* (London 1960); J. L. Mays, *Hosea* (London-Philadelphia 1969). SZS

Ḥoseyn Vā'eze Kāshefī, real name Ḥoseyn b. 'Alī Beyhaqī (d 1505 Herat), Persian poet of the Herat circle in the late fourteenth century. He spent his youth in Sabzavar and Mashhad and came to Herat in 1456, where he knew 'Abdorraḥmān Jāmī and 'Alīshīr Navā'ī (qqv). He was a theologian, astrologer (eight works on this subject), and general writer. Of his literary works we know *Sayāhatnāmeye Hāteme Tā'ī* (The Travels of Hāteme Tā'ī, 1486), fairy-tales about that legendary hero; the didactic *Fotūḥatnāmeye solṭānī* (Royal Book of Victories); works on literature and model letters; the important *Akhlāke Moḥsenī* (The Morals of the Beneficient, 1496), comprising forty chapters of prose and verse dedicated to Abū 'l-Moḥsen, the son of Ḥoseyn Bāyqarā. His best known work is *Anvāre Soheylī* (The Shining Star Canopus, 1504), a version of the story of *Kalīle o Demne* (qv) in the highly artificial style favoured in his day.

Trans.: H. G. Keene, *The Morals of the Beneficient* (Herford 1850); A. N. Wollaston, *The Anwar-i-Suhaili, or Lights of Canopus* (London 1877).
RHIL 313–14. JB

Hovhannisian, Mikhayel, Armenian novelist, see **Nar-Dos.**

Hüsäin, Mahti (b 1909 Shikhly, d 1965 Baku), Azerbaijan writer, dramatist and literary historian. The son of a village teacher, he graduated from the Kazakh Teachers' Training College, the Historical Faculty in Baku and the Scenario Faculty in Moscow, and became a journalist. He was active in political life and in the Union of Azerbaijan Writers. Writing for the Azerbaijan press he began with stories of village life during the collectivization campaign. After writing novels about the civil war in Azerbaijan, village life and the heroes of World War II, he achieved outstanding success with his novel, *Apsheron* (Apsheron Peninsula), about the oil-field workers. His plays deal mainly with historical subjects, and he also wrote a number of critical articles and studies. Novels: *Kin* (Hatred); *Färyad*; *Tärlan* (1937); *Dashgyn* (Flood, 1934–36); *Ölkäm* (My Land, 1947); *Vätän chichäkläri* (Blossoms of the Fatherland, 1943); *Sähär* (Dawn, 1956); *Komissar* (1953); *Gara dashlar* (Black Stones, 1958); dramas, *Nizami* (1940); *Javashir* (1943); etc.

PHTF II 690. ZV

Husrav I Kavātān U Rētakē (Husrav, son of Kavāt, and his Page), a Middle Persian seventh century text, light reading which presents a conversation between the king and his young high-born attendant. The youth answers all the king's questions faultlessly, on many different subjects, and then performs a heroic deed to show his valour; he thus gains a high position at court. The unknown author gives us a picture of the ideal courtier, as well-informed about religious dogma as about the art of cooking, not to speak of the knowledge possessed by every true knight.

RHIL 46; TMSLZ. OK

I

Iashvili, Paolo (b 1894 Argveti, d 1937), Georgian poet. He began publishing verse as a student; in 1913 he went to Paris to study at the art school attached to the Louvre. He got to know poets and painters round Picasso and Apollinaire, and began translating French poetry. On his return to Georgia he and V. Gaprindashvili formed the Blue Horns group. After Soviet rule was established in Georgia (1921), a development he welcomed heartily, he held various official posts until in 1937 he was accused of nationalism and activities against the interests of the people, and committed suicide. His work is not voluminous; no volume of verse appeared in his lifetime, and his poems are scattered in periodicals. Even so he was of great importance for the development of modern Georgian poetry. After his posthumous rehabilitation in 1955 several selections of his poetry were published, both in the original and in Russian translation.

Trans.: UAGP. VAČ

Ibn 'Abd Rabbihi, Aḥmad b. Muḥammad (b 860, d 940 Córdoba), Arabic Andalusian writer and poet. He was court poet to the Spanish Umayyads, excelling in love poetry rather than in panegyric. His real fame is based on his extensive anthology of Arabic poetry, prose and scientific knowledge entitled *al-'Iqd* (The Necklace), later called *al-'Iqd al-farīd* (The Unique Necklace), divided into 25 chapters called after precious stones. It is an encyclopaedia, as it were, of the knowledge which might be considered useful to a well-educated man, containing all that the standards of the time designated as general culture.

BGAL I 154, S I 250–1. IH

Ibn Bājja, Abū Bakr Muḥammad b. Yahyā b. aṣ-Ṣā'igh, known as Avempace (b Saragossa, d 1138 Fez), the philosopher of Muslim Spain; he was also a physician, held public office, and studied astronomy. The most important of his works, *Tadbīr al-mutawaḥḥid* (The Regimen of the Solitary), was translated into Hebrew in the Middle Ages. The main idea of the book is the communion of the human reason with the divine cosmic reason (*Intellectus agens*), through which (as all Arab philosophers thought) man could develop his rational ability to the utmost. Ibn Bājja describes in this context a utopian and absolutely perfect society formed by individuals who have achieved the highest level of reason, but who, in our imperfect world, remain isolated and alien.

Miguel Asín Palacios, *Avempace: El regimen del solitario*, edición y traducción (Madrid-Granada 1946); S. Munk, *Mélanges de philosophie juive et arabe* (Paris 1955); CHPI 317–25, 361–2. VSD

Ibn Baṭṭūṭa, Muḥammad b. 'Abdallāh (b 1304 Tangiers, d 1368/9 Morocco), Arab traveller. In 1325, having acquired the usual legal and theological education, he set out on a pilgrimage to Mecca, in the course of which he determined to devote his life to exploring lands near and far. In the following two years he travelled extensively in Arabia, Mesopotamia and Azerbaijan, and then spent three years in Mecca. From here he sailed down the Red Sea to Yemen and Aden, covering the east African coast as far as Kilwa, and returned via Oman and the Persian Gulf to Mecca. In 1332 he set out through Egypt and Syria to visit Asia Minor, and after travelling back and forth throughout the land he crossed to the Crimea and the kingdom of the Golden Horde in southern Russia. In the suite of a Byzantine princess he arrived in Constantinople, and then returned to the capital of the Golden Horde. After travelling across Central Asia and Afghanistan, he arrived at the valley of the Indus in 1333 (according to his own, not very reliable, dating); from there he went to Delhi. There Ibn Baṭṭūṭa spent seven years as a judge in the services of the Sultan; in 1342 he was appointed ambassador to China, but it was several years before he got there; he was travelling all over southern India, and spent some time on the Maldive Islands and visiting Ceylon and Bengal. He then sailed to Sumatra, Java, and southern China, travelling overland to Peking. From China he did not return to India but to Baghdad, via the Persian Gulf; arriving in 1347, he went on a pilgrimage to Mecca, and thence via Tunis and Sardinia back home to Morocco (1349). That same year he visited southern Spain, and later crossed the Sahara to the kingdom of Mali in the western Sudan, where he remained for a year. Returning to Fez late in 1353, he spent the rest of his life somewhere in Morocco as a judge.

In Fez he dictated his memoirs to a local man of letters, Ibn Juzayy, who adorned the simple travel notes with poetic interpolations and descriptions in rhymed prose, and gave the work the pompous title *Tuḥfat an-nuẓẓār fī gharā'ib al-amṣār wa-'ajā'ib al-asfār* (A Gift to Those Interested in the Curiosities of the Cities and Marvels of the Ways), but it is generally known as *Riḥla* (Travels). Ibn Baṭṭūṭa was the greatest traveller of the Middle Ages, the extent of his journeying being surpassed only in modern times. His descriptions of the countries visited, as well as the record of his adventures and experiences, make his book an important document for the social and cultural history of the Islamic world in the 14th century, and a first-class source for the history of India, the Maldives, Asia Minor and West Africa. His information is reliable; some inaccuracies and contradictions are to be found in his chronology, but they are not difficult to resolve. His *Travels* enjoyed great popularity in the Islamic world as well as in Europe in modern times, having been translated (either in full or in selections) in most of the major languages.

Trans.: H. A. R. Gibb, *Ibn Battúta, Travels in Asia and Africa* (London 1929); Sir H. A. R. Gibb, *The Travels of Ibn Battuta*, vols. I–III (Cambridge 1958, 1961, 1971).
H. F. Janssens, *Ibn Batoutah, Le Voyageur de l'Islam* (Brussels 1948). IH

Ibn al-Fāriḍ Sharaf ad-Dīn'Umar (b 1181 Cairo, d 1235 ibid.), Arab poet and mystic. He lived as an ascetic, spending long periods at the holy places (about 15 years in Mecca). His mystical verse made full use of the symbolism of physical love, and the praises of wine; his originality lies in this ambiguity. The allegories which fill his poems are not always clear. The language is difficult, and full of poetical tricks. The best known of his poems is the *qaṣīda* (qv) *at-Tā'īya al-kubrā* (Long Poem Rhyming in the Letter *Tā'*), describing his mystical experiences. Numerous commentaries have been written on this poem.

Trans.: R. A. Nicholson, *Studies in Islamic Mysticism* (Cambridge 1921), 199–266.
BGAL I 262, S I 462. SP

Ibn Danyal, Turkish writer, see **Karagöz.**

Ibn Fāris, Abu'l-Ḥusayn, known as al-Qazwīnī ar-Rāzī (d c1004 Rayy), Arab philologist. He studied in Qazwīn and became famous as a teacher in Hamadān, where the famous author of *maqāmāt* (qv), al-Hamadhānī, was among his pupils; later he lived at the court of a Buwayhid prince in Rayy. Although he was probably of Persian descent himself, he was convinced of the superiority of the Arabic language and defended it against Persian critics, with many effective arguments. His work dealt with lexicography, grammar, poetry, jurisprudence, the lives of the Prophet and other matters. His interest in linguistics was roused particularly by semantics and a broad field of linguistic speculation. His work comprises about forty texts, of which only roughly half have survived, and only some of which have been printed. His place in the history of Arabic philology is secured principally by his dictionary and his writings on Arabic.
LK

Ibn Ḥazm, 'Alī Aḥmad (b 994 Córdoba, d 1064 near Casa Montija), Arab Andalusian poet, historian and scholar, one of the greatest thinkers of mediaeval Muslim civilization. As a young man he began a political career and became vizier, but after spending several years in prison he led a life of retirement, devoted to poetry and various fields of Islamic scholarship. He is the author of many works on Islamic law, interpretations of the *Qur'ān* (qv), works on psychology, practical ethics, history, philosophy, politology and comparative religion. His most important work is *Ṭauq al-ḥamāma* (The Dove's Necklace), a treatise on love and lovers, full of deep psychological insight, with much autobiographical detail and examples taken from Arabic love poetry.

Trans.: R. A. Nykl, *A Book Containing the Risala Known as the Dove's Neckring About Love and Lovers* (Paris 1931); Léon Bercher, *Le collier du pigeon, ou, de l'amour et des amants* (Alger 1949); A. J. Arberry, *The Ring of the Dove* (London 1953).
I. Friedlaender, *The Heterodoxies of the Shiites according to Ibn Hazm* (New Haven 1909); EI[2] III 790–9. IH

Ibn Hishām, Arab writer, see **maghāzi.**

Ibn Khaldūn, Abū Zayd 'Abdarraḥmān b. Muḥammad (b 1332 Tunis, d 1406 Cairo), Arab historian, philosopher of history and sociologist. Born of a family of Andalusian emigrés to Tunis, he received a thorough education in literature, law and theology, and early took an active part in political life, holding the office of minister at various North African courts. He retired in 1374 to devote himself to the writing of his historical work, which he completed in 1382, making additions later. In that year he went to Egypt, where he alternated the profession of teacher with that of supreme Malikite judge. In 1401 he met the conqueror, Tamerlan, in Syria, but refused to enter his services. Ibn Khaldūn left a full account of his active and varied life in an autobiography.

His principal work, *Kitāb al-'Ibar wa-dīwān al-mubtada' wa-l-khabar fī ayyām al-'Arab wa-l-'Ajam wa-l-Barbar* (Book of Instructive Examples and Register of the Beginning and the Account of the History of the Arabs, Persians and Berbers), is a universal history with special emphasis on North Africa and the Near East; particularly the chapters devoted to the history of the Berbers are remarkable for their originality. The long Introduction to this work (*al-Muqaddima*) represents the first systematic attempt to discern the laws governing the evolution of society and culture.

Ibn Khaldūn's approach is strictly objective and scientific, being based on the social and economic factors inherent in social organization; he also points out the influence of climatic and geographical factors. He differs from his contemporaries in rejecting the teleological concept of history and divine intervention in its course. He stresses the deterministic factors in the historical process, showing that similar conditions produce similar results. The rise of a civilization and culture in its various forms (Ibn Khaldūn distinguishes nomad culture, settled agricultural society, and urban civilization) is dependent on man and the human activity necessary for feeding and defending the group, and for aggressive action. A crucial role is played also by group solidarity, which, at its most developed stage, leads to the emergence of dynasties and states; this solidarity he found most marked among the nomads and inhabitants of the steppes, who periodically overflow into more fertile regions and found dynasties, but in the course of three generations succumb to the softer settled life and disappear, being superseded by a new wave of invaders from the steppes. Yet Ibn Khaldūn was well aware that the harmonious all-round progress of civilization is only possible among settled peoples.

Ibn Khaldūn further presented a penetrating analysis of state power, warfare, the branches of science, and various forms of economic activity; here it is noteworthy that he was the first to stress that human work is the only producer of value. His comments on education, language and literature are also full of interest. His concept of history was naturally restricted to his experience of the Muslim world, but a number of his ideas are of a wider and universal validity. Ibn Khaldūn was one of the greatest figures of Islamic and world culture, and his genius anticipated many ideas that only became apparent in the 19th century; he is considered the founder of sociology and of the philosophy of history. In the Islamic world, already in decline in his time, his ideas were not followed up; Europe only became acquainted with them as late as the 19th century.

Trans.: F. Rosenthal, *The Muqaddimah—an Introduction to History*, 3 vols. (New York 1958).
N. Schmidt, *Ibn Khaldun, Historian, Sociologist and Philosopher* (New York 1930); W. J. Fischel, *Ibn Khaldun and Tamerlane* (Berkeley 1952); M. Mahdi, *Ibn Khaldun's Philosophy of History* (London 1957); M. A. Enan, *Ibn Khaldun, His Life and Work* (Lahore 1941). IH

Ibn Khallikān (1211–1282), Arab biographer, of a respectable Arbela family. After studying there, in Aleppo and in Damascus, he was a lawyer in Cairo until his nomination as Chief Qadi in Damascus. Losing this position later, he returned to Cairo and became a professor. Shortly before his death he was again nominated Chief Qadi for Syria. His main work, *Wafāyāt al-a'yān wa anbā' abnā' az-zamān* (The Book of the Deceases of Notabilities and of the Reports on the [Illustrious] Son of Time), comprises over 800 biographies of well-known and famous men from the political, cultural and literary life in Islam. It is one of the most widely used sources of Arab cultural and literary history. Ibn Khallikān was most fortunate in his choice of the wealth of material at his command; written in simple classical Arabic, his selection is most informative and well-balanced as a source of knowledge and understanding of the Abbasid period in Islamic society.

Trans.: M. G. de Slane, *Ibn Khallikān's Biographical Dictionary* (Paris 1842–71, published in several editions).
BGAL²I 398ff, S I 561; EI² III 832–3. RS

Ibn al-Muqaffa' (720–757), Arab writer and translator. From a highborn Persian family of the western province of Fars, he was called 'Son of the cripple' after his father, a tax collector, whose hands had been severely injured by torture. Like many other influential Persian families they seem to have sympathized with the Abbasid and Alid factions against the Umayyads; in any case immediately after the latter's fall (750) we find him in the post of government secretary in Iraq. He was one of those in positions of influence who tried to make a contribution of his own to bring the Arabs and Persians closer together. An educated, cultured Persian, he was aware that his own people were at a higher stage of civilization than the Beduin Arabs in the past, and therefore, with the consent, if not at the instigation, of the second Abbasid Caliph, al-Manṣūr, he wanted to preserve at least the memory and evidence of those great days before the theocratic Islamic Arabo-Persian state.

He therefore began to translate from Pahlavi (Middle Persian) into Arabic; most important was his translation of *Kalīla va Dimna* (see *Kalīle o Demne*), which was translated from Arabic into the most important European as well as Oriental languages in the Middle Ages. In his translation Ibn al-Muqaffa' treated the text freely, enlarging the introduction and in places the text. His version soon became popular reading, and was often issued in enlarged or abbreviated form, and even in verse. In view of the relatively recent date of the existing manuscripts, it is difficult to restore his original text.

Of great significance for later works on Arabian and Persian history is Ibn al-Muqaffa''s translation of a chronicle of a pre-Islamic Persian king (*Khvatāy Nāmak*, qv), known only in fragmentary form. With similar works, it served as material for histories of the period, and was also used by the poet Ferdousī (qv) for his famous *Book of Kings* (*Shāhnāme*). Besides these and other Persian works made accessible to the Arab-Islamic world for the first time in masterly translations, Ibn al-Muqaffa' himself wrote in Arabic, among other works a 'great' *adab* (qv), dealing not only with aristocratic ethics but also with the moral principles and rules to be followed by different social classes; and an extremely interesting pamphlet *ar-Risāla* (The Epistle) dedicated to the Caliph al-Manṣūr, setting out for discussion religio-political and social questions and the ideas current in his time. This example of his free-thinking may even have contributed to his tragic early death; as a dualist he had been converted to Islam, and the Caliph, under the pretext of heresy, had him tortured to death for base personal and political motives.

BGAL² I 158, S I 233–7; EI² III 883–5. RS

Ibn al-Mu'tazz, Abu'l-'Abbās 'Abdallāh (b 861 Sāmarrā, d 908 Baghdad), Abbasid Arab poet, rhetorician and anthologist. The son of the thirteenth Abbasid caliph, he was orphaned early in life and was brought up by his grandmother, who entrusted his education to leading scholars including the philologists al-Mubarrad and Tha'lab. Political intrigues dictated that he lived at court in retirement. Finally he allowed himself to be proclaimed Caliph under the name of al-Muntaṣif B'illāh, but was murdered within the day. Extremely well versed in ancient models, he was also in his own poetry a leading exponent of the 'new style', showing a preference for shorter and lighter metres, for elegance and artifice. His compositions include dutiful encomia and elegies about Abbasid princes, and haughty rebukes of Shī'ah sympathizers. An account of the achievements of al-Mu'taḍid in 417 couplets of *rajaz* is worth noting because of the paucity of narrative verse in Arabic, but it does not achieve the stature of an epic.

Most highly prized were his descriptive pieces, which abound in striking and pretty, if sometimes over-ingenious, imagery. The same aestheticism is reflected in his prose. In his *Kitāb al-badī'*, he set out to demonstrate that the rhetorical devices which his contemporaries called 'new' occurred in sacred texts and in Bedouin language; he then listed and illustrated seventeen such 'embellishments' of style. Although his exposition is scarcely systematic, the work is nevertheless an early landmark in the imposing development of Arabic rhetoric, in which the word *badī'* eventually became a technical term for 'tropes'. His *Ṭabaqāt ash-shu'arā' al-muḥdathīn* consists of biographical notices and anecdotes about, as well as quotations from, poets who had praised his dynasty. It is a valuable source-book for the period, and one of the earliest examples extant of the *Ṭabaqāt* type of work, which is largely anthology but also, in part, embryonic literary criticism. PC

Ibn Qutayba (828–889), Arab philologist and scholar. Coming from Kufah, he studied philology, theology and the traditional sciences under the best scholars of the time in Iraq, before taking up the post of *qadi* (state inspector) in northern Persia and in Basra. Favoured by the court, he moved to Baghdad, where he was considered the founder of the 'mixed' or eclectic school of grammar. His works show the many sides of his talent; first among them is his handbook for secretaries (*adab*, qv) based on his own practice, and prefaced by a famous cultural and political introduction. He compiled two extensive anthologies, *Kitāb ash-shi'r wa'sh-shu'arā* (The Book of Poetry and Poets), *Kitāb al-ma'ānī al-kabīr* (The Great Book of Poetry), which are informative for the poet's biography and for aspects of rhetoric. His compendium of literature and science, *'Uyūn al-akhbār* (The Sources of Information) and his history book, *Kitāb al-ma'ārif* (The Book of Learning) had a lasting influence on the *adab* form of literature. His works on the *Qur'ān* (qv), *Tafsīr gharīb al-Qur'ān*, *Ta'wīl mushkil al-Qur'ān* (Qur'ānic Exegesis) and on theology, *Ta'wīl mukhtalif al-ḥadīth* (Theological Interpretations), written from the orthodox Muslim standpoint, are highly informative about the early Abbasid period.

I. M. Huseini, *The Life and Works of Ibn Qutayba* (Beirut 1950); G. Lecomte, *Ibn Qutayba, l'homme, son oeuvre, ses idées* (Damascus 1965, with a bibliography); BGAL² I 124–7, S I 184–7; EI² III 844–7. RS

Ibn Rushd, Abu'l-Walīd Muḥammad b. Aḥmad b. Muḥammad, known as Averroes (b 1126 Córdoba, d 1198 Morocco), the philosopher of western Islam whose work marks the culmination of Aristotelian philosophy written in Arabic. He was a legal authority and physician in Andalusia and in Morocco, but in the end his teachings were condemned and he was sent into exile. His works have survived in the original and in Hebrew and Latin translations. Ibn Rushd regarded Aristotle as the greatest authority in philosophy. His main works are commentaries on Aristotle, but he also wrote a number of original philosophical works, among them *Tahāfut at-tahāfut* (Destructio destructionis), refuting the anti-philosophical notions of al-Ghazzālī (qv). Ibn Rushd's philosophy follows Aristotle's teachings very closely; it embraces a number of theses which could hardly be held to be compatible with the principles of Islam, such as the belief in the perpetual duration of matter, life after death seen as the survival of the reason only, and in a collective but not individual form (monopsychism). Ibn Rushd explained the relation of religion to philosophy by the theory of 'two truths' and the existence of a hidden philosophical truth in religious texts. Philosophy and religion do not conflict; the latter is for the common people, while philosophy is for the scholar, whose duty it is to elaborate his philosophical ideas. The theory of 'two truths' was radicalized by the European Averroists, who believed there were two separate and even conflicting realms, that of theology and that of philosophy. Opposition to scholasticism grew up from these ideas in France and Italy, attributing to Averroes many sharp attacks on religion (eg *De tribus impostoribus*). The Christian scholastics rejected the teachings of Averroists but revered Ibn Rushd as a scholar versed in Aristotle, and a commentator of his work. Ibn Rushd also exerted an influence on mediaeval Jewish philosophy.

E. J. J. Rosenthal, *Averroes' Commentary on Plato's Republic* (Cambridge 1956); G. F. Hourani, *Averroes on the Harmony of Religion and Philosophy* (London 1961); S. van den Bergh, *Averroes's Tahāfut al-Tahāfut* (The Incoherence of the Incoherence) (London 1954); S. van den Bergh, *Die Epitome der Metaphysik des Averroes* (Leiden 1924); E. Renan, *Averroës et l'averroïsme* (Paris 1925); L. Gauthier, *La théorie d'Ibn Rochd sur les rapports de la religion et de la philosophie* (Paris 1909); M. Horten, *Die Hauptlehren des Averroes nach seiner Schrift 'Die Widerlegung des Gazalī'* (Bonn 1913). vsd

Ibn Sīnā (Ebne Sīnā, Avicenna) Abū 'Alī (b 973, 975 or 979 Afshana, d 1037 Hamadan), Central Asian philosopher and universal scholar writing in Arabic and Persian. In his autobiography of the first forty years of his life he writes that he was born in Afshana, Transoxania, where his father who had come from Balkh was a government official. Later his parents took him and his brother who was five years younger to Bukhara where his remarkable intelligence and unusual powers of memory enabled him to complete his schooling (the *Qur'ān*, qv, literature) at the age of ten. He then studied law, the natural sciences, and above all philosophy; his genius was already apparent, and at seventeen he was so famous that the doctors attending the sick ruler of Bukhara recommended him to the young scholar's care. This gave Ibn Sīnā access not only to court, but also to the unique court library which included a large number of works translated from the Greek. On the death of his father Ibn Sīnā entered government service for a time, leaving it when the death of the ruler brought disturbed times to the country. For years he then moved from one princely court in Persia to another, as doctor, adviser, or even minister. In this changeful life he was not always admired and respected. A man of genius, but certainly not easy to get on with, he often brought anger and hatred down on his head; he had to flee overnight more than once and he even experienced imprisonment. Not until he settled in Isfahan did he find the peace and quiet necessary to devote himself almost entirely to scholarship. On a short visit to Hamadan, with his admirer the prince of Isfahan, he died of a colic which had been troubling him for many weeks.

The work of Ibn Sīnā, the first bibliography of which comes from one of his disciples, determined for centuries the whole system of Islamic scholarship. He distinguished between essence and existence; what is possible and varied is found in the essence, which logically precedes existence. His view of the world, with its forms and elements, and with its substance arising from the movements of the spheres, is based on the neo-Platonic tradition. Regarding revelation as allegory expressing intelligible truths, and the prophets as essential monitors in a universal determinism, Ibn Sīnā managed to combine his philosophy with Quranic theology.

His extensive works have not come down to us complete, nor free from pseudepigraphs. They reveal him primarily as a philosopher and doctor, but also as a scholar versed in mathematics, astronomy, physics, chemistry and music. His treatment of political and economic questions shows that he was no mere theorist, but also a man of broad practical experience. His principal philosophical works are *Shifā'* (The Book of Healing, ie the soul), a philosophical encyclopaedia; *Najāt* (The Book of Salvation, ie from error) and *Ishārāt* (The Book of Directives and Remarks). His book of *Impartial Judgment* between the Easterners and the Westerners and his *Eastern Philosophy* are known only in fragmentary form. Of his medical writings, the *Qānūn* (Canon of Medicine) became a famous textbook in the West; under the Latinized form of his name, Avicenna, Ibn Sīnā was regarded as the 'Arabic Galen', and his work was translated up to the seventeenth century. His minor works deal with exegesis of the *Qur'ān* (qv), metaphysics, religious questions and Sufism (qv); it is of interest that some subjects, eg logic and medicine, he attempted to treat in verse. In Persian Ibn Sīnā wrote *Dāneshnāme* (The Book of Knowledge), a treatise on logic, metaphysics and physics which was translated (partly) into French in 1955, and (wholly) into Russian in 1957. From Ibn Sīnā's lyrics only a few *robā'īs*, *ghazals* and a *qaṣīda* (qqv) are extant, but he exercised influence on the whole of Persian mysticism by his Sufi writings.

BGAL[2] I 589–99, S I 812–28; EI[2] s.n.; A. M. Goichon, *Introduction à Avicenne: Son épitre des définitions* (Paris 1933); A. M. Goichon, *La philosophie d'Avicenne et son influence en Europe médiévale* (Paris 1944); *Avicenna, Scientist and Philosopher, a Millenary Symposium* (London 1952); RHIL 179–81. RS

Ibn Taymiyya, Taqī ad-Dīn Aḥmad (b 1263 Harran, d 1328 Damascus), Arab Islamic theologian and jurisconsult. He taught theology and law and was active in public life in Syria and Egypt during the time of the Mongol threat from the East. He was a strict adherent to the Ḥanbalī school of legal thought; he systematically denounced the rationalist position of the theological school of al-Ash'arī (qv) as well as a number of innovations in religious life and thought, such as the cult of saints, some of the ideas and practices of the mystics and Greek logic, all of which he regarded as heretical. His strict ideas and sharp attacks on his opponents landed him in prison more than once, in Cairo, Alexandria and Damascus. Although he was persecuted and indeed died in prison, his influence remained considerable for many generations. His doctrinal position might be defined as conservative reformism. He emphasized the need to follow the *Qur'ān* and *ḥadīth* (qqv) and found the ideas of the Prophet's Companions preferable to the later formulations of the legal schools. He did not accept the view current in his time that independent interpretation of the Sacred Law was not permissible since the establishment of the legal schools. These leading ideas in his complex system became important again later, and gave rise to the Wahhābī revivalist movement in 18th century Arabia. They are not without echo in the main lines of reformism in modern Islam in the Arab countries today. Ibn Taymiyya's work was very extensive; its greater part has been published in modern times. His chief theological work, *al-Wāsiṭiyya; Kitāb as-siyāsa ash-shar'iyya*, a treatise on juridical policy, was translated into English by Omar A. Farrukh (Beirut 1966).

H. Laoust, *Essai sur les doctrines sociales et politiques d'Ibn Taymiyya* (Cairo 1939), and *Contribution à une étude de la méthodologie canonique* (Cairo 1939). LK

Ibn Ṭufayl, Abū Bakr Muḥammad b. 'Abd al-Malik (b Cadiz, d 1185 Morocco), Arab philosopher in Andalusia and Maghreb. He also practised medicine, held official positions in the state, and was a friend of Ibn Rushd (qv). Ṭufayl's most important work was the philosophical romance, *Ḥayy b. Yaqẓān* (Alive the Son of Awake), modelled on the allegorical philosophical writings of Ibn Sīnā (qv), from whom he also borrowed the title of his romance and the names of his characters. The hero of the book lives alone on a deserted island, achieving metaphysical and scientific knowledge by his own efforts. Towards the end the allegory passes from an apotheosis of the power of human reason to a more pessimistic note; leaving his island to preach his philosophy to mankind, the hero returns disillusioned by failure, and convinced that only the few can understand philosophy; for most people traditional religion is adequate (see Ibn Bājja). Ibn Ṭufayl was known to the scholastics as Abubacer (from Abū Bakr) and his romance was translated into Latin as early as the 17th century (*Philosophus autodidactus*) by Pocock, and soon after by Ockley into English.

A. S. Fulton, *The History of Ḥayy Ibn Yaqzan* (London 1929); L. Gauthier, *Ibn Thofaïl, sa vie, ses oeuvres* (Paris 1909); *Hayy ben Yaqdhân roman philosophique d'Ibn Thofaïl*, texte arabe et traduction française (Paris [2]1936). VSD

Ibn Yamin, Persian poet, see **Ebne Yamin.**

Ibrāhīm, Ḥāfiẓ (b 1870 Dayrut, d 1932), Egyptian poet. After graduating from the Cairo Military Academy, Ḥāfiẓ saw war service in the Sudan under Kitchener. His military career ended under a cloud after his involvement in an army mutiny. After some years of unemployment he became director of the literary section of the Egyptian National Library in 1911, a post which he held until shortly before his death. Ḥāfiẓ is remembered primarily as a poet, but was also a prose writer and translator: his book *Layālī Saṭīḥ* (Evenings with Saṭīḥ) is similar in form and content to the writings of Ibrāhīm and Muḥammad al-Muwayliḥī (qv), and he translated most of Victor Hugo's *Les Misérables.* As a poet he belongs to the generation of Shawqī and Muṭrān, and the work of al-Bārūdī was a major formative influence. His verse is famous for its fervent patriotism, and passionate concern for social and political issues. One of his most famous poems is that written about the Dinshawāy incident in 1906.

M. A. Khouri, *Poetry and the making o, modern Egypt* (Leiden 1971). RCO

al-Ibrāhīmī, Bashīr (1889–1965), Algerian thinker. He studied in the Arab East, and was influenced by Jamāluddīn Afghānī and Muḥammad 'Abduh (qqv). One of the founders of the Algerian reform movement, he led the '*ulamā*' association in Algeria from 1940. Apart from his organizational work, he set up schools and wrote for the press. His articles and speeches were famous for their brilliant style, poetic improvisation, subtle humour and irony. He was one of the foremost scholars in the renaissance of Arabic culture in Algeria.

SP

Ibrahimov, Mirzä (b 1911 Eva in Iranian Azerbaijan), Azerbaijan writer, dramatist and literary critic. An orphan, he worked as a servant, and later in a factory. After the revolution he studied in Baku and became a journalist. Graduating in literature at Leningrad University, he became director of the Baku opera, President of the Union of Azerbaijan Writers, Member of the Soviet, and holder of many state awards. He began writing in the thirties. His poems, stories and plays describe the life of the oil workers in peace and war, their fraternal feelings towards other Soviet peoples and towards Spain in her fight against fascism, the transformation of Azerbaijan women and life in collective farm villages. His novel, *Gyälädjäk gyün* (The Day Will Come), about life and the struggle in Iranian Azerbaijan is one of the finest works of Azerbaijan Soviet literature, and has been translated into many languages. He also writes on Azerbaijan literature and on the theory of literature, and publishes translations from Russian and world literature. Volumes and stories, *Azad* (1949); *Salam* *sänä, Rusiya* (Greetings, Russia!, 1950); novels, *Böyük dayag* (Great Help, 1957); *Pärvanä* (1971); dramas, *Häyat* (Life, 1935); *Madrid* (1938); *Mähäbbät* (Love, 1942); *Kändchi gyzy* (Village Girl, 1963) etc.

PHTF II 690. ZV

al-'Īd, Muḥammad 'Alī Khalīfa (b 1904), Algerian poet, who studied at the traditional az-Zītūna school in Tunis. After 1926 he turned to journalism, putting his poetic talent at the service of the Algerian reform movement. He taught for many years, and became a headmaster; in 1955 he was imprisoned by the French authorities. His main themes are the heroism and faith of the Arabs and their future, many poems being directly inspired by specific events. Al-'Īd uses many traditional themes and formal elements, but introduces new ones (eg drops the monorhyme, gives new poetic meanings to words, etc). One of the leading Algerian poets, he helped to maintain the continuity of Arabic literature in Algeria. In 1967 he was awarded a literary prize. He also wrote a historical play *Bilāl* in verse form. Works: poems, *Dīwān M.al-'Īd M. 'Alī Khalīfa* (1967). SP

Idrīs, Yūsuf (b 1927), Egyptian short-story writer, novelist and dramatist. He graduated in medicine, and worked for some years as a doctor before turning to writing. The first phase (from 1953) produced several volumes of short stories in a critical realist style (eg *Arkhaṣ layālī*, The Cheapest Night, 1954); they deal with ordinary people and their worries, and criticize bureaucracy and social injustice. Idrīs condemns evil prejudices and gives a colourful picture of rural folk lore. The stories in *al-Baṭal* (The Hero, 1957) stress the author's patriotic, anticolonial ideas. The two novellas *al-Ḥarām* (Guilt, 1959) and *al-'Ayb* (Sin, 1962) are also deeply committed; the former depicts the miserable life of seasonal agricultural labourers, the latter shows how a woman deciding to go out to work is made the victim of the social and moral ills of urban society. In the sixties his work began to turn away from critical realism to general human themes, and to concentrate on psychological analysis. Some of the stories in *Lughat al-āy āy* (The Āy-āy Tongue, 1965) could be called irrational literature. His plays deal mainly with the theme of social responsibility of the individual, and the dramatic conflict between egoism and the moral duty of patriotism (*al-Laḥẓa al-ḥarija*, The Critical Moment, 1958). Other works: volumes of stories, *Ḥādithat sharaf* (A Matter of Honour, 1958), *A-laysa kadhālik?* (Is that not so?, 1957); play, *Jumhūrīyat Farḥāt* (The Republic of Farḥāt, 1957). JO

al-Idrīsī (1100–1165), Arab geographer from Ceuta. He studied in Córdoba and travelled widely in Spain and France, reaching the shores of England and visiting North Africa and Asia Minor. He finally settled at the court of the Norman king, Roger II, in Palermo, where he probably died. His principal work is the famous *Book of Roger*, *Nuzhat al-mushtāq fī 'khtirāq al-āfāq* (The Stroll of One Desirous of Crossing the Horizons of the Globe), a detailed geographical work which the king not only commissioned, but also helped to write. It is based on Idrīsī's own observations, on a wealth of information drawn from earlier Arab geographers, and from reports, sketches and notes provided by experts sent out by the king, or collected from sailors and merchants.

Starting from Ptolemy's theory of seven climates in the world as then known, the *Book of Roger* described countries and provinces with their towns and villages, their trade, economy, manners and customs, languages and dialects, as well as their rivers and seas, plains and mountains, islands, straits and deserts. Of great interest are the reports on central, northern and eastern Europe and on the interior of Africa, for which there are no other contemporary sources. Idrīsī imagined the world as a sphere in space, surrounded by water like the yolk of an egg; the sphere is firmly held in the centre of the universe, and is inhabited only at its northern part, the south being uninhabitable. The ocean encircles the sphere like a belt, and the lower hemisphere rests on the sea. The text of the *Book of Roger* is illustrated by 70 maps; one large map is based on Ptolemy.

Idrīsī engraved a large map of the world on a silver sheet, completing it, together with the book, only a few weeks before the king's death in 1154.

Trans.: in *Opus geographicum* (Leiden 1970). BGAL[2] I 628f; EI[2] III 1032–5 (with bibliography); K. Miller, *Mappae arabicae 1–2* (Stuttgart 1926–7). RS

Ikromi, Jalol (b 1909 Bukhara), Tajik writer and dramatist. He lost his mother in childhood and was brought up by his father, a Muslim judge, and his two wives. At 14 he lost his father and entered the teachers' training college in Bukhara. In 1927 he met S. Aynī (qv), under whose influence he wrote his first story, *Shabi dar Registoni Bukhoro* (Night in the Square of Bukhara). He learned more of the craft from translations of Chekhov. Going to the Tajik capital, Dushanbe, he worked in the Institute for Tajik Language and Literature, then as editor of a literary journal, and finally as director of the Tajik theatre. His stories reflect the problems of the day: agricultural reform, and the fight against the Basmachis (*Du hafta*, Two Weeks, 1933). In 1936 he wrote his first drama, *Dushman* (Enemy), followed by others. Sholokhov's *Virgin Soil Upturned* inspired his first novel, *Shodī* (1940), to which he later added the story of the village after Shodi's return from fighting in World War II. During the war he also wrote stories, reportages and plays about the heroism of the Soviet army, the best known being the play *Khonai Nodir* (Nodir's House), about the defence of Stalingrad. The psychological novel, *Man gunahgoram* (I am Guilty, 1958), on problems of family life, love and ethics, is one of the finest in Tajik literature. The three-part *Dukhtari otash* (Daughter of Fire, 1962) gives a vivid picture of life in Bukhara in the early 20th century, up to the revolution, and a striking portrait of a Tajik woman; the long novel *Duvozdah darvozai Bukhoro* (Twelve Gates of Bukhara, 1969) has a similar historical theme. His other works include: novellas, *Tirmor* (Viper, 1935), *Javobi muhabbat* (Love's Answer, 1947), *Tori ankabut* (Cobweb, 1960); volumes of stories, *Hayot va ghalaba* (Life and Victory, 1934), *Dostoni mardi tankshikan* (Legend of the Destroyer of Tanks, 1944); plays, *Dili modar* (A Mother's Heart, 1942), *Khor dar guliston* (Thorn among Roses, 1964).

RHIL 574–5. JB

Illuyanka, Hittite word for a snake or dragon, used in literature for the myth of the struggle between the storm god and the dragon. The myth was recited annually at one of the most important Hittite religious festivals, but the names of some of the gods show that the myth originated much earlier. The myth has been preserved in two quite distinct versions. In the first, the storm god fights the dragon (*illuyanka*) and loses; he plans his revenge, asking the goddess Inara to prepare a feast, and to fill the goblets well. Inara invites a human to help her, Hupashiya. The dragon comes to her feast, and gets so drunk that he cannot get back into his lair; Hupashiya ties him up and the storm god kills him. Later Inara built a house, and setting Hupashiya in it, ordered him not to look out of the window. For twenty days he obeyed, but at last opened the window; seeing his wife and children, he felt homesick and begged Inara to let him go. The text is very damaged at this point, but it seems that Inara killed Hupashiya instead of freeing him.

In the second version, the victorious dragon took the eyes and the heart of the storm god away. In this deprived state the god married and had a son; growing up, the boy married the dragon's daughter and in obedience to his father, the storm god, asked the dragon for the eyes and heart. His wish granted, he brought them to the storm god who thus regained his strength and set out to do battle with the dragon; the victorious storm god then killed not only the dragon, but his own son as well.

O. R. Gurney, *The Hittites* (London 1969); A. Goetze, in ANET[3], 125; H. G. Güterbock, in *Mythologies of the Ancient World* (New York 1961), 150–2. JS

Ilumilku (14th century BC), a priest in Ugarit. According to the alphabetical cuneiform tablets he wrote down myths and epics, but it is not clear whether he merely wrote down existing oral versions or whether he was a poet using earlier material for works of his own. See also Baal, Aqhat, Keret.

O. Eissfeldt, *Sanchunjaton von Berut und Ilumilku von Ugarit* (Halle 1952). SZS

'Imrāni, Jewish-Persian poet, see **'Omrāni.**

Imru' (Imra') al-Qays (6th century), Ancient Arabian poet, the most famous of the pre-Islamic period. A member of the ruling family of the Kinda tribe, he tried for a long time to regain the lost power of his tribe and power for himself. Most of what is told of his life is invention, like the story of his visit to the Emperor Justinian in Constantinople. The poetry of this era bears the features of tribal poetry, composed both by professional bards and by folk poets; the work of the former is of course more cultivated, but both are written in poetic diction. It is a well-developed poetry, with metres based on quantity, with stress, in a variety of metrical schemes and with a single rhyme throughout. The poetic elements are established (topoi, imagery) and the themes set.

Arabic literary criticism ascribes an important place in pre-Islamic poetry to Imru' al-Qays, a place which is confirmed by the negative attitude towards him adopted by Muhammad and the emergent forces of Islam. Imru' al-Qays is also supposed to have developed the love prologue in the *qaṣīda* (qv), as well as other innovations, for which there is no evidence. His famous poem, *mu'allaqa* (see mu'allaqāt), which Arab critics accept as authentic, is in all probability not genuine. Many of the poems in his *dīwān* (qv) are not of his authorship, and only between 20 and 25 can safely be attributed to him. In these poems Imru' al-Qays shows himself the master of the common Ancient Arabian genres, especially those of description and of love poetry. He reveals potential power of thought, but for the most part he did not go beyond the ideas of ancient Bedouin Arabia. His *dīwān* has survived and been published (Baron de Slane, *Le diwan d'Amro'l kais*, 1837, and new Arabic editions).

BGAL I 24, S I 48; EI[1] II 506–5; NLHA 103; BHLA 261. KP

Indian style (*sabke hendī*) came into being in Persian literature in the 14th–15th centuries as the logical development of the Iraqi style (qv). It arose in India, where Persian poetry flourished most abundantly, and then spread to the other lands where Persian was the language of poetry, carrying with it the influence of Indian culture. It was the predominant style up to the 19th century and still has many adherents in the eastern sphere of Persian literature. In language and expression this poetry is simple, but the philosophical and social ideas it conveys, and the involved allegory and metaphors cryptic to the point of absurdity, make it difficult to understand, especially the examples from the 17th and 18th centuries. The best known poets writing in the Indian style are Mīrzā Moḥammad 'Alī Ṣā'eb, 'Orfī (qqv), Nazmī and others. The style was brought to its highest level by the Indo-Persian poet 'Abdulqādir Bēdil (qv), who was also its greatest exponent, inventing the particular form known as *sabke bēdilī*, which predominated in Central Asian poetry in the 18th century. Ideas as to when and where the Indian style arose vary greatly, Homāyūn Farokh believing it to be a combination of the Khorasani, Iraqi (qqv) and Herat styles, for which he proposed the name 'Isfahani style'. The Uzbek literary historian Z. Rizoyev derives the Indian style from Alīshīr Navā'ī (qv) and his pantheism.

RHIL 112ff, 469f, 515–20. JB

Ioane-Zosime (d probably 978), Georgian church writer, a simple monk of the St Saba Georgian monastery in Palestine, who led a group of monks to Sīnai in 973, to found a Georgian colony and monastery. He died there. He is the author of the *Kebay da didebay kartulisa enisay* (Praise of the Georgian Tongue), in which he compares Georgian and Greek and concludes that, for its beauty and rich vocabulary, Georgian will be the official language of the Last Judgment. He also drew up a calendar, *Krebay ttuetay tselitsdisay* (The Months of the Year), and wrote a number of hymns and comparative liturgical studies, in which he gave a detailed description of the Georgian liturgy; in this way he helped to establish the Georgian church. His chief influence on Georgian literature was through his patriotic enthusiasm. VAČ

Ipuwēr, ancient Egyptian sage whose utterances are preserved on a fragmentary papyrus in the Leyden Museum (No. 344). It is a nineteenth dynasty copy (13th century BC) of a work originally written probably under the twelfth dynasty (20th or 19th century BC). It describes the social upheaval and administrative deterioration in the land under an ageing king who is unaware of what is going on beyond the palace walls. It is not clear whether the writer is describing the contemporary scene or prophesying what will happen. The work seems to belong to the decadent period between the Old and the Middle Kingdoms (c2200 BC), which would make it a prophesy written after the event.

Trans.: A. H. Gardiner, *The Admonitions of an Egyptian Sage* (Leipzig 1909); J. Wilson, in ANET, 441–4.
J. Spiegel, *Soziale und weltanschauliche Reformbewegungen im Alten Ägypten* (Heidelberg 1950). JČ

Īraj Mīrzā, called Jalāl ol-Mamālek (b 1874 Tabriz, d 1924/5 Teheran), Persian poet. A Qajar prince, great greatgrandson of Fatḥ ʻAlī Shāh, he was educated at the Polytechnic school in Tabriz and, besides the languages of Islam, spoke French and Russian well. He began his career as the court poet of the crown prince Moẓaffarod-dīn with the title, Ṣadr oshshoʻarā (Head of Poets), but he resigned this and held other posts at court. He travelled in Europe, and spent the last years of his life in Mashhad, returning to Teheran not long before his sudden death. Unlike most of the important Persian poets at the turn of the century (ʻĀref, Amīrī, Bahār, Neshāpūrī, qqv, and others), he did not belong to the nationalist revival in literature, and indeed politically committed work became the target for his poems. He wrote on the theme of motherhood and motherly love, which was quite new in Persian literature. He also wrote poems for children in which he simplified his idiom, using words and phrases from colloquial speech and even French words which were already assimilated into the vernacular. In this respect he was an important forerunner of the modernist poets. These experiments were probably influenced by his knowledge of French literature, from which he made a number of translations (or free paraphrases) especially of the *Fables* of La Fontaine.

MLIC 120–9; RHIL 384–5; MR 61–7 with examples in English or French translation. VK

Iraqi style (*sabke ʻerāqī*) developed from the Khorasani style (qv) in the 11th century in western and southern Iran, and soon spread throughout Persian literature, which it dominated up to the 14th century. It was marked by a greatly increased influence of Arabic, with old Persian words and expressions being rarely used. The formal procedures and poetic ornament became more and more elaborate and subtle. The frequent use of astronomical, mathematical, philosophical and similar terms made this poetry more difficult to understand. Sufi terms were also introduced (see Sufism). Anvarī (qv), Ẓahīre Faryābī and others wrote in this style, which gave rise in 14th-15th century India to the Indian style (qv).

RHIL 112ff. JB

Isahakian, Avetikh (b 1875 Ghazarapat, d 1957 Yerevan), Armenian writer. After studies in Vienna, Leipzig and Zurich, he was a member of the literary group, *Vernatun* (see Thumanian), was imprisoned for liberation activities and emigrated in 1911. Except for 1926–30 (visiting USSR), Isahakian lived and published in Paris, for a time in Venice, and Geneva, until 1936 when he returned to USSR. His first volume of love and nature lyrics *Erger u verk'er* (Songs and Wounds, 1898) reflects folk sources; after 1905 social and revolutionary themes appear in his works. In

1911 he published his best work and one of the best romantic lyrical-epic poems in modern East Armenian literature, *Abulala Mahari* (English trans. by Z. Boyajian, 1958, French by J. Minassian, 1952, and many other languages). This poem expresses the author's rejection of contemporary social norms, institutions and relationships, and his personal solution of the eternal conflict between the hierarchy of values of the individual and the society. Writing in exile, Isahakian drew most on homesick reminiscences of childhood and early years; this is the period of the first version of his paraphrase of the folk epic *Sasma Mher* (1936, see Sasuntsi Davith). He also wrote stories, prose poems, legends and fables (often from myths or folk epic of other nations), and, during World War II, anti-fascist lyrics and articles. With Demirtchian, Thumanian (qqv) and others, he links early 20th century Armenian literature with that of the Soviet period.

Further trans.: P. Gamarra, in ELA; H. Salakhian, in SLM.
HO 7; LACC. LM

Isaiah (8th century BC), Israelite prophet, and the book called after him. He admonished the king and the people of Israel not to abuse their religion or ally themselves with heathen peoples. The *Book of Isaiah* includes the prophecy of eternal peace among men and in nature. To the original book (chapters 1–39) a late 6th century prophecy was subsequently added (the *Second Isaiah*) predicting in poetic terms the fall of Babylon and the deliverance of Israel. With their poetic imagery, energetic reproaches and words of comfort, both parts of the book are among the finest early Hebrew writing. See also *Bible* and *Prophets*.

J. L. McKenzie, *Second Isaiah* (Garden City 1968). SZS

al-Iṣfahānī, Abu'l-Faraj (897–967), Arab writer and biographer from Isfahan. Of an Arab family related to the Umayyad caliphs, he studied in Baghdad; here he gained access to the favour of the powerful Buwayhid family, but he was also well received at the court of the Hamdanid, Sayfaddawla, in Aleppo. He was in contact with the Umayyad dynasty in Córdoba, in Spain. After years of travelling he died in Baghdad. Iṣfahānī's chief work, *Kitāb al-aghānī* (Book of Songs), consists of over 20 volumes, on which he claimed to have worked almost 50 years. It is based on songs which the Caliph, Hārūn ar-Rashīd, had written down by the musicians of his time. Iṣfahānī not only increased their number considerably, but put them into some sort of historical or traditional literary context and contributed a wealth of information on the melodies, poets, singers and composers, as well as about innumerable figures mentioned in thousands of traditions. This has made his work an invaluable source, as yet not exhausted, for the history of culture and literature in the pre-Islamic and Islamic period, up to the 10th century. His biographies of the descendants of Abū Ṭālib (*Maqātil aṭ-Ṭālibīyīn*), the uncle of the Prophet Muhammad, are of great importance for the history and theology of early Islam, and especially for the followers of 'Alī, the Shiites.

BGAL[2] I 152f, S I 225 f; EI[2] I 118. RS

Isfahani style of Persian literature, see **Indian style.**

Ishtar's Descent to the Nether World. Ishtar (Sumerian: Inanna), 'Queen of Heaven', was the goddess of love in Sumer and Babylonia; her astral form was the planet Venus. The subject of the poem on the *Descent of Ishtar to the Nether World* was probably inspired by the period during which the planet is not visible, and by the arrested fertility of nature during the winter. The goddess is said to dwell for a time in the kingdom of the dead, the poem thus providing important knowledge of Babylonian ideas on the after-life. It is known in a more extensive Sumerian version and a shorter, slightly different Akkadian version. According to this myth, Ishtar, bearing her insignia of power, sets out for the 'Land of no Return', to rule there as well as in the sky. Her sister,

Ereshkigal, ruler of the underworld, orders the guardian of the gate to deprive Ishtar of the insignia of her power one by one as she passes the seven gates, so that she arrives before her sister's throne naked and helpless. In her absence the earth becomes barren. To free Ishtar from the underworld the god Ea creates a eunuch who is to sprinkle her corpse with the water of life. When the guardian of the underworld leads Ishtar out again into the world, he restores her insignia one by one at each gate, but she can only leave under one condition, that she sends someone else to take her place in the kingdom of the dead. This substitute is the lover of her youth, Tammuz, who in the earlier, Sumerian version is said to spend half the year on the earth and the other half in the underworld. The Sumerian version says that Enki (Babylonian Ea) despatched to the nether world two divine sexless beings created from the dirt of his finger-nails.

Trans.: E. A. Speiser, in ANET³ 106–9.
S. N. Kramer, *The Sacred Marriage Rite* (Bloomington 1969), 107 ff; ANET³ 52–7; LRPOA 258–65. LMa

I'tiṣāmī Parvīn, Iranian poetess, see **Parvīn E'teṣāmī.**

İzzet Molla Keçecizade (b 1785 Istanbul, d 1829 Sivas), Turkish poet. Of a craftsman's family, he studied Islamic law and became a judge active in political life; a supporter of reform of the Ottoman regime, he was exiled twice, to Keşan and to Sivas. His verse is notable for its wit and laconism. The arrangement of his *dīwān* (qv) in chapters, according to the subject and the type of poem, is unusual. Most important work: *Mihnet Keşan* (A Sufferer, or: Several Sufferers, or: Suffering in Keşan), which is 7,000 distichs of dialogue with the poet's other self, in various poetic forms; it wryly describes life in a little provincial town and sadly remembers Istanbul. Other works: *Gülşeni Aşk* (Delightful Place for Love), a romantic epic; *Layhalar* (Reflections), prose treatise on Ottoman politics.

GHOP IV 304 ff; PHTF II 447 ff. ZV

J

Jabbarly, Jäfär (b 1899 Hyzy, d 1934 Baku), Azerbaijan poet, writer and dramatist. From a poor peasant family, he was only able to attend the school of drama in Baku after the establishment of the Soviet regime. He later became a film and theatrical producer and writer. His first plays and poems were published in 1915, influenced by Turkish literature and the ideal of pan-Turkish unity; later he began to depict the Azerbaijan national awakening, and then the new life of Soviet Azerbaijan. He also wrote stories, film and opera librettos. Jabbarly paid great attention to language and style, trying to perfect the literary language. He is considered the founder of Soviet Azerbaijan drama. The richness of his language has not yet been surpassed. His main works are the stories, *Gülzar* (1924) and *Dilbär* (1927), and the dramas, *Od gyälini* (Fire's Bride, 1928), *Ogtay Eloghli.*

PHTF II 661, 685. ZV

al-Jāḥiẓ, Abū 'Uthmān 'Amr b. Baḥr (b c776 Basra, d 868 Basra), Arab prose writer and philosopher. He was given the nickname al-Jāḥiẓ (goggle-eyed) on account of his protuberant eyes, and became known by it; his family was said to be descended from East African slaves. He was educated in his native Basra, completing his studies in Baghdad, where he was sent for by the Caliph al-Ma'mūn (813–833); he was indirectly commissioned to write books justifying the Abbasid seizure of power. In Baghdad he supplemented his already wide learning in Arabic by reading Greek works in translation. He was the author of almost 200 works, of which only a small part survives; they are *adab* (qv) in style, endeavouring to educate through entertainment. The innumerable anecdotes, amusing stories and jokes scattered through the text are meant to please the reader and lead him on to study further. He had religious and political aims in view; in his popular writings and in the great seven-volume *Kitāb al-ḥayawān* (Book of Animals) he propagated the ideas of the Mu'tazilites.

The latter book describes the special features of animals, presents varied religious information and adds the author's own observations and conclusions; it is a work of apologetics, explaining the miracles of creation, glorifying Islam and confuting the heretical superstitions of the non-Arabs.

Kitāb al-bayān wa't-tabyīn (Book of Eloquence and Exposition) is a political work directed against the Shu'ubites, who proclaimed the superiority of the non-Arabs over the Arabs; it is both a treatise on rhetoric and an anthology compiled to prove the literary ability of the Arabs. His anti-Shu'ubite stand is even clearer in *Kitāb al-bukhalā'* (Book on Misers; French trans.: C. Pellat, Paris 1951), with its description of certain social classes of his day; he attempts to show the generosity of the Arabs in contrast with the miserliness of the urban Persian middle class. In defiance of tradition, al-Jāḥiẓ was concerned with contemporary reality in his work, which is a partial synthesis of the Arabic heritage in thought and culture, enriched by elements assimilated from the Greeks. His significance in the history of literature is immense, for he was the first really good Arabic prose writer, attempting to treat in prose subjects hitherto reserved for verse. He was a master of irony, which he developed to a degree unknown before. Al-Jāḥiẓ proved that Arabic is a subtle language, rich, and capable of expressing the development of ideas. His own idiom was indeed rich in all its aspects. In spite of various criticisms raised against him, he is rightly considered one of the greatest of Arab writers. Further works: *Manāqib al-atrāk* (Outstanding Qualities of the Turks); *Fakhr as-sūdān 'ala'l-bīdāb* (In Praise of Black against White); *Dhamm akhlāq al-kuttāb* (Blame of the Morals of Officials).

Trans.: Charles Pellat, *Le livre des avares de Ǧāḥiẓ* (Paris 1951); Charles Pellat, *Le livre de la couronne, sur les règles de conduite des rois* (Paris 1954).

BGAL I 158–60, S I 239, 421; C. Pellat, *Le Milieu Baṣrien et la formation de Gāḥiẓ* (Paris 1953). SP

Jalil, Musa (b 1906, Mustafa, Orenburg district, d 1944 Berlin), Tatar poet; of a poor family, he studied at the *madrasa* and later at Moscow University. Jalil was editor of the Tartar papers published in Moscow, *Kechkene ipteshler* (Young Comrades) and *Oktyabr balosy* (Children of October). He joined the Soviet army in 1941, was severely wounded, taken prisoner and executed in Berlin. His first volume of verse appeared in 1925, *Barabyz* (We are Marching); his subsequent work celebrated the building of socialism. The finest is a volume of poems written in prison in Berlin, *Moabit törmesende yazygan shigyrlar* (Poems Written in Moabit Prison, 1953). NZ

Jalil, Rahim (b 1909 Khujand, now Leninobod), Tajik writer. His father was a craftsman; both parents were illiterate. He studied at the first Tajik training school for teachers, and began teaching at 18. He then turned to writing and became a leading figure in the Leninobod group of the Tajik Writers' Union. He published his first short story, and first poems under the pseudonym Dahrī, in 1931. The volume of verse, *Mavjhoi muzaffariyat* (Waves of Victory) appeared in 1933, about new people and a new, happier life. He was influenced by S. Aynī and A. Lāhūtī (qqv). Jalil wrote many stories about the emancipation of the women of Central Asia, the fight against enemies of the Soviet system, and respect for work. His satires ridicule laziness, religious prejudice and superstition. His first collection of stories was *Orzu* (Wish, 1937), followed by *She'rho va hikoyaho* (Poems and Stories, 1939). He also wrote plays, like *Sanavbar* (1938), on the struggle between the old and the new way of life in Tajik villages; *Du vokhurī* (Two Meetings, 1942) is a war play, while *Dili sho'ir* (The Poet's Heart, 1963) is a historical play about the life of Kamāl Khujandī, the poet. In 1941 Jalil published his first novel, *Gulrū*; it was republished after the war in extended form as *Odamoni jovid* (Immortals). It is a historical novel, like the later trilogy *Shūrob* (1959), of the revolution as it affected the Tajiks, the birth of a

Tajik proletariat, and the brotherhood of the Tajiks, Russians, Uzbeks and other nationalities. In 1969 Tajik newspapers published excerpts from a historical novella, *Ma'voi dil* (The Refuge of the Heart). Further works: stories, *Hissa az qissa* (Tales, 1941); *Hikoyahoi zamoni jang* (Tales of Wartime, 1944); *Umri dubora* (Double Life, 1949); *Hamida* (1961); poems from various periods, *Nasimi Sir* (The Syrdarya Breeze, 1969).

RHIL 575–6. JB

Jamāluddīn Afghānī (b 1838 As'adābād in Afghanistan or Asadābād in Iran, d 1897 Istanbul), philosopher, Islamic reformer, politician and writer, on whose origin Afghan and Persian scholars cannot agree. As a young man he lived in Kabul; after a visit to India in 1856 and to Mecca, he entered the services of the Emir Dost Muhammad, serving as minister for a time. He started what was probably the first newspaper in Afghanistan, *Kābul.* After the death of the Emir he lived in India, Egypt (where he profoundly influenced the national liberation movement), Iran, Russia, and for several periods in Turkey, where he died. In 1941 his remains were brought to Kabul and laid to rest in the mausoleum. A pan-Islamic thinker, he wrote mainly in Arabic, but also published in Persian. Most of his writings are scattered throughout the journals of his time, like the lecture on 'Islam and Science' which he gave in Paris in 1883, or his attack on Darwinism and materialism. One of the few books he published is the *Tatimmat ul bayān fī tārīkh ul-afghān* (A Satisfying Explanation of Afghan History), a brief history of Afghanistan.

Elie Kedourie, *Afghani and 'Abduh: An Essay on Religious Unbelief and Political Activism in Modern Islam* (London 1966); A. A. Kudsi-Zadeh, *Sayyid Jamāl al-dīn al-Afghānī, An Annotated Bibliography* (Leiden 1970); Nikki R. Keddie, *Sayyid Jamāl ad-Dīn al-Afghānī* (Berkeley, Los Angeles, London 1972). JB

Jamālzāde, Moḥammad 'Alī (b c1895 Isfahan), Iranian writer, founder of the modern short story in Persian. His father, Seyyed Jamāloddīn, was a leading figure in the Iranian national revival. A progressive spirit, he gave his son a modern education at the French lycée near Beirut, after which Jamālzāde studied in Dijon and Lausanne, where he completed his law studies. He then joined the patriotic emigrées in Berlin, whose aim was to fight, under the leadership of Ḥasan Taqīzāde, against the Anglo-Russian occupation of Iran during World War I. An important medium for their views was the paper, *Kāve*, named after the legendary smith in Ferdousī's (qv) *Book of Kings*. Founded in 1916, it turned from a purely political journal into an excellent literary magazine. Jamālzāde published his first stories here. For a time he worked at the Iranian Embassy in Berlin, and since 1931 has lived in Switzerland, at first in the International Labour Organization and from 1956 as a writer and critic.

His magazine stories attracted attention, but his first book of six stories (published in 1921) marked a revolution in literature, enthusiastically greeted by some and utterly condemned by others. The title, *Yekī būd va yekī nabūd* (There Was Once—Or Was There?), is the phrase with which Persian fairy tales begin, but these particular tales were not fairy stories. They gave a critically realistic view of life in Iran, not in the traditional abstract allegorical manner open to various interpretations, but portraying real life situations and characters, and with a clear and undeniable point. In this Jamālzāde was an absolute pioneer in Persian writing, although in style and language he had predecessors in Malkom Khān Mīrzā, Dehkhodā (qqv), and other scholarly translators of European prose. In addition, his *Yekī būd va yekī nabūd* is a literary work which is still very much alive, and its characters unforgettable.

For the next twenty years he published little, in magazines, and it was not until 1941 that he began publishing books again; longer works like *Dār ol-majānīn* (House of Fools, 1942), *Qaltashan Dīvān* (Dictator of the Imperial Office, 1946), *Sar o tahe yak karbās yā Eṣfahānnāme* (All of a Pattern, or the Book of Isfahan, 1956) and collections of stories and essays, like *Talkh va shīrīn* (Bitter and Sweet, 1956), *Kohne va nou* (The Old and the New, 1959) and many others. They are all of a high literary standard and deal with social themes, in latter years with a tendency to philosophical and mystical speculation. They also inevitably throw light on the problems and specific features of Iranian life. Jamālzāde also writes critical and polemical studies for the Persian press (eg for *Sokhan*, *Rāhnamāye ketāb* and others), and is an excellent translator of European prose (Schiller, Oscar Wilde, Anatole France). Some of his own stories have appeared in English, German or French translations in various periodicals.

Trans.: R. Gelpke, *Persische Meistererzähler der Gegenwart* (Zürich 1961); S. Corbin and H. Lotfi, *Choix de Nouvelles* (Paris 1959).
AGEMPL 134–45; KMPPL 91–112; RHIL 389–90. VK

Jambul Jabayev, Kazakh folk singer, see **Zhabayev,** Zhambyl.

Jāmi, Moulānā 'Abdorraḥmān (b 1414 Jam, d 1492 Herat), Persian poet and writer who spent most of his life at the Timurid court of Herat, with the exception of his pilgrimage to Mecca and a short stay in Baghdad. A close friendship existed with Alīshīr Navā'ī (qv), the vezīr and inaugurator of Chaghatay literature. Jāmī was 'too exalted for there to be any word of praise for him', as Bābur (qv) says in his memories. In fact, Jāmī was a very prolific writer, and is considered the last great classical poet in Persian who excelled in every form of literature, from lyrical poetry to epics. We admire him most as an erudite and versatile author who influenced, for a while, both Indo-Persian and Chaghatay literature. His work extends over all fields of knowledge: theology, poetics and rhetorics, grammar and letter-writing. As a member of the Naqshbandiya order he composed the *maṣnavī* (qv) *Toḥfat ol-aḥrār* in honour of Shaikh 'Obaydollāh Aḥrār; but at the same time he expressed ideas close to Ibn 'Arabī's essential monism, wrote a commentary on Ibn 'Arabī's *Fuṣūṣ ol-ḥekam* as well as on 'Irāqī's *Lama'āt*. He commented likewise upon famous Arabic verses, eg of Ibn al-Fāriḍ (qv). His mystical inclinations are shown in full in the collection of 616 biographies of saints, *Nafaḥāt ol-ons*, relying upon earlier hagiographic works.

Jāmī's fame mainly rests upon his achievements as a poet. He left three lyrical *dīwāns* (qv) which display an amazing amount of rhetorical refinement (not in vain did he compose a book on logogryphs). These *ghazals* (qv) are charming though after a while they become monotonous despite their ornate wording. Faithful to existing tradition, he wrote a collection of *maṣnavīs* under the title *Haft ourang* (The Seven Thrones), three of the topics being taken from Neẓāmī's *Khamse* (Leylā Majnūn, Khosrou Shīrīn, Kheradnāmeye Eskandarī), and three connected with mysticism (the *Sobḥat ol-abrār*, *Toḥfat ol-aḥrār* and the *Selselat oẕ-ẕahab*). The allegory of the *maṣnavī Salmān o Abṣāl* goes back to hermetic ideas elaborated by Naṣīroddīn (qv). The most famous work in the *Haft ourang* is *Yūsof and Zoleykhā*, which is considered the best poetic version of the famous story contained in Sura 12 of the *Qur'ān* (qv). It has often been illustrated by miniaturists. Jāmī's graceful *Bahārestān* (Orchard of Spring) is modelled upon Sa'dī's (qv) *Golestān*; its eight chapters with anecdotes, biographical notes and poetry was widely read. Although Jāmī's excellence in all forms of literature is unquestionable, the modern reader will probably admire his erudition and skill more than his poetry, in spite of some beautiful lyrical passages.

Trans.: Ralph D. H. Griffith, *Yūsuf u Zulaikhā* (London 1881); A. Rogers, *Yūsuf u Zulaikhā* (London 1889); E. Fitzgerald and F. Falconer, *Salaman and Absal* (ed. A. J. Arberry, 1956); E. H. Whinfield and Muḥ. Qazwini, *Lawā'iḥ. A Treatise on Sufism* (London 1906). AS

Janayev, I. V., Ossetic poet, see **Niger.**

Javakhisvili, Mikheil (b 1880 Tserakvi, d 1937), Georgian writer. Of a peasant family, in 1901 he graduated from the fruit-

and wine-growers' school in Yalta. After a few years in an office, he edited the journal *Iveria* from 1904, but emigrated when the attacks on the Tsarist regime published in the paper were laid to his account. He studied at the Sorbonne and travelled in Europe and North America, returning home illegally in 1909 to lead the patriotic group round the journal *Eri* (People). For this activity he was arrested and jailed, and was not allowed to live in the Caucasian area. He returned after Soviet Georgia was established (1921) and held various responsible posts until he was accused in 1937 of activities against the interests of the people, arrested and executed.

Besides a number of short stories and novellas, Javakhishvili wrote several novels of which the most significant is *Kvachi Kvachantiradze* (1924), a brilliant satire on certain aspects of the Georgian national character, the story of a swindler. The novel *Jaqos khiznebi* (Jaqo's Favourites, 1925) is also satirical while *Arsena Marabdeli* (1930) is a historical epic tale of the famous rebel of the early 19th century. The slight novel, *Tetri saqelo* (White Collar, 1924), contrasts the comfortable 'alienated' life of the city with the romantic idea of patriarchal village life. He nevertheless comes to the conclusion that, with all its drawbacks, city life is the life of the future, while idyllic dreams of village life are a poetic but nevertheless vain attempt at escape.

Trans.: Ruth Neukomm, *Georgische Erzähler der neueren Zeit* (Zurich 1970); M. Tougouchi-Gaiannée and D. Renet, *Les invités de Jako* (Liège 1946); I. Petrova, *Too Late* in *Mindia, the Son of Hogay and Other Stories by Georgian Writers* (Moscow 1961). VAČ

al-Jawāhiri, Muḥammad Mahdī (b 1900 Najaf), Iraqi poet, the greatest representative of the traditional line in Arabic poetry. He grew up and studied in Najaf, then went to Baghdad where he edited the periodicals *al-Furāt* (Euphrates), *al-Inqilāb* (Revolution) and *ar-Ra'y al-'āmm* (Public Opinion). Victimized for his political activities and the progressive opinions expressed in his verse, he left Baghdad for Cairo; on his return he again worked in editorial offices. In 1956 he went to live in Syria until the Revolution of 1958, when he became the chairman of the Union of Iraqi Writers and was given his due as a writer. When the Qāsim regime failed to fulfil the hopes placed in it by progressive opinion in Iraq, al-Jawāhirī again left the country and lived for a time in Prague. Returning home he once more held office in the Union of Writers. Now he lives in Baghdad.

Al-Jawāhirī is generally considered the best of the contemporary poets in the traditional manner, and is greeted with lively interest at all international gatherings of poets. He is primarily a progressive political poet, profoundly committed to the struggle against foreign domination, feudalism, and political reaction. Besides his verse on these themes he writes love and nature poetry. The significance of his political verse has extended far beyond the boundaries of Iraq and become a symbol for Arabs in their struggle for emancipation. His poetry is vividly expressive and possesses a musical quality which justifies the unusual choice of vocabulary; the composition of his *qaṣīdas* (qv) is truly remarkable. Both his language and his poetics are based on the old classical foundations, but speak in the idiom of a new context, the modern age and the struggle of the Arabs to forward their own progress. It is precisely this balance of traditional (in form) and modern (in thought) that has brought al-Jawāhirī his reputation; he can perhaps be called the Emir of poets.

NTALAC 124. KP

Jeremiah (late 7th–early 6th century BC), Israelite prophet who bitterly reproached Judah for social and religious evils, and prophesied the destruction of Jerusalem and of the Temple. He was taken to Egypt in captivity after the capture of Jerusalem by the Babylonians in 586 BC. In the *Book of Jeremiah* his friend Baruch the scribe preserved the prophet's reproaches to his own and other peoples. The prophesies of Jeremiah are remarkable for their depth of feeling. *The Book of Lamentations* over the destruction of Jerusalem and of the

Temple is wrongly attributed to him. See also Five Festal Scrolls and Bible.

J. Bright, *Jeremiah* (Garden City 1965). SZS

Jibrān, Jibrān Khalīl (b 1883 Bisharrī, d 1931 New York), Lebanese-American poet; author of narratives, articles, poetic prose-pieces, parables and aphorisms; also a painter. At the age of eleven, he emigrated to Boston with part of his family; but he returned to the Lebanon to attend al-Ḥikmah school 1897–99, and again briefly in 1902 as guide to an American family. His painting brought him the patronage of Fred Holland Day and Mary Haskell. The latter in particular sent him to study art in Paris from 1908 to 1910. In 1912 he finally settled in New York. He maintained an amorous and literary correspondence with the poetess Mayy Ziyādah over many years, but he never married. His body was eventually taken back to the Lebanon.

Although not a submissive son of the church (his bold early articles on social regeneration did not spare the clergy) Jibrān was a Christian living in a Western environment, yet nostalgically attached to his Arab homeland. Christian ideals mingled with the literary influences of Rousseau, Blake, Nietzsche and others to forge a creed in which universal love and communion with Nature were uppermost. The poetic prose pieces which form the peak of his literary achievement have the majesty and rhythm of Biblical language and, although after 1918 it was mostly in English that he wrote, he was the leading light in the Syro-American literary Society called *ar-Rabiṭah al-qalamiyyah,* founded in New York in 1920, which did much to familiarize Arab readers with Romantic motifs and symbols. He is often excessively sentimental, sometimes even guilty of inflated sententiousness; but his impulses were genuine, and in his time seminal. Chief works: autobiographical love story, *al-Ajniḥah al-mutakassirah* (Broken Wings, 1912, trans. by A. R. Ferris, 1959); verse, *al-Mawākib,* (The Processions, 1919); miscellaneous prose, *'Arā'is al-murūj* (Nymphs of the Valley, 1906, trans. by H. M. Nahmad 1948), *al-Arwāḥ al-mutamarridah* (Spirits Rebellious, 1908, trans. by A. R. Ferris, 1947); *Dam'ah wa ibtisāmah,* (Tears and Laughter, 1914, trans. by A. R. Ferris, 1949; A Tear and a Smile, trans. by H. M. Nahmad, 1950). In English: *The Prophet* (1923); *Jesus, the Son of Man* (1928).

K. S. Hawi, *Kahlil Gibran,* 1963; Mikhail Naimy *Kahlil Gibran,* 1964; Suheil Bushrui, *An introduction to Kahlil Gibran,* 1970. PC

Job, Hebrew book of the Old Testament (qv) called after the chief character, a wise man of Edom. In a narrative in the folk manner the book tells how Satan asks God to test Job's faith by the loss of his wealth, his family and his own health, all of which is returned to the sufferer manyfold in the end. The poetic discussions between Job and the friends who pity him but do not agree with his ideas are an attempt to work out the purpose and the cause of human suffering. The profound thought in these discussions surpass religious convention both in their sincere faith and in the descriptions of nature as an expression of God's omnipotence; it is this which gives the book its literary worth. The story and individual passages have exercised an influence on both mediaeval and modern poetry. See also Bible.

M. H. Pope, *Job* (Garden City 1965). SZS

Jonah (perhaps 5th century BC), Hebrew narrative about the prophet Jonah. Fleeing by ship, Jonah was thrown into the sea and rescued after three days in the belly of a giant fish, whereupon he prophesied the fall of Nineveh. See also Bible. SZS

Joel (perhaps c500 BC), Hebrew book of prophecies, and apocalyptic vision of the descent of the enemy, their destruction, and the salvation of Jerusalem. See also Bible. SZS

Joshua, Hebrew book of the Old Testament (qv), giving the traditional version of the conquest of the Promised Land (Palestine) by the tribes of Israel led by Joshua, in about 13th century BC. It includes the story of the fall of Jericho to the sound of trumpets. See also Bible.

J. A. Soggin, *Joshua* (London-Philadelphia 1972). SZS

Judges, a book of the Hebrew Old Testament (qv), dealing with the life and battles of the tribes of Israel after the conquest of Palestine under Joshua (qv; probably 13th–11th centuries BC), before the institution of the monarchy. It includes the song of Deborah (qv) celebrating the victory of the Isralites (probably late 12th century BC). See also Bible.

Trans.: HBKJV; HBRSV; NEB. SZS

K

Kadiri, Abdullo, Uzbek writer, see **Qodiriy,** Abdullo.

kalām (Arabic: speech; argument, scholastic theology), the designation of Islamic theology (*mutakallimūn, loquentes,* scholastic theologians). The first phase was the Mu'tazilite school, centred in Basra and Baghdad, and proclaimed the state religion by the Caliph al-Ma'mūn in 827. Under the Caliph al-Mutawakkil (847–861) the opponents of the Mu'tazilite school were victorious. The Mu'tazilite school came into conflict with those thinkers who accepted the *Qur'ān* (qv) literally and denied man's free will. It revealed some rationalist and philosophical elements, which have gained it recognition as the forerunner of later Islamic philosophy (*falsafa*). The fundamental beliefs elaborated by the Mu'tazilite school were: man's free will; emphasis on the role of human reason; elaboration of an abstract conception of God, shorn of positive attributes and necessarily acting in the spirit of divine justice; and denial of the eternal existence of the *Qur'ān.*

In opposition to the Mu'tazilite school, and as a reaction to its teachings, the *kalām* of the school of al-Ash'arī (qv) developed; originally a follower of *mu'tazila* al-Ash'arī later opposed the school. His views prevailed and later formed the main theological line of Sunnite Islamic thought (see al-Ghazzālī). Abū Manṣūr al-Māturidī (d 944 Samarkand) elaborated a similar system. Atomist conceptions (similar to those of Indian atomism) also penetrated into *kalām* and served to demonstrate *creatio ex nihilo* and to disprove the principle of natural causality.

H. Corbin, *Histoire de la philosophie islamique* (Paris 1964), 152–78, 352–3; L. Gardet-M.-M. Anawati, *Introduction à la théologie musulmane* (Paris 1948); A. S. Tritton, *Muslim Theology* (London 1947). VZD

Kalīle o Demne or Kalīla va Dimna is a collection of moralizing, humorous and erotic fables; their history goes back to Buddhist Sanskrit models and particularly to the Pañcatantra (see Vol. II). About the first half of the sixth century the work was freely translated into Pahlavi, perhaps by order of the Iranian ruler Khosrou Anoshirvān; it was entitled *Kalilak u Dimnak* after one of the fables, about two jackals whose names in the Sanskrit original were Karataka and Damanaka. This Pahlavi version has not survived, but an Arabic translation of it by Rūzbeh, an Iranian better known as Ibn al-Muqaffa' (qv), provided the original for a number of versions in other oriental languages. Rūdakī (qv) was the author of a 10th century poetic version of which only about a hundred *bayts* (see 'arūḍ) have survived. There was a prose version by Abū 'l-Ma'ālī Naṣrollāh in the twelfth century, and the prose *Anvāre Soheylī* (The Shining Star Canopus) by Ḥoseyn Vā'eze Kāshefī (qv) at the turn of the 15th–16th centuries, in the intricate style then fashionable. Works in imitation of *Kalīle o Demne* were produced by various authors from the 10th to the 13th centuries, entitled *Marzbānnāme.* A Latin translation by John of Capua (about 1270) gave rise to a number of versions in European languages, eg English in 1570, Czech in 1528 AO.

Trans.: André Miquel, *Le livre de Kalila et Dimna* (Paris 1957).
RHIL 222ff, 660. JB

Kamāloddīn Esmā'īl Eṣfahānī (c1172/3–1237), Persian poet. Mainly a panegyrist, he played an important part in the development of the *qaṣīda* (qv). He seems to have inherited literary talent from his father Jamāloddīn Moḥammed b. 'Abdorrazāq Eṣfahānī (d 1192), who is usually classed among the outstanding poets of the *ghazal* (qv) before Sa'dī (qv). His thematically varied work is less sensitive in feeling than that of his father, but the epithet Khallāq ol-ma'ānī (Creator of original thoughts) given to him in Persian literary history is evidence of outstanding intellect and poetic invention. HS

Kamberdiev, M. B., Ossetian poet, see **Qamberdiaty,** Mysost.

Kanî, Abu Bekir (b 1712 Tokat, d 1792 Istanbul), Turkish poet; his varied life as a poet, singer and official took him from place to place in the Ottoman Empire. At one time interpreter to the Rumanian Alexander in Bucharest, later at the sultan's court in Istanbul, he was forced to leave because of his disloyal views. He died in poverty, in an order of whirling dervishes (see Sultan Veled). He wrote easily and much; his *dīwān* (qv) embrace only a part of his diverse work, verses in praise of the Prophet and also ironic descriptions of contemporary society and amusing episodes. He also left an interesting collection of letters in which real epistles are mingled with imaginary ones like the *Hirre Namesi* (Cat's Letter), a satire in the mouth of a cat.

GHOP IV 159 ff; PHTF II 72, 450. ZV

Kanık, Orhan Veli (b 1914 Istanbul, d 1950 Istanbul), Turkish poet. He did not finish university, and after wartime military service entered the translation department, Ministry of Education, where he edited translations of world classics (including some of his own) such as works of Molière, Musset, Gogol and Anouilh. He later turned to journalism, and a year before his death published the literary journal *Yaprak* (Paper) which attracted attention primarily for his personality. His early verse was exceptionally talented; it was published in *Varlık* from 1936 on. Friendship with the poets Melih Cevdet Anday and Oktay Rifat (qqv) led to publication of a joint volume, *Garip* (Strange, 1941), and the establishment of a literary group. Fundamentally they rejected traditional forms and literary clichés in content and in the language of poetry. Unlike his early verse Kanık turned to the everyday life of ordinary people for his themes; his style became almost that of aphorisms, and his language close to the colloquial. The novelty of this programme attracted attention and ensured an important place in Turkish literature for Kanık. Other works: *Vazgeçemediğim* (What I Cannot Give up, 1945); *Destan Gibi* (As a Destan, 1946); *Yenisi* (New, 1947); *Karşı* (Against, 1949).

E. Heister, *Orhan Veli* (Köln 1957). LH

Kaputikian, Silva (b 1919 Yerevan), Soviet Armenian poet. She studied philology in Yerevan and published her first volume of verse *Oreri het* (With the Days) in 1945. Since then there have been many volumes: *Im harazatner* (My Kinsfolk, 1951), *Srtabac zruyc* (Sincere Discourse, 1955), *Mtorumner čanaparhi kesin* (Meditation Midway, 1961) and others. Her verse is strict in form, dealing sympathetically with life's difficulties, especially those facing women. It is well known in Armenia, the Soviet Union, and throughout the Armenian diaspora. Her *K'aravanner der k'aylum en* (Caravans Are Still Marching, 1964) gives an impassioned picture of the life of Armenian minorities in the Near East.

Trans.: D. Rottenberg, in SLM; J. Gaucheron, in ELA; L. Mardirossian, in OeO. LM

Karacaoğlan (17th century), the greatest of the Turkish folk poets (see âşik). Nothing definite is known of his origin or life; he was probably from south-east Anatolia, where he spent most of his life, in a semi-nomad environment. He passed through other regions, perhaps serving in the Ottoman army. Legends soon overlaid the real facts of his life. His poems (originally sung) were handed down orally or preserved in manuscript collections, but all those attributed to him are probably not his, while others underwent changes in the course of oral transmission. Even so his work reveals an unusually strong poetic personality, deeply rooted in Turkish folk tradition and little affected by Arabo-Persian influences, unlike some folk poets of his time. His verses are composed in syllabic folk forms, and in unadulterated Turkish. His themes are taken from nomad and village life, inspired by the beauty of women and of the mountainous landscape. His style is vivid, dynamic and realistic, without artificiality. His influence is traceable down to the present day, and was particularly felt in the formative years of modern literature and the modern idiom. His work still enjoys a wide popularity. JZ

Karagöz, the principal character in Turkish shadow plays. Karagöz (the 'black-eyed') is usually thought to be a gipsy, and the actors in the shadow plays use gipsy elements in their language, just as the plays themselves use a certain measure of gipsy idiom. This may prove the part played by the gipsies in handing on this genre. The shadow plays probably originated in China, for the brightly-coloured transparent figures, made of leather, are strongly reminiscent of the figures used in Chinese shadow plays. It was formerly thought that the shadow play had come to Turkey by way of Central Asia and Persia, but in these areas only puppet plays were known. That this was an erroneous assumption can be seen, too, from the fact that there is evidence of the existence of shadow plays in Egypt as early as the 12th century. The texts of plays by Ibn Danyal (d 1311 in Egypt) are known to us, and we know, too, that in 1517 shadow plays were performed before the Sultan Selim I in Egypt. It would appear that the shadow plays came to Turkey by this route, there to develop further.

The Turkish guild of shadow actors derives its craft from Şeiyh Küşteri in the 14th century, but this is no more than a legend. The shadow theatre is often mentioned by poets and mystics, who attribute symbolic significance to it, as an expression of the ephemeral nature of life and the vanity of the world. The theatrical *ghazal* (qv), *Hey Hakk*, with its invocation and praise of God could refer to mystical ideas. All shadow plays follow a set scheme; after the introductory scene (*gösterme*) has disappeared and the psaltery music (*velvele*) comes to an end, Hacivat appears and sings, praising God and trying to get Karagöz to come out. When Karagöz appears there is a thrashing scene in which he gets the worst of it. Then the *muhavere* begins, the dialogue between Hacivat and Karagöz in which the play itself is built (*fasl*). The plots of the plays come either from folk tales (*Ferhad ile Şirin*, *Tahir ile Zühre* and others) or more frequently from the life of the ordinary people, with Karagöz carrying on a trade or doing something that is not permitted. As H. Ritter has shown, the acts consist of a series of dramatic scenes in which different characters find themselves successively in the same comic situation, which is finally resolved by a character who can rise above the situation.

Besides the two main characters, Karagöz the rough, uneducated gipsy and Hacivat the educated, better-off man, different social and linguistic types appear: *çelebi* (religious title) Zenne, Matiz Bekri Mustafa, an opium smoker, and others, as well as representatives of the different nations of the Ottoman Empire, ie an Armenian, Jew, Persian, Arab, Albanian and others. The comic effect of the play, orally, depends on the innumerable plays on words, the witty jokes and sayings; in the subject matter, comic effect derives from references to politics and local conditions. The political satire involved was the reason why the shadow plays were often prohibited in the Ottoman Empire. The shadow plays were also widely performed in the provinces of the old Ottoman Empire, and that is why they are found in North Africa, in the Balkan countries, and in Greece. Although they are dying out in these regions, as in Turkey itself, in Greece the shadow theatre underwent an important and specific development. This can be seen in the Greek texts collected by H. Ritter (Karagöz II.XII–XIV).

Trans.: H. Ritter, *Karagös* (I Hannover 1924, II Leipzig 1941, III Wiesbaden 1953).
KLL V 2863–8; KLL IV 327; PHTF II 158–64; EI II 783–6. OS

Karaosmanoğlu, Yakup Kadri (b 1889 Cairo), Turkish writer. His childhood was spent partly in the aristocratic milieu of Cairo, partly in the provincial town of Manisa, the home of the noble family of the Karaosmanoğlu. He attended a French school in Alexandria, but since 1908 his home has been Istanbul. Here he made his living as a journalist, as a writer (member of the Fecri Ati group, 1909), and as a teacher. In 1924, he became a deputy in the National Assembly, later ambassador (to Albania, 1934–35, Czechoslovakia, 1935–39, Holland, 1939–40, Switzerland, 1942–49, Iran, 1949–51). Retiring from the Foreign Service, he again took up writing and journalism. Once more the staunch 'Kemalist' served as a member of the National Assembly, in 1961–65.

Karaosmanoğlu is the most representative writer for the period of the Greco-Turkish war, in which he actively participated, and of the early Republic, under Kemal Atatürk. The years between the early 1920's and the middle 30's were his most productive, although he started his literary career in 1912 and was still actively writing in the late 1950's. A man of broad culture, great energy, and strong convictions, his novels are remarkable not by the style of writing, which is rather conservative and even somewhat pedestrian, but by the perceptiveness and fullness with which the complex realities of life are unfolded before the reader's eye: the impoverishment of an upper-class family through the war and the collapse of the Empire (*Kiralık Konak*, Palace to Let, 1922), the twilight world of a fashionable representative of a mystical order and of his admirers (*Nur Baba*, 1922), the political circles and wrangles of the corrupt Young Turk period (*Hüküm Gecesi*, Night of Judgment, 1927), the strange, unreal atmosphere in the capital under foreign occupation (*Sodom ve Gomorrah*, 1928), an idealistic intellectual's disillusioning encounter with the Anatolian village during the Greco-Turkish war (*Yaban*, Stranger, 1932), the demoralizing effects of rootless life in emigration (*Bir Sürgü*, Exile, 1937). All of these pictures are painted in bold, apt strokes, and there is a definitive unforgettable quality about them. Further works: novels, *Ankara* (1934); *Panorama* (1935–54); *Hep o Şarkı*, 1956; short stories collected in 3 volumes (1913, 1923, 1947); lyrical essays (prose poems), *Erenler Bağından* (From the Garden of the Blessed, 1922); *Okun Ucundan* (From the Tip of the Arrow, 1940). Mémoirs: *Zoraki Diplomat* (Diplomat against his Will, 1953; *Anamin Kitabı* (My Mother's Book, 1957); *Vatan Yolunda* (On the Road of the Mother-Country, 1958).

PHTF II 589 ff. AT

Kārnāmak I Artakhshēr I Pāpakān (Book of the Deeds of Artakhshēr, Son of Pāpak), a Middle Persian tale, by an unknown author, dating from c600. It tells of the founder of the Sassanian dynasty, Artakhshēr I, and his successors, Shāhpuhr and Ōhrmadz I (224–273). The greater part deals with Artakhshēr, recounting his birth and childhood, his humiliating time at the Parthian court and his escape, helped by a slave-girl who was in love with him; it tells of his subsequent victory over the Parthians, and his Parthian wife's plots against his life. This is followed by the birth of his son Shāhpuhr, his recognition of him, and the story of his grandchild Ōhrmazd. A fundamental element of the composition is the ineluctable action of supernatural forces, determining the fate of man by the constellation of the heavenly bodies and by dreams. This account of the royal fortunes aims to show models of chivalry on the throne of Persia, the wisdom, far-sightedness, courage, heroism, and noble pursuits of the rulers, with their human feelings. There is very little historic fact in the work, whose original form was that of a folk narrative.

RHIL 44 f; TMSLZ. OK

Kartlis Tskhovreba (The Life of Georgia), Georgian chronicles from the earliest times to the fifteenth century, with additions to the end of the 18th century. The basic parts are the works of Leonti Mroveli; *Tskhovreba kartvelta mepeta* (Lives of the Kings of Georgia), *Tsameba tsmidisa da didebulisa motsamisa Archilisi* (The Martyrdom of the Holy and Noble Martyr Archil), and *Juansher's Tskhovreba da mokalakoba Vakhtang Gorgaslisa* (Life and Reign of Vakhtang Gorgasal), all date from the 11th century. *Tskhovreba da utseqeba Bagrationianta* (Lives of the Bagrations) by Sumbat Davitisdze belongs to them, while *Tskhovreba mepet-mepisa Davitisi* (Life of the King of Kings, David), by an unknown author, deals with the period of David the Builder (1089–1125). Other single works deal with the reign of Queen Tamar (1184–1213), the Mongol invasion, etc, up to the decay of the Georgian monarchy in the fifteenth century. During the Georgian national revival (16th–18th centuries) further additions were made. There are several copies of the work, first edited and completed by King Vakhtang VI at the beginning of the eighteenth century. Only the earlier parts have any literary value of their own; the remaining sections are for the most part royal chronicles.

C. Toumanoff, *Medieval Georgian Historical Literature*, Traditio, I, 1943. VAČ

Kāshefī, Ḥoseyn Vā'eẓ, Persian poet, see **Ḥoseyn Vā'eẓe** Kāshefī.

kaside, Turkish poetic form, see **qaṣida.**

Kateb, Yacine (b 1929), Algerian writer and poet, writing in French. He studied at a French lycée, was imprisoned in 1945 and expelled for taking part in a demonstration. He then travelled in Europe and the Middle East. By 1946 he had already published a slim volume of verses, *Soliloques.* His travels inspired sketches and poems published in newspapers. In his *Nedjma ou le poème ou le couteau* (Nedjma, Poem or Knife, 1948), the symbol of Nedjma, the motherland or the beloved, appears for the first time. The same symbol is stressed in the play about the Algerians' fight, *Le cercle des représailles* (1955 in journal, 1959 in book form, *Circle of Reprisals*), influenced by classical drama, and successfully produced on the stage. The novel *Nedjma* (1956) presents a hallucinatory vision of Algeria, and the novel *Le polygone étoilé* (Starry Polygon, 1966) develops a similar set of symbols. Kateb's work is involved, the ideas progressing as it were in a spiral, presenting certain analogies with oral literature. The critics are sharply divided in their assessment of his work. Kateb believes in the writer's freedom not to commit himself and write to order. Other works: play, *L'homme aux sandales de caoutchouc* (The Man in Sand-shoes, 1970).

DLM 173–209. SP

Kātebī Torshīzī (d 1434), Persian poet who was considered by Alīshīr Navā'ī, Doulatshāh and Jāmī (qqv) to be one of the great masters of poetry in the early 15th century. Born in Torshīz or Nishāpūr, he spent some time at the Timurid court of Herat; he also visited different courts in Iran and eventually settled in Astarabad, where he died. His *ghazals* (qv) always consist of seven verses; but his poetry is highly sophisticated and full of rhetorical tricks and is therefore uneven. His *maṣnavī* (qv), *Dah bāb* (Ten Chapters), consists of chapters rhyming in homonyms; another *maṣnavī* of his, called *Majma' al-bahrayn* (The Meeting-point of the Two Oceans, or, of the Two Metres) can be scanned in two different metres. It deals with a mystical love-story in the fashion of those days. Kātebī also wrote a *Khamse* (Quintett) of which only two parts survive. His *Sīnāme* is a collection of 30 love-letter-poems in imitation of the *Dahnāmes,* letter-like poems which were fashionable in early Timurid days. Some satirical verses show Kātebī's wit and technique very well. AS

Katib Çelebi, see **Haci Halifa,** Katib Çelebi.

al-Kawākibī, 'Abd ar-Raḥmān (1849–1902/3), Arab nationalist writer of Syria. Born in Aleppo, he served the Ottoman government there in various capacities and contributed to various Arabic periodicals.

In 1877 he was a candidate for the Ottoman parliament, but was not elected. Friction with the authorities compelled him to move to Egypt in 1898. There, he started writing in the Arabic press against the despotic rule of the Sultan Abdul Hamid II. His two main books, both serialized at first in Arabic periodicals in Egypt, are *Ṭabā'i' al-istibdād* (The Characteristics of Tyranny) and *Umm al-qurā* (Metropolis, a reference to Mecca). The first is a frontal attack on political and religious tyranny and lays the groundwork for Arab nationalism. Translating from V. Alfieri's *Della Tirannide*, al-Kawākibī maintains that tyranny is incompatible with religion, and holds that Islam is a liberal religion. His second book argues that tyranny has forced stagnation on Islam, and that this deprives the Ottoman caliphate of any right to rule the Arabs. He calls on the Arabs themselves to regenerate Islam and to have a spiritual caliph of their own in Mecca, someone from the tribe of Quraysh, on the assumption (an axiom for him) that the Arabs were the better people.

Trans.: S. G. Haim, *Arab Nationalism: An Anthology* (Berkeley 1962).
Kh.S. al-Husry, *Three Reformers: A Study in Modern Arab Political Thought* (Beirut 1966).

JML

Kāẓim Khān Khaṯak, 'Shaydā' (b c1725, d c1780), Pashtun poet. A younger son of Afẓal Khān Khaṭak (qv), Kāẓim spent most of his life in northern India. His poetry is within the Sufi tradition of mysticism, his pen-name signifying 'Lovesick (for God)'.

RSPA 305–25. DNM

Kemal, Orhan (real name: Mehmet Raşit Kemali Öğütçü, b 1914 Ceyhan near Adana, d 1970 Sofia), Turkish writer. His father, a lawyer and politician who in 1930 founded a short-lived oppositional party in Adana, lived in exile in Syria for many years. Kemal received only an elementary education. He worked in a cotton mill, as a social security employee, and in many other jobs; he spent five years in prison (1938–1943), finally moved to Istanbul (around 1950) and devoted himself completely to writing as a freelance journalist, and movie script writer. Starting with poetry in 1937, Kemal later wrote short stories upon the advice of Nazım Hikmet (qv) with whom he spent several years in prison. In the 1950's he began writing full-length novels. His prose is characterized by an ease and accuracy of expression which convince the reader that he has lived everything he writes. His heroes are sharply observed types of Gorkiesque realism, truthfully portrayed in their way of acting and speaking, workers in a cotton mill or migrating farm workers, kicked around, but virile and unbroken. Kemal's poems are to be found scattered in magazines (1939–1946); his short stories are collected in 11 volumes (1949–1968). Novels set in Adana area: *Bereketli topraklar üzerinde* (On the Blessed Soil, 1954), *Vukuat Var* (There Are Events, 1959), *Hanimin Çiftliği* (The Mistress' Farm, 1961), *Eskici ve Oğulları* (The Old-Clothes Man and His Sons, 1962), *Kanlı Topraklar* (The Bloody Soil, 1963). Novels set in an Istanbul proletarian milieu: *Murteza* (1952), *Suçlu* (The Culprit, 1957), and many others. Autobiographical novels: *Baba Evi* (The Father's House, 1949), *Avare Yıllar* (Wasted Years, 1950), *Cemile* (1952), *Dünya Evi* (The House of the World, 1960), *Nazım Hikmetle Üc Buçuk Yıl* (Three and a Half Years with Nazım Hikmet, 1965). Altogether he wrote 27 novels. One play: *Ispinozlar* (Finches, 1965). AT

Kemal, Yaşar (real name Kemal Sadik Göğçeli, b 1922 Hemite, village in Osmaniye County, Province of Adana), Turkish writer. He received only rudimentary education in schools in the Adana area. After having worked in many jobs of varied character he moved to Istanbul (1951) where he worked as a freelance writer and journalist. Several of his novels were made into plays or films. In the 1965 elections he was a candidate of the Workers Party. Kemal is the writer of the largely under-populated and under-developed plains and mountains of south-eastern Anatolia, with their Turkish and Kurdish population of poor peasants, ruthless landowners, and helpless local administrators. His heroes are the people, poor but endowed with superior moral standards and an inherited folk culture. Persons and scenes are perfectly blended into the surrounding landscape. Images and language are dramatic, local speech and folklore elements abound. Since 1955, five novels have appeared, especially: *İnce Memed* (Memed the Hawk, 1955); *Orta Direk* (The Wind from the Plain, 1960); *Yer Demir, Gök Bakır* (1963). Other works include short stories, essays, reportages, poetry.

Trans.: Thilda Kemal, *Anatolian Tales* (New York 1969); Edouard Roditi, *Memed, my Hawk* (New York 1961); Guzine Dino, *Memed le mince* (Paris 1961); Guzine Dino, *Le pilier* (Paris 1966); Thilda Kemal, *The Wind from the Plains* (London, Toronto 1964). AT

Kerbabayev, Berdi (b 1894 Kouki-Zeren), Turkmen writer; educated in the *madrasa*, he later worked in the Leningrad Oriental Institute and edited the poetry of Mahtumkuli (qv). His own work draws on Soviet as well as pre-revolutionary themes; his best-known novel *Aigytly ädim* (Determined Step, 1940) describes the revolutionary years (1915–20) in Turkmenia. Kerbabayev is a classic writer of Soviet Turkmen literature. He has translated Russian classical works. His other writings are collections of poems; *Adatyn gurbany* (Victim of Customs, 1929) and *Amyderya* (1931); stories, *Khekayalar* (Tales, 1941); novellas, *Batyr* (1935), *Aysoltan* (1949); plays, *Kurban Durdy* (1942); *Abadan* (1941); novel, *Nebit-Dag* (1957). LH

Kerem, A., Turkish folk singer, see **âşık.**

Keret, an Ugaritic epic preserved on alphabetical cuneiform tablets of the 14th century BC (see Baal). It tells of King Keret, who lost his wife and children and then, ordered to do so by the god El in a dream, gathered a large army and went to the city of Udum. The king there gave him the beautiful Hurriya as a wife, without putting up resistance. They had seven sons and eight daughters, but only the youngest daughter took care of her sick father and saved his life. The sons used their father's illness to fight for his throne. The text as preserved ends with Keret's curse on his ungrateful sons. The same theme, of a military expedition to win a wife for the king, occurs in the Greek Iliad.

Trans.: ANET 142–9.
DCML; GT. SZS

Kesā'ī, Abū 'l-Ḥasan(?) (b 953, d after 1002), Persian poet, who came from Merv and wrote panegyrics in honour of the Samanids, whose court was the centre of the neo-Persian renaissance. Later, at about 50 years of age, he turned to Shiism and extolled the Twelve Imams, thus introducing a topic which was to become very common with later Persian poets. Ethé stressed the fact that the Ismailite missionary Nāṣere Khosrou (qv) disliked Kesā'ī, probably for religious and political reasons; whereas E. G. Browne regards Nāṣer's relevant verses rather as an expression of a sort of hidden admiration. Kesā'ī belongs to the early masters of elegant style; in his lyrics one finds forms and expressions comparable to that 'New Style' which was then fashionable in Arabic poetry. Descriptions of flowers and objects in partly far-fetched, but always poignant, comparisons are reminiscent of verses in 10th century Arabic poetry.

H. Ethé, *Die Lieder des Kisā'ī*, SB Bayr. Akademie der Wissenschaften, philos-philol. Kl. (München 1874). AS

Ketābe Anūsī (The Book of the Forced Conversation), a Jewish-Persian chronicle of great historical importance. It describes, in the classical Persian poetic form, the persecutions of the Jews under Shah Abbas I and II by Bābā'ī b. Luṭf from Kashan, in the 17th century.

W. Bacher, *Les juifs de Perse au XVIIIe et au XVIIIe siècles d'après les chroniques poétiques de Babaï b. Loutf et de Babaï b. Farhad* (Strassburg 1907). JPA

Khādim, Qiyāmuddīn (b 1912 Kāma), Afghan poet writing in Pashto. He has worked as a journalist, heading the department for the abolition of illiteracy; he was also editor of the journal, *Hēwād.* Khādim writes poems, stories and sketches, and translates. His work is informed with profound patriotism and religious feeling. In a series of quatrains he praised learning and science, and declared that literature must seek new subjects. His most important volumes of verse are *Rūhī gulūna* (Spiritual Blossoms, 1947) and *Khiyālī dunyā* (The World of Thought, 1960); *Bāyazīd Rōshān* (1944) is an interesting study in literary history (see Bāyazīd Anṣārī). *Pashtunwālay* (Pashtunism or a Pashtun code of morality, 1952) shows him as an essayist. JB

al-Khansā' (6th–7th century), Ancient Arabian poet, author of the finest elegies. Her work was bound up with the life of her tribe, the Sulaim, whose fallen heroes she lamented. Her two brothers, Mu'āwiya and Ṣakhr, were among her chief subjects. Towards the end of her life she lived in Medina, close to the prophet Muhammad. Since it was their duty to weep for the dead and those fallen in battle, women are found particularly among the elegiac poets; al-Khansā' was the greatest of them, with a depth of emotion and passionate feeling, and a striking, simple style. Her *dīwān* (qv) has survived and was first published in Beirut, in 1888 (R. Cheikho).

BGAL I 40, S I 70; BHLA 290; EI[1] II 901–2; NHLA 126; ANAP 38. KP

Khalili, Khalīlullāh (b 1907 Kabul), Afghan poet and historian writing in Dari. Of a high official's family, he became a teacher. After holding various official posts he became the monarch's adviser on cultural matters in 1954, and later entered the diplomatic service. A poet in traditional manner, his model is the poet Farrokhī (qv). He writes lyrics, often on philosophical subjects such as the ephemeral nature of life, or its burden. He also introduced new themes such as the struggle against backwardness, or the idea of world peace, but not peace at any price; peace must be based on justice. His favourite forms are the *qaṣīda* (qv) and the quatrain; a number of robā'ī (qv) have appeared, for example, in the volume *Barghāi khazān* (Autumn Leaves, 1957). He has published a number of volumes of verses, eg *Muntakhabāti ash'āri ustād Khalīlī* (Selected Poems by the Master Khalīlī, 1954), and *Dīwāni Khalīlī* (1962, Teheran). A work of serious scientific and historical interest is his *Āṣāri Herāt* (Herat Monuments, 1930–31), a history of the literary figures connected with Herāt from the Anṣārī (qv) period (11th century) to the present day. His historical works include *Salṭanati Ghaznaviyān* (The Ghazna Sultanate, 1954), *Ārāmgāhi Bābur* (Bābur's Tomb), and *Fayẓi quds* (The Consummation of Holiness) which deals with Bēdil (qv). JB

al-Khamisi, 'Abdarraḥmān (b 1920), Egyptian poet, writer and journalist. Lacking the means to finish university, he worked as shop-assistant, conductor, teacher, radio announcer and proof-reader. He developed from romantic poetry to realistic psychological tales, concerned with the ethics of society and the forces working against natural human happiness. He contrasts the longing for happiness of his heroes with the untenable social order which engenders poverty and proves destructive of human lives. He often deals with ethical questions, attacking hypocrisy. Psychological analysis of his characters in the context of their sexual life is characteristic of his stories; his desire to penetrate the human mind sometimes draws him to pathological and perverted types. As a journalist he always supported progressive trends in Egyptian culture. Main works: volumes of short stories, *Qumṣān ad-dam* (Bloodstained Shirts, 1953), *Lan namūt* (We shall not Die, 1953), *Dimā' lā tajiff* (Blood which Will not Dry); essays, *Al-fann alladhī nurīduhu* (The Art we Want, 1966). JO

Khāni, Aḥmade (b 1650 Bayazid, d 1707 Bayazid), Kurdish poet writing in the Kurmānjī dialect. He belonged to the Khāniān tribe which moved from Hakkari to Bayazid at the end of the 16th century; he probably spent most of his life there, but there is little reliable information about his life, and scholars are not agreed on his exact dates. He was the spokesman of the educated, patriotically minded feudal strata. His broad knowledge of Persian, Arabic and Turkish literature, with his poetic talent, enabled him to realize his aim of arousing the consciousness of his people in a work which became the foundation of the Kurdish classical school of poetry. His work is marked by the endeavour to give artistic form to Kurdish folk literature and show that the Kurdish language can produce literature of the highest value. His aim can be seen in the many *ghazals* (qv) on poetics and philosophy, in his versified Arabic–Kurdish dictionary *Nūbār* (First Fruit) and from his greatest work, *Mam u Zīn*. The Nūbār dictionary was written to help Kurdish children learning Arabic, and comprises sixteen chapters each in a different metre. *Mam u Zīn* is based on the popular Kurdish folk tale of the tragic love of Mam and Zin, handed down by itinerant singers (*dangbezh*) under the title *Mame Alan*. The literary version, particularly in its formal aspects, bears clear traces of Arabic and Persian influence. This poem of 2655 couplets became very popular; it is rich in poetic imagery and lyrical scenes. It first appeared in book form in Istanbul in 1919.

Trans.: Jemal Nebez published a prose version of *Mam u Zīn* in German (Munich 1969); Roger Lescot edited the text of the folk tale *Mame Alan* with a French translation (*Textes kurdes II*, Beyrouth 1942).

The International Society Kurdistan (Amsterdam 1969) published the folk version *Mam ū Zīn* with an English translation by A. Ward.

AK

Khāqānī, Afẕaloddīn Badīl b. ʻAlī (b 1121/2 Shirvan, d 1199 Tabriz), Persian poet; he had the *takhalloṣ* (ie pseudonym), Ḥaqā'eqī. His father was a carpenter; his mother was converted from Christianity to Islam. He was educated in the sciences by his uncle, a physician. He was introduced to the court of the Shirvān Shāh Manūchehr by the poet laureate, Abulʻalā Ganjevī, his teacher and father-in-law, but quarrelling with him almost brought Khāqānī's poetical career to an end. After Abulʻalā's death, Khāqānī succeeded him as court poet. His upright character made him unhappy with conditions at court, but attempts to gain a position with the Seljuqs and Khvarazm Shāhs failed. A characteristic feature of his life was steady contact with Christian and Byzantine affairs. He wrote poems of praise of the Byzantine emperors, Manuel Komnenos and Andronikos Komnenos, and the latter intervened in his favour when he was imprisoned by Manūchehr's successor, Akhsitan. Finally disgusted with court life, he went to Tabriz, where he spent the rest of his life in retirement.

With Anvarī (qv), Khāqānī's work represents the culmination of the *qaṣīda* (qv) genre. His *qaṣīdas* are characterized by a didactic element. The introductory parts lay more stress on a description of nature, in contrast with Anvarī's work. Both have in common the frequent use of scientific terminology and obscure allusions, which render their verse difficult and make commentary necessary. Both a religious spirit and a keen awareness of the beauties of nature permeate Khāqānī's poems, while national feeling and a philosophical spirit are characteristic for those of Anvarī. Allusions to Christian concepts are peculiar to Khāqānī; this is exceptional in Persian poetry, in which direct knowledge of Christianity is rare, and stereotyped Qur'ānic misunderstandings about it prevail. The Arabic influence is less strong, although he imitates Buḥturī (qv) and others occasionally. He was influenced by such Persian predecessors as ʻOnṣorī, Sanā'ī and Manūchehrī (qqv). He gives free expression to personal feelings, for instance in the poems mourning his wife and son, without rhetoric devices. Biographical data are also frequent in his works. Besides his *dīwān* (qv), in which he dedicated some poems to ordinary people from the lower walks of life, a work named *ʻAjā'eb ol-gharā'eb* (Curious Rarities), and letters, his fame rests on the first travel-account in verse, the *maṣnavī* (qv) *Toḥfat ol-ʻErāqeyn* (The Gift from the Two Iraqs), which contains autobiographical material and a description of his pilgrimage to Mecca and Medina in 1156/7.

BLHP II 391–9; RHIL 202–8. WH

Khayyām, ʻOmar, Persian poet, see ʻOmar Khayyām.

Khetaegkaty, Kosta (Russian Khetagurov, Konstantin Levanovich, b 1859 Aul Nar, d 1906 Georgiyevsko-Ossetinskoye), famous Ossetian poet and revolutionary democrat, also one of the first Ossetian painters. The son of a poor officer, from 1881 he studied at the Petersburg Academy of Arts, but could not finish his studies because of financial difficulties. In 1885 he returned to the Caucasus and worked in Vladikavkaz until 1891. In the same year he was expelled from the Terek region and banished to Karachay for protesting against the government. In 1893 he moved to Stavropol, one of the cultural centres in the northern Caucasus. There he began writing poetry and working as a journalist. In 1896, after his expulsion had been cancelled, he was allowed to visit his mother country occasionally. He settled at Piatigorsk. In 1899 he was again banished for five years for sedition and lived for some months in Kherson, until he received permission to return to the northern Caucasus except Ossetia. From 1900 to 1902 he lived in Piatigorsk, and from 1902 to 1904 he worked in Vladikavkaz again. In 1904 the poet, seriously ill with progressive paralysis, was taken to Georgiyevsko-Ossetinskoye, where he died.

Khetaegkaty is the greatest Ossetian poet, although the bulk of his work is written in Russian. In his poetry, his plays and journalism he describes the hard life of the Ossetians, denounces Czarist oppression and calls upon the people to fight for freedom and prosperity. Among his best known poems in Ossetian published in the anthology *Iron faendyr* (The Ossetian Lyre, 1899) are *Saldat* (The Soldier), *Sidzaergaes* (The Mother of the Orphans), *Chi dae?* (Who Are You?). In *Uaelmaerdty* (At the Cemetery) he uses a *Nart* theme (qv). His historical poem in Ossetian, *Khetaeg*, remained unfinished. Besides *Stikhotvoreniya* (Poems, 1895) he also wrote great poems in Russian: *Fatima*, *Pered sudom* (At the Trial), *Komu zhivyotsya veselo?* (Who Lives Well?), a satiric imitation of Nekrasov's poem (*Who Lives Well in Russia?*) and a drama, *Dunya* (1890–93). Among his best known journalistic works (in Russian) are *Gorskiye shtrafnyye summy* (Fines in the Mountains, 1898), *Neuryaditsy severnogo Kavkaza* (Confusion in the Northern Caucasus, 1899) and *Osoba. Etnograficheskiy ocherk* (The Ossetian Way of Life. Ethnographical Sketch, 1902). By his mastery of style and form Khetaegkaty has set an example for all future Ossetian poets.

G. I. Kravchenko, *Kosta Khetagurov* (Ordshonikidze). ML

Khety, Ancient Egyptian writer, see **Akhtoy.**

Khorasani style, also known as the Turkestani style (*sabke khorāsānī* or *torkestānī*), in Persian literature it arose in Central Asia (the region now part of Soviet Central Asia and northern Afghanistan) and predominated from the outset to the eleventh century. It was used, for example, by Daqīqī, Rūdakī, Ferdousī, 'Onṣorī, Farrokhī, Gorgānī (qqv) and others; in spite of its name the style was not confined to Khorasan. Its principal feature was simplicity of expression, poetic images taken from real life, few Arabic words and expressions, and the use of old Persian words and expressions from the vernacular. In 19th century Iran there was a movement to return to this simple style (*bāzgasht*) which was highly significant for Persian literary history in Iran. In the eleventh century the Khorasani style evolved into the Iraqi style (qv).

RHIL 112 ff. JB

Khosrow, Khusraw, Indo-Persian poet, see **Nāṣere Khosrou.**

Khraïef, al-Bashīr (b 1917), Tunisian writer. He went to Khaldūnīya school, followed many professions including secondary school teacher, and is now in broadcasting. His first novella (1938) introduced colloquial Arabic to Tunisian literature, rousing great criticism. His first success was the novel, *Iflās aw ḥubbuk darbānī* (Decline, or Your Love has Destroyed me, 1957), again using colloquial Arabic in dialogue; it depicted the Tunisian petty bourgeoisie feeling the impact of modernism; it is highly autobiographical. Equally successful was *Barq al-layl* (1961), an historical novel of 16th century Tunis, aiming not at historical precision but at creating a symbolical folk hero. His novel *ad-Dagla fī 'arājīnihā* (Dates in the Tree, 1969) describes the local customs of a southern Tunisian oasis. The works of Khraïef have been awarded several literary prizes. SP

Khrakhuni, Zareh (real name Artho Tchiumpiushian, b 1926 Istanbul), Armenian poet. Educated at a Mekhitarist monastery, he studied philosophy, psychology and sociology at the University of Istanbul. He now works as an editor and contributor to the Armenian press in Turkey. His verse follows the tradition of modern West Armenian poetry: *K'ar kat'ilner* (Stony Drops, 1964), *Es ew urišner* (I and Others, 1965), *Lusnapartēz* (Moon Garden, 1968). He has published many translations into Armenian, and adaptations from classical Armenian literature (H. Thumanian, qv) for children.

Trans.: S. Boghossian, in ELA. LM

Khushḥāl Khān Khaṯak (b 1613 Akoṛa, Peshawar, d 1689 Cherat), Pashtun national poet. Khushḥāl was hereditary chieftain of a branch of the Khaṯak tribe living on the west bank of the river Indus, below the confluence of the Kabul river. He succeeded his father *Shāhbāz Khān*, killed in battle against the neighbouring Yūsufzays, in 1641. He was a faithful dependant and march-warden of the Mughal emperor Shāhjahān, serving him in peace and war, but soon fell into disfavour under his successor, Aurangzēb. In 1664 he was taken in chains to Delhi and then imprisoned in Jaipur. Finally released in 1669, he resigned the chieftaincy to his son Ashraf (qv) in 1672 and by degrees Khushḥāl abandoned his loyalty to the ungrateful Mughal ruler. His last years were spent in open rebellion, exiled from his home by his treacherous son Bahrām. Khushḥāl fathered a very large family and his poetic gift was inherited through several generations (see 'Abdulqādir Khān, Afẓal Khān, Kāẓim Khān).

Pashto literature had, in its infancy, produced only a handful of poets of whom the best known were either exponents or opponents of the heretical doctrines of Bāyazīd Anṣārī (qv). Khushḥāl outshone this dim clerical light like a new sun. A man of action in many fields, huntsman, soldier and local patriot, moralist and philanderer, throughout his life he expressed his feelings in a series of lyrics and odes. Some of his poems written in prison and exile have a poignancy, and his descriptions of nature a freshness, unsurpassed in either Pashto or Persian. The verse forms of Persian, the *ghazal*, *qaṣīda*, *robā'ī* (qqv), were the model for poetry in Pashto, but the language does not lend itself to the rules of Persian quantitative metre. Instead rhythmic stress patterns were adapted from the national folk-song and combined with the borrowed rhyme schemes to produce new forms. Khushḥāl's large output at least consolidated this new style, if it did not actually formulate much of it. Besides his *dīwān* (qv) of shorter poems, he also wrote a number of *maṯnavī* (qv) poems, including a *Bāznāma*, on his favourite sport of falconry, and a description of the country of Swat.

D. N. MacKenzie, *Poems from the Diwan of Khushâl Khân Khattak* (London 1965); E. Howell and O. Caroe, *The Poems of Khushal Khan Khatak* (Peshawar 1963). DNM

Khvājūye Kermānī, Kamāloddīn Abo'l-'Aṭā Maḥmūd Morshedī (b 1281 Kerman, d 1352 Shiraz), Persian poet, a contemporary of Ḥāfeẓ (qv). Apparently the son of a wealthy family (khvājū means 'little master'). He wandered throughout Iran earning his livelihood as panegyrist at the courts of various dynasties, chiefly the Muzaffarids, and the court of the ruler of Shiraz Sheykh Abū Esḥāq. Khvājūye was attached to the Sufi Order of the Kāzarūniyya and received his spiritual education from Sheykh 'Alāoddoule Semnānī. His work is influenced by Sufi concepts (see Sufism) and is typical of the Iraqi style in Persian poetry. His poems bear a strong resemblance to those of Sa'dī and Ḥāfeẓ (qqv), in form as well as content, but without achieving their force and originality. His *dīwān* (qv) consists of *qaṣīdas*, *ghazals* and *robā'īs* (qqv) but the *ghazals* form the central and most typical part. Khvājū wrote a *khamse* (collection of five epics) of lesser importance in imitation of Neẓāmī and Amīr Khosrou (qqv). It bears testimony to his Sufi convictions in the choice of themes for three mystic epics, *Kamālnāme*, *Gouharnāme*, and *Rouẓat ol-anvār*, while *Gol o Nourūz* and *Homāy o Homāyūn* are romantic *maṣnavīs* (qv). All follow the traditional form.

BHLP III 222–9; RHIL 260–2. WH

Khvatāy Nāmak (Book of Kings, finished probably 635), an extensive Middle Persian historical work, on the history of Iran from the creation of the world. The original form has been lost, but it was translated into Arabic and rewritten and supplemented. It was used by Ferdousī (qv) for his epic, *Shāhnāme*.

RHIL 58, 66; TMSLZ. OK

al-Khwārizmi, Abū ʻAbdallāh Muḥammad b. Mūsā (first half of the 8th century), Arab mathematician, astronomer and geographer. Originally from Khwārizm in Central Asia, he lived in Iraq under the Caliph al-Maʻmūn (813–833), well known for his encouragement of the exact sciences, who commissioned al-Khwārizmī to prepare excerpts from the Indian astronomical tables of Siddhānt. His mathematical and astronomical writings show strong Indian influence; the best known, translated into Latin in the Middle Ages, is the *Mukhtaṣar min ḥisāb al-jabr wa-'l-muqābala* (Manual of the Calculation of Integration and Equation). The term *al-jabr*, in its Latinized form of *algebra*, has found its way into the languages of Europe, while the mathematical term *algorithm* is a distortion of al-Khwārizmī's name. He also translated and edited Ptolemy's geographical tables, supplementing them with current knowledge, under the title *Kitāb ṣūrat al-arḍ* (Picture of the Earth); the map he drew for this work has been lost, but can be reconstructed from the co-ordinates given. His mathematical work was the chief textbook used in European universities up to the 17th century.

G. Sarton, *Introduction to the History of Science* Vol. I (Baltimore 1925), 563 ff. IH

al-Kindi, Abū Yūsuf Yaʻqūb b. Isḥāq (b Kufa, d after 870 Baghdad), Arab philosopher of the Kinda tribe, known as the 'Philosopher of the Arabs'. He lived in Basra and in Baghdad, and suffered persecution under the Caliph al-Mutawakkil. His work was marked by exceptional polyhistoric breadth and universality, dealing with practically all the branches known to science in his day. Only a fragment of his vast work has survived (he wrote over 250 works), partly in Latin (*Tractatus de erroribus philosophorum*, *De Quinque essentiis*, *De somno et visione*). Kindi was the first philosopher of the Arab philosophy (ie written in Arabic) based on Aristotelian and neo-Platonist doctrine. He influenced philosophical thought particularly by his classification of the degrees of reason (*De intellectu*), a theme elaborated further by mediaeval philosophers.

A. Nagy, *Die philosophischen Abhandlungen des Ja'qub ben Ishaq al-Kindi* (Münster 1897); CHPI 217–222, 355. VSD

Kings I and II, Old Testament (qv) books giving the history of King Solomon (qv), the division of his kingdom into those of Israel and Judah, and the history of these kingdoms down to their destruction in 722 and 586 BC respectively. They also recount the deeds of the prophets Elijah and Elisha. See also Bible.

J. Gray, *I & II Kings* (London–Philadelphia 1970). SZS

Kırk Vezir Hikayesi (The Stories of the Forty Viziers), a Turkish work comprising 80 tales in the framework of the story of a faithless queen. The queen tells a story to rouse the king's anger against his son (who has refused her advances), which is followed by a story told by a vizier, proving the prince's innocence by a tale of feminine cunning. After forty pairs of tales the prince is saved and the queen executed. This Turkish collection of stories is a translation from an Arab original, now lost. It was probably translated by Şeyhzade Ahmed Misrî who added numerous rhetorical ornaments, Turkish sayings and poetic insertions. Nothing is known of him, but he probably lived under the Ottoman Sultan Murad II

(1421–1451), to whom the work is dedicated in the majority of the manuscripts. The name Misrî (the Egyptian) suggests that he lived and probably also wrote the book in Egypt, which is also the scene of most of the stories. The work spread rapidly throughout the Turkish cultural sphere, and became popular reading in the broadest sense of the words. It has therefore survived in a great many manuscripts, differing both in length and in the inclusion of several episodes. Extracts from the cycle, and the whole work, have several times appeared in popular Turkish editions. ZV

Kldiashvili, Sergo (b 1893 Simoneti), Georgian writer. He is the son of the writer Davit Kldiashvili (1862–1931) and studied law in Moscow, graduating in 1917. His works consist largely of short stories, of which he has published several volumes, eg *Provintsiis mtvare* (Provincial Moon, 1924); *Svanuri motkhrobebi* (Svan Tales, 1935). He has written a novel about World War I, *Perpli* (Ashes, 1932), and several plays, including *Gmirta taoba* (Generation of Heroes, 1937) and *Irmis khevi* (Valley of Deer, 1944).

Trans.: *The Little Imps*, in *Mindia, the Son of Hogay and Other Stories by Georgian Writers* (Moscow 1961). VAČ

Koriun, known as Skantcheli (the Beautiful, 5th century), Armenian historian, preacher, teacher and translator. Being a disciple of Mesrop Mashtots, the creator of the Armenian script, Koriun wrote Mesrop's biography, probably the first work to be written in that script. On grounds that a nation whose existence is threatened has urgent need of its own alphabet, Koriun preached the right of all peoples to use their mother tongue. Koriun's *Vark' Maštoci* (Life of Mashtots, trans. by B. Norehad, New York 1964) gives an interesting picture of Armenian society in the 4th–5th centuries. His style is at times laconic, at times elevated and emotional, with abundant rhetorical ornament. The difficulty presented by the text made it necessary to produce a more popular edition, the *Pseudo-Koriun*, with additional material from the work of Movses Khorenatsi (qv). Koriun inaugurated a pleiad of 5th century Armenian historians and founded the tradition of eulogist biography in Armenian literature.

AAPJ; NACL; HO 7; THLA. LM

koshma, strophic form, see **Süleyman of Stal.**

Kostandin Yerznkatsi (b 1250–60, d after 1336), Armenian poet. A monk of a monastery near Erznka, one of the main authorities in the development of Armenian mediaeval love and nature lyric poetry. His early verse was religious and didactic; however, it gradually turned to the pleasanter side of life and the beauties of nature, in the manner of Persian poetry and the preceding Armenian *ashughs* (see Frik and *âşik*), only tenuously connected with Christian symbolism.

Trans.: TTA, TRA II, III. LM

Kōyi, Hājī Kādyr (b 1817 Koysanjaq, d 1897 Istanbul), revolutionary Kurdish poet. The dialect in which he wrote was close to Mūkrī. He gained an elementary education in a church school, but was early orphaned and lived in poverty. He travelled widely through Iraqi and Iranian Kurdistan, using his position as a mullah to spread very progressive, revolutionary ideas which more than once got him into trouble. The *dīwān* (qv) of his verses is extensive. His work shows the influence of A. Khānī (qv) to some degree. His poetry expresses the longing of his people for their rights, for independence and freedom. He attacked both secular and religious leaders sharply, and believed that religious fanaticism was one of the greatest dangers for the nation. He was outspoken against sects, and against the abuse of religion for political ends. He exhorted the Kurds to preserve their own language and attacked Kurdish poets for writing nothing but love lyrics; for him education and technical progress were more important. The Turks destroyed many of his poems, and his *dīwān* did not appear until 1925, in Baghdad.

FKH xxxiii–xxxvii. AK

Kulakovskiy, Aleksey Yeliseyevich (b 1877 Zhekhsogonsk, Alexeyev district, d 1926 Moscow), the first Yakut poet, writer and scholar. In 1897 he graduated from the Yakut Secondary School. At first a nationalist in his attitudes, he soon found his way to support for the Soviet regime and was active in forming a new cultural life for the Yakut people. He was a philologist and ethnographer, and a tireless propagator of culture among his people. He founded the nation's literature; his poem *Bayanay algyha* (The Enchantment of Bayana, 1900) was the first literary work ever to be written in the Yakut language by a Yakut, and was the foundation stone of the national literature. In his work, and especially in the ballad *Oyuun tüüle* (The Shaman's Dream, 1910), he criticized life and society in Yakutsk under the Czarist regime, when the rich dominated everything. The publication of his works in 1924–5, under the title *Yrya khohoon* (Songs, Verse), was the most significant cultural event of the time in Yakutsk. He wrote in Russian from 1897 and in Yakut from 1900. MM

Kumarbi, Hurrian god, the main character in a number of myths preserved in a Hittite version, dating from mid-second millennium BC. The first tells of violent changes among the ruling gods; in the third generation Kumarbi seized the throne. Overthrown by the storm god, Kumarbi becomes his revengeful enemy. There is a loose connection between this text (which is very fragmentary, but nevertheless can be said to be influenced by Babylonian mythology, and to show several parallels with the later theogony of Hesiod) and the myth of Hedammu and the song of Ullikummi, describing Kumarbi's revenge. In the myth of Hedammu, Kumarbi allows a great dragon (Hedammu) to grow in the sea, terrible mainly because of his insatiable ravenous hunger. He is not only the enemy of the storm god, but of the whole pantheon. The assembly of the gods sends the goddess Ishtar, sister of the storm god, against him; her songs and her womanly charm enchant the dragon, seduce him, and thus turn the menace away from her brother and his fellow-gods.

In the song of Ullikummi, Kumarbi lies with a mountain and begets a heartless stony monster, Ullikummi, to wreak his revenge on the storm god; Ullikummi rises to the very home of the gods, and is immune even against the charms of Ishtar. At last the wise god Ea, friend of the other gods, helps to defeat Ullikummi by suggesting that he be cut off from the earth, in which he is rooted, and thus rendered helpless. Although the song of Ullikummi is only fragmentary, it is the greatest work of Hittite literature; three tablets have been reconstructed. The term 'song' is taken from the Hittite; the text appears to be rhythmical, and was probably intended for instrumental accompaniment.

O. R. Gurney, *The Hittites* (London 1969); A. Goetze, in ANET[3] 120 ff; H. G. Güterbock, in *Mythologies of the Ancient World* (New York 1961), 156 ff. VS

Kunanbayev, Abay (1845–1904), classical national poet of Kazakhstan, founder of Kazakh written literature. He was educated in the madrasa and came into contact with Russian and European culture through Russian revolutionary democrats exiled in Semipalatinsk. His lyrics took traditional folk forms and enriched them with new poetic techniques. Kunanbayev translated Pushkin, Lermontov, etc; he composed melodies for many of his own and many translated lyrics, which then became national songs (including one from Goethe). He was a propagator of Russian culture, which is one of the themes of his philosophical-didactic volume, *Ghakliya* (Instruction).

PTF 751–3. LH

L

Labid b. Rabi'a (b mid 6th century, d c660), Ancient Arabian poet, one of the 'long-lived' poets (*mu'ammarūn*) and one of those born before Islam and surviving into the Islamic era. He is reported to have accepted Islam. Although his gnomic verse on religious themes is valued by later Arabic tradition, it is mainly forged. On the other hand the pagan poems are a faithful reflection of Bedouin life. Most of them are in *qaṣīda* (qv) form and are inspired by tribal life. His *dīwān* (qv) has survived and been published (Khālidī 1880, C. Brockelmann 1892).

BGAL I 36, S I 65; EI[1] III 1; NLHA 119; WILA 31. KP

Lāhūtī, Abolqāṣem (Tajik: Abulqosim Lohutī, b 1887 Kermanshah, d 1959 Moscow), Persian and Tajik poet and revolutionary. His father was a craftsman with an interest in poetry. Lāhūtī went to school in Teheran, where he came into contact with revolutionary ideas, and took part in the Iranian revolution (1905–11) as a fighter as well as a poet. After the failure of the revolution he was condemned to death, but escaped to Baghdad. Returning in 1915, he emigrated again in 1917 to Istanbul. He returned to Iran illegally in 1921, and led the second Tabriz rebellion in 1922; it was also known as 'Lāhūtī khān's rebellion'. After the city was surrounded by a Cossack force sent by the central government of Iran, Lāhūtī and his companions fled to the Soviet Union, where he lived until his death. From 1925 he spent some years in Tajikistan, helping to organize a modern cultural life there.

His first verses, in the traditional manner, date from 1903; later his poetry was closely bound up with his revolutionary activity. The poem *Be ranjbar* (To a Worker) was written in 1909, and the *ghazals* and *robā'ī* (qqv) written in Baghdad and Istanbul resounded with patriotism. His first long poem written in the Soviet Union, *Kreml'* (1923), celebrates the victory of the oppressed against tyrants; it is in classical ode form. He introduced new strophe forms, created new dimensions and also made changes in the prosody of folk poetry. Several volumes of *ghazals* appeared, and of poems with philosophical, political and love themes. Some of his lyrics have been put to music and are popular among the Tajiks as folk songs. In *Tāj va bayraq* (The Crown and the Flag, 1935) he tried to write a heroic epic on the building up of Soviet society, in the metre of Ferdousī's (qv) *Shāhnāme*. In 1940 he wrote the libretto for the Tajik opera *Kāveye āhangar* (Tajik: *Kovai ohangar*, Kave the Smith).

During the war his themes included the brotherhood of nations, resistance to fascism and the heroism of Soviet soldiers. The volumes published at this time include *Surūdhāye mosallaḥ* (Song in Arms, 1942), *Hadīye ba front* (Gift to the Front, 1943), and *Mardestān* (Land of Heroes, 1944). *Parīye bakht* (Fairy Good Fortune, 1948), about the pre-revolutionary struggle of subjugated peoples against Czardom, was condemned as dangerous, and for some time his poetry rarely appeared. His reputation was restored, however, and his work is held in high esteem especially in Tajikistan, where he is considered the founder of Soviet Tajik poetry. Most of his later books have appeared in Tajik, and some in Persian as well (in Arabic script). His work as a translator is also important, introducing to Persian and Tajik readers the works of Pushkin, Gorky, Mayakovskiy, Shakespeare, Lope de Vega, and others.

RHIL 564–6. JB

Lamiî, Şeyh Mahmud b. Osman (d 1532/3 Bursa), Turkish writer and poet. Of an official's family he early resigned office and devoted himself to literature, as a Sufi supported by the Sultan. Twenty-four volumes of his extensive works have survived, the prose being mainly translation or paraphrase of Persian and Arabic texts, highly appreciated in the literature of the time. Some of them may only have pretended such originals in order to achieve recognition more readily. In the approach used, his *Şeref ül-İnsân* (Man's Honour), a theme from the teachings of the Muslim Brothers of Purity, and the worldly *Latâif-nâme* (Anecdotes), are close to popular literature. Besides his long *dīwān* (qv) and nine allegorical epics in praise of the most earnest submission to the will of Allah, *Şehrengiz* (He who Rouses the Town) is his most striking work; about beautiful lads in

Bursa, it gives a unique picture of life in the Anatolian town. It was this work which made Lamiî the model for later poets.

GHOP III 20 ff; PHTF II 436 ff. ZV

landey, the typical Pashto folk poetry form. The word means 'short', ie a poem of two unrhymed lines, the first of nine and the second of thirteen syllables, ending in *-na* or *-ma*. Each *landey* is a complete unit (like the Japanese *haiku*, see vol. I), usually on a love theme, but often a description of nature, a social subject, or the poet's longing for his birthplace. Many *landey* were composed by women. JB

Leonidze, Giorgi (b 1899 Patardzeuli, d 1966), Georgian poet. He studied theology and then philosophy in Tbilisi (Tiflis). At twenty he joined the symbolist group, Blue Horns, but soon left. Besides his poetry he studied Georgian literature, and was a member of the Georgian Academy of Sciences. He was a lively, temperamental lyric poet of exceptional talent, writing as easily in the old style (love poetry of the oriental type) as in the modern manner (intricate reflective lyrics). In several long poems on the past of Georgia he revealed an epic talent (eg *Samgori, ambavi Tbilisisa da Samgoris*, Samgori, the Story of Tbilisi and Samgori, 1950). This can also be seen in his prose, which combines a poetic imagination with scientific analysis and an essayist's approach. His verse is remarkable for the wealth and vividness of his images, drawing as much on traditional oriental imagery as on the inventions of the symbolists.

Trans.: UAGP. VAČ

Lohuti, Abulqosim, Tajik and Persian poet, see **Lāhūti,** Abolqāṣem.

Lortkipanidze, Niko (b 1880 Chuneshi, near Kutaisi, d 1944), Georgian writer. He belonged to a family well known for its learning, with whom the writers Tsereteli (qv), A. Qazbegi and others were friendly. He attended the classical secondary school and Kharkov University, from where he was expelled for activity in the student revolutionary movement. He went abroad, and studied mining engineering in Loeven, Austria, but did not graduate. Returning home in 1907 he taught German in Kutaisi secondary schools and 1928–39 at the Tbilisi (Tiflis) Technical College. After the establishment of Soviet rule in Georgia (1921) he became Chairman of the Union of Arts Workers and a member of the Executive Committee of the Georgian Writers' Union.

His work falls into several genre categories. One comprises poetical descriptions and tales of village life (gathered mainly in the cycle *Imeri*) and includes also the novel *Bilikebidan liandagze* (From Footpath to Railway Line, 1928), describing life in western Georgia round the revolutionary year, 1905. The second important category comprises tales and novellas called by their author 'non-historical'; although their themes are taken from Georgian history, the approach varies from that of the historical novel (fragments of *Davit Aghmashenebeli*, David the Builder, 1912–44), to parodies on knights 'sans peur et sans reproche' (eg *Raindebi*, Knights, 1923), and to novellas so packed with cruelties (which correspond to the period concerned and taken singly would be realistic) that they come close to grim 'black' humour (eg *Mriskhane batoni*, The Angry Gentleman, 1911, and *Zhamta siave*, The Evil of Time, 1919). The realism of these works does not lie in a truthful account of Georgian history, but rather in their critical attitude to the usual Georgian interpretation of that history.

Lortkipanidze also wrote several cycles of prose poems and symbolical stories (*Panashvidi*, Funeral Song, 1914, etc.). He also tried his hand at drama (*Keto*, 1914; *Sheutankhmebelni*, Those Who Do Not Agree, 1929). Many of his short and longer stories recall film librettos in their construction. This is so particularly in the gradual focusing of attention, and the juxtaposition of details which are apparently unconnected, and also by the careful treatment of the dialogue which in many places is the vehicle of the action, and not the author's own account. In his last completed work, the 'non-historical' story, *Tqved qopilis dabruneba* (Return from Captivity, 1944), Lortkipanidze dropped his ironical tone and achieved a classical simplicity and purity. His work is among the finest ever produced in Georgian literature, and is a fitting model for the younger generation of Georgian writers.

Trans.: Ruth Neukomm, *Die Frau mit dem Kopftuch*, in *Georgische Erzähler der neueren Zeit* (Zurich 1970); *The Woman in Black*, in *Mindia, the Son of Hogay and Other Stories by Georgian Writers* (Moscow 1961). VAČ

Ludlul Bel Nemeqi (I shall Praise the Lord of Wisdom), the title and opening words of a poem in Akkadian wisdom literature. It has been given an analogous title in modern times, the *Babylonian Job*, from the hero of the Old Testament book (qv). The suffering hero of the Akkadian poem is called Shubshi-meshrê-Sakkan. The poem covers four clay tablets and dates from the Kassite period (c1500–1200 BC). In a monologue of 400–500 lines the hero tells his life story; prosperous, in favour with the gods and the king, in good health, fulfilling all his civil and religious duties to the full, he was suddenly plunged into sickness, poverty, misery, and the disfavour of the gods and the king. Rites and incantations were of no avail. Then the gods reveal their decisions to him in three dreams, and the supreme Marduk restores him to his former prosperity and favour. The sufferer passes through the twelve gates of the great temple of Marduk in Babylon; his health is restored to him, for Marduk the supreme, just god rewards the pious and punishes evildoers. Together with the *Babylonian Theodicy* (qv) this poem expresses the religious thought of the Kassite period, when men began to doubt the justice of the divine order and to ask whether misfortune and suffering was really a punishment for neglect of religious duties. The human mind cannot know the true intentions of the gods, and even in the deepest despair man must trust in the benevolence of the gods.

W. G. Lambert, *Babylonian Wisdom Literature* (Oxford 1960), 21–62; LRPOA 328–41. LJKČ

Lugalbanda, mythical Sumerian ruler of the early dynastic period (28th century BC), hero of the two important poems, *Lugalbanda in the Depths of the Mountains*, and *The Epic of Lugalbanda*. Written during the Third Ur dynasty (2111–2003 BC), they survive in several variants. Both poems are loosely connected with the epic, *Enmerkar and the Lord of Aratta* (see Enmerkar), linked by the fragmentary poem *Lugalbanda and Enmerkar*.

Born of the gods, Lugalbanda, ruler of the city of Uruk-Kulaba, leads his army to conquer enemy lands in far-off mountains, accompanied by his seven friends at the head of picked bands of warriors. Far from home he is seized by a strange fever and paralysis, which his friends try to cure with food and drink and healing rites. Failing, they decide to return home, first lamenting their fate before the chief Sumerian gods. The moon god, Su'en, and the goddess of love, Inanna, refuse to help, but the sun-god Utu restores Lugalbanda's strength. In the meantime his army has forsaken him. Lugalbanda then sets out alone to find his companions, among other adventures meeting the mythical bird, Anzu, who offers him wealth and prosperity. Lugalbanda only asks for better luck and for the strength to find his friends, which Anzu helps him to do in return for various services. The army is in terrible straits at the gates of Aratta, whose defenders will not submit. Lugalbanda decides to go back to Uruk alone to bring help, but Inanna tells him how his predecessor, Enmerkar, once conquered that same city, and gives him magic powers. Lugalbanda and his seven friends fight on with their army, and win a famous victory over Aratta, seizing the magnificent treasures of the city.

C. Wilcke, *Das Lugalbandaepos* (Wiesbaden 1969); S. N. Kramer, *The Sumerians* (Chicago 1963), 269–76. BH

Luṭfi as-Sayyid, Aḥmad (b 1872 in Lower Egypt, d 1963), one of the most famous disciples of Muḥammad 'Abduh (qv), studied in Cairo at the School of Law. He was a founder member of the so-called *ḥizb al-umma* or 'People's Party', which developed from the former group of 'Abduh's supporters known as *ḥizb al-imām*. The *ḥizb al-umma* began to issue a periodical known as *al-Jarīda* (The Review), and Luṭfī as-Sayyid became its editor. Through the numerous articles which he wrote in *al-Jarīda*, there emerges a carefully thought-out concept of the Egyptian nation, expressed in terms which derive very much from European thinkers such as Rousseau, Comte, Mill and Spencer. Although the theory and articulation of Egyptian nationalism owed him a great deal, he was never at the centre of Egyptian political life. After the foundation of the Wafd Party, he devoted most of his working life to the Egyptian University, where he was Professor of Philosophy.

A. H. Hourani, *Arabic Thought in the Liberal Age* (Oxford, 1962); J. M. Ahmed, *The Intellectual Origins of Egyptian Nationalism* (London, 1960). RCO

M

al-Ma'arrī, Abu 'l-'Alā (b 973 Ma'arrat an-Nu'Mān, north Syria, d 1058 ibid.) Arab poet and thinker of the Abbasid period, the finest writer of philosphical poetry of the time. Blind from the age of four, he tried to establish himself as a panegyrist, and succeeded at the Hamdanid court of Aleppo. He returned to his home-town, spent some time in Baghdad, for unknown reasons, and returned again to withdraw from the world and live as an ascetic until his death, surrounded by disciples and sought out by scholars. In his early poems (*Siqṭ az-zand*, The Spark of the Kindling Stick), his manner was that of those poets who were attempting a synthesis of the traditional and the new style in panegyric, elegy and occasional poetry; he was particularly influenced by al-Mutanabbī. In the volume of mature verse, *Luzūm mā lā yalzam* (The Necessity of that which is not Necessary, the term for his self-imposed limitation to difficult rhymes), the tone is pessimistic and critical. Here he reveals himself as a bold philosopher touching on all the great problems of humanity without regard for Islamic dogma. He does refrain from criticism of evil in all strata of society. His scepticism and free thinking in religious matters reached its peak in *Risālat al-ghufrān* (Treatise on Forgiveness); here he saw God as impersonal Fate, rejected the dogma of revelation and that of life after death and attacked the leaders of Islam; his view of religion was an historical one, regarding it as the work of man, his training and customs. He is said to have written a blasphemy (*al-Fuṣūl wa 'l-ghāyāt*, Book of Chapters and End Rhymes on the Qur'ān qv). His letters have been preserved (*Mukātabāt*).

BGAL I 254–5; S I 449–54; EI[1] I 79–81; AAP 112; NLHA 313–24; WILA 189–92. KP

maghāzi, numerous traditions of battles and military expeditions made by the Muslims during the Prophet's lifetime; also important Arabic folk epics, some of proved historical basis, often drawing on the Arabic historians (esp. Ibn Hishām, the author of the biography of the Prophet). Most are legendary in character, however, and the heroes (especially Imām 'Alī, later Caliph), are presented with some heroic idealization. The narrative is characterized by the repetition of certain clichés and schematic epic treatment. The prose narrative is interspersed with sections in verse. The classical Arabic used shows traces of the vernacular. The *maghāzī* provide interesting material for folklore study, containing many themes and motifs typical of the Arab heroic epic. They are important for the understanding of the history and developments in Islam, since they show the popular religiosity of mediaeval Muslims. The *maghāzī* cannot be dated; it is assumed that the epic cyclization lasted over a long time. They are linked with the name of the learned Abu'l-Ḥasan al-Bakrī, late 13th and early 14th centuries, who may have had a hand in the partial compilation of the *maghāzī*.

R. Paret, *Die legendäre Maghāzi-Literatur* (Tübingen 1930). JO

Maghrebi, Moḥammad, Persian writer, see **Neẓāmolmolk,** Ḥasan b. 'Alī Ṭūsī.

Mahari, Gurgen (real name Atchemian, b 1903 Van, d 1969 Yerevan), Soviet Armenian writer. Son of a teacher, when his parents were killed in the genocide he was brought up in an orphanage. A member of the literary groups *Hoktember* and *Noyember* (see Tcharents), he was victimized in the thirties and did not return to public life and to literature until the fifties. His early lyrics on the sorrows of his childhood were followed by emotional poems on the building of socialism (*'1920–23?!'*, 1923; *Titanik*, 1924; *Širaki kanal*, The Shiraki Canal, 1925). His short stories combine lyrical and epic elements in the style of Bakunts (qv) (*Otnajayner ayguc*, Steps in the Orchard, 1935; *Lrut'yan jayn*, The Voice of Silence, 1962). In the last period of his work he completed his autobiography (*Mankut'yun*, Childhood, 1929; *Patanekunt'yun*, Boyhood, 1930; *Eritasardut'yun šemin*, Early Youth, 1955), and wrote memoirs and literary criticism.

Trans: J. Champenois, in OeO. LM

Maḥfūẓ, Najīb (b 1912), Egyptian novelist and short-story writer. He studied at Cairo Arts Faculty and has been in responsible cultural posts for some years. His first period (1939–44) gave three novels using ancient Egyptian history as allegory to criticize conditions in monarchic Egypt. In his second ('social') period (1945–57) he turned to contemporary social themes, using methods of critical realism; he aptly described the mentality and environment of the Cairo petty-bourgeoisie between the wars. His finest work from this period is the trilogy *Bayna al-qaṣrayn*, *Qaṣr ash-shauq*, *as-Sukkarīya* (1956, 1957, 1957), a broad well-balanced canvas showing the changes wrought by political and social developments in the life of three generations of a traditional Muslim family in the period 1917–44. In his third (philosophical) period his work changed sharply in form and content, deserting descriptive realist methods for new modes. Plot disintegrates, allegory and symbol appear, character and events represent ideas or spiritual values. Past and present interwoven take the place of chronological sequences; the stream of consciousness, dreamlike visions and broken dialogue are made use of. Maḥfūẓ turns to philosophical subjects and ideological problems; the relation of reason to faith, science to religion, the meaning of life, the new social order, socialism, revolution, etc. The same involved development can be traced in the short stories which began to predominate after 1968. His work reflects the conflicting trends in the Arab world today and the critical points in Arab thought; it is informed by passionate feeling, a relentless search for truth, and profound humanism. Other works: novels, *al-Qāhira al-jadīda* (New Cairo, 1945), *Bidāya wa nihāya* (Beginning and End, 1949), *al-Liṣṣ wa 'l-kilāb* (The Robber and the Dogs, 1961), *aṭ-Ṭarīq* (The Way, 1964), *ash-Shaḥḥādh* (The Beggar, 1965), *Mīrāmār* (1967); volumes of stories, *Dunyā 'llāh* (Allah's World, 1963), *Bayt sayyi' as-sum'a* (House of Ill Fame, 1965), *Taḥta al-miẓalla* (Under the Umbrella, 1969).

Trans.: Trevor Le Gassick, *Midaq Alley, Cairo* (Beirut 1966). JO

Mahmud of Kahabroso (b 1873 Kahabroso, d 1919), Avar poet. He came from a poor family in the Avar region of Dagestan, and became a disciple of the poet Chanka, whom he met when he attended the Arabic school in the mosque. At first he sang Chanka's songs, but later he composed his own. Mahmud incurred the hostility of the district governor (*nayib*) of Hotsatl, Nazhmudin, by falling in love with his daughter. He went away to Azerbaijan, where he stayed until 1905. In World War I he served in the Dagestan cavalry, fighting with the Russian army in the Carpathians. After the Revolution he returned to Dagestan and took part in the revolution there as a soldier; he was murdered in 1919. His surviving works consist entirely of love poetry, passionate confessions and complaints of faithless women, and colourful erotic dreams; his lyrics are full of allusions to classical Arabic literature, which he read first in school and later from block-books. He was thus one of the last 'oriental' poets in Dagestan, at the same time taking his stand alongside the native artists. His best known poem is the narrative *Mariam;* although a Muslim, he compares his lady to the Virgin Mary, having seen a coloured print of her in some Catholic Carpathian home. In his mastery of language and form, Mahmud is one of the greatest Avar poets. VAČ

Maḥmūd Qāri of Yazd, Persian poet, see **Bosḥāq,** Aḥmad.

Mahtumkuli (b c1733), Turkmen poet. Little is known of his life; his father was also a poet. Mahtumkuli was taken prisoner with his family by the Persians, and later travelled widely. His works include secular as well as mystic verse, reflecting the social reality of the day. He favoured the unification of the Turkmen tribes, and was the first Turkmen poet to write not in the Central Asian literary language (Chaghatāy) but in a language close to the Turkmen dialects.

PHTF II 726–7. LH

Majnūn, term used in Arabic, Persian and Turkish literature linked with the late 8th century Arabic poet, Qays b. Mulawwaḥ Majnūn (= Madman), who was famed for his unhappy love for Laylā. He was regarded as the prototype of the lover-poet, destroyed by his passion. The verses attributed to him and the tales told of his love are close to folk fantasy. His historical authenticity is doubtful. His unhappy love became a favourite theme in Islamic literature, particularly in Persian (Neẓāmī, Amīr Khosrou, Jāmī, qqv). In mystical poetry Majnūn became the symbol of the soul, striving through devotion, self-denial and repentance to achieve union with God.

BGAL I 48, S I; EI² I 103; BLHP II 406. KP

Makal, Mahmut, (b 1933, Demirciköyü, Aksaray-Niğde), Turkish writer. He graduated from a Teachers' Training College in 1947 and began his career as a writer with letters and articles sent to the review, *Varlık*, from the Anatolian village where he taught. Published as a book, *Bizim Köy* (Our Village), these writings aroused great interest amongst the Turkish reading public and were soon afterwards translated into many foreign languages. The works of Makal portray the crucial problems of backwardness and social change in the Turkish village; they were the initial inspiration to a whole group of writers who, like him, had graduated as school-teachers. (Of these the best known are Talip Apaydm and Fakir Baykurt.) On the model of *Bizim Köy*, their first works were naturalistic descriptions of present-day life in Turkish villages. Following the success of his first works, Makal went to the university and graduated. Since 1967 he has taught at a secondary school for deaf-mutes in Istanbul. Works: *Köyümden* (From My Village, 1952, under the title of *Hayal ve Gerçek*, Vision and Reality, 1957, 1965), *Memleketin Sahipleri* (The Masters of the Land, 1954, 1958), *Kuru Sevda* (Idle Dreams, 1957, 1964), *17. Nisan* (17 April, 1959), *Köye Gidenler* (Those Who Go to the Village, 1959), *Kalkınma Masalı* (The Lullaby of Progress, 1960).

Trans.: Wyndham Deedes, *A Village in Anatolia* (London 1954); O. Ceyrac and G. Dino, *Un village anatolien* (Paris 1963). DL

Makhdūm Kalla, see **Donish,** Ahmad.

Malachi (c5th century BC), Hebrew book of prophecies. It reproves Israel for wickedness in its religious and private life. See also Bible, *Twelve Lesser Prophets.* SZS

Malāyē Jizrī, Shēkh Aḥmad (b c1570 Jazīrat ibn 'Umar, d 1640 ibid.), Kurdish poet. He wrote in the Kurmānjī dialect. He received an elementary education at church schools in Diyarbakir and Amadiya and probably became court poet to the ruler of Jazirat (now Jizre), but there is practically no reliable information about his life. The authorities diverge considerably in their ideas on when the poet lived. His *dīwān* comprises mainly *ghazals* (qqv), lyric poems with a rich imagery and many philosophical ideas, clearly influenced by Sufism (qv). He had many followers who signed their work '*Malā*' after his death, and much of that work is therefore wrongly attributed to him.

M. Hartmann, *Der kurdische Diwan des Schēch Ahmad von Geziret ibn 'Omar genannt Māl'i*

Gizri (Berlin 1904); D. N. Mackenzie, *Malâ-ê Jizrî anf Faqî Tayrân* (reprinted in *Yādnāme-ye Īrānī-ye Minorsky*, Teheran 1969); FKG 116–117. AK

Malkom Khān Mīrzā (b 1833/4 Isfahan, d 1909), Iranian scholar and politician, an important figure in the Iranian national revival. Of an educated Armenian family, he gained most of his education in Europe and on his return became a teacher at the Polytechnic school recently established in Teheran and run on European lines. The situation in Iran at the time included a despotic system of administration, which was leading the country into dependence on the European powers; this and Iran's general backwardness drove Malkom Khān to devote himself entirely to enlightenment and political work. He was called to court as an adviser more than once, and then was sent into exile for his radical opinions. He is generally considered to have been one of the foremost instigators of the struggle for a constitutional regime. He was the author of political pamphlets and of many articles published in journals furthering the cause of revival, and particularly in his own model paper, *Qānūn* (Law), published in London. He also wrote several plays on social themes. His original prose style, with its clear straightforward sentences shorn of rhetorical ornament, its precise definitions and logical modern way of argument, was an innovation of far-reaching significance at a time when modern Persian prose was coming into existence. Many of his neologisms became the accepted terms in the civil administration.

Trans.: A. Bricteux, *Les comédies de Malkom Khan* (Paris 1933).
AGEMPL 64–9; KMPPL 14–16. VK

Mämmädguluzadä, Jälil (b 1869 Nahichevan, d 1932 Baku), Azerbaijan writer and dramatist. Son of a tradesman, he became a teacher, and early began writing short stories in the realist manner of M. F. Akhundzadä (qv). He also turned Mämmädguluzadä towards translation from Russian and world classics through the medium of Russian. This influence is clearly seen in his stories and plays, ridiculing religious fanaticism, ignorance and social backwardness. From 1906 he published the satirical paper, *Molla Näsräddin*, attracting the leading Azerbaijan writers of the day; it played an important social and literary role in Czarist times and later became the medium for Soviet ideology in Azerbaijan literature (1922–30). Mämmädguluzadä had a lively style and the characters in his stories and plays were well-drawn, whether simple Azerbaijan peasants, their cunning or stupid wealthier countrymen, or Czarist officials, ignorant and alien in their new surroundings. His works are still republished and widely read, for the lively language and the veracity of his presentation, which is unsurpassed by later writers. Stories: *Pocht gutusu* (The Post Box, 1903); *Usta Zäynal* (Master Zäynal, 1928); *Gurbanälibeg* (1907); *Ässäyin itmäkliyi* (The Lost Ass); plays: *Ölälär* (The Deads) and others.

PHTF II, 684. ZV

Mammeri, Mouloud (b 1917), Algerian novelist writing in French. A Kabyle countryman, his native tongue is Berber. He studied in Rabat, Algiers and Paris, and fought in Europe in World War II. Teaching at a *lycée* in Algiers, then Morocco, he is now a professor at Algiers University. His novels mark three phases in the social evolution of Algeria. *La colline oubliée* (Forgotten Hill, 1952) expresses longing for a new life to replace grim reality; it depicts Kabyle village life in World War II and social conflict on various levels. Evidence of a talent for observation, it is written with a feeling for mounting tension. *Le sommeil du juste* (The Sleep of the Just, 1955), also on a war theme, vents social, national and religious problems and condemns colonialism and the senseless traditions of the old world. *L'opium et le bâton* (Opium and a Stick, 1965) depicts the revolutionary war of liberation. Mammeri also writes plays, translated from Kabyle, and writes on Kabyle poetry.

J. Déjeux, *La Littérature maghrébine d'expression française* (Algiers 1970), 143–171. SP

Manas, Kirghiz folk epic, unequalled in

length in any other literature: one version consists of 500,000 lines. It is in three parts, one about Manas, the second about his son, Semetey, and the third about his grandson, Seytek; it includes myths and legends which reflect actual events. The first part is an heroic epic while the second and especially the third tend to romanticize. The epic, which is entirely in verse, is sung by *manaschi* singers, with no instrumental accompaniment.

Trans.: Pertev Boratav, *Aventures merveilleuses sur terre et ailleurs de Er-Töshtük, le géant des steppes. Epopée du cycle Manas* (Paris 1965).

LH

Mandūr, Muḥammad (1907–1965), Egyptian critic. He graduated in Arts and Law, Cairo (1930) and studied classical philology and French literature for nine years at the Sorbonne. In 1939 he returned to lecture at Alexandria and Cairo University, taking a doctor's degree in 1943; he became a journalist in 1944. His first critical work, *an-Naqd al-minhajī 'inda 'l-'Arab* (Arab Methodical Criticism), comprises studies on the methods of classical Arab critics. Mandūr started from the aesthetics of G. Lanson and stressed the soundness of method based on personal impressions and cultivated taste. This phase of his development is represented by *Fī'l-mīzān al-jadīd* (Weighed on New Scales, 1944). In his next phase, while acknowledging the role of subjective impressions in the perception of literature, he stressed the need for rational analysis as the basis for objective conclusions. In the mid-50's Mandūr developed a conception of 'ideological' criticism, according to which literature has an important social function and must be based on the needs of society and of modern man, and committed to the struggle for social progress. He consistently defended literary criticism as an independent discipline and resisted the introduction of methods from other sciences, particularly psychology. Other works: *Fī 'l-adab wa 'n-naqd* (Literature and Criticism, 1949), *An-naqd wa 'n-nuqqād al-mu'āṣirūn* (Criticism and Contemporary Critics).

JO

al-Manfalūṭi, Muṣṭafā Luṭfī (b 1876 Manfalūt, d 1924 Cairo), Egyptian narrative- and essay-writer, and journalist. At the Azhar, he became a devoted disciple of Muḥammad 'Abduh (qv). He occupied a number of minor governmental posts, but only when his patron Sa'd Zaghlūl was in office; at other times he lived modestly by his pen. His most substantial works were free adaptations (at second hand, for he knew no foreign language) of 19th century French novels (Bernardin de St Pierre's *Paul et Virginie*, 1923, and Alphonse Karr's *Sous les Tilleuls*, 1923 under the titles of *al-Faḍīlah* and *Majdūlīn*). Otherwise he wrote essays on social questions and short narratives depicting pathetic situations, but inspiring rectitude and compassion. His shallow philosophy, repetitive stories and cardboard characters were acceptable in his time; his style, pure yet supple, shorn of verbal artifice yet emotionally charged, has influenced many. Collected pieces: *al-Naẓarāt* (Glances, 3 vols., 1902–10); *al-'Abarāt* (Tears, 2 vols., 1916–22).

PC

Māni (b 216/7 Mardinū or Awrūmyā in Babylonia, d 276 Gundēshāhpuhr in Khuzestān), a Babylonian scholar and religious thinker, of Persian origin. Brought up among a sect of Baptists, he proclaimed his own religion at the age of 24. He gained the favour of the Persian king, Shāhpuhr I, but did not succed in making his teaching the state religion of Iran. The ruling circles in the country, fearful of the social consequences of his teaching, imprisoned him, and he died in jail. He was one of the learned men of his time, studying all the sciences then known. The religious theosophical system he elaborated during his travels combined the teachings of Zarathushtra with those of Christianity and Buddhism, and was based on the idea of conflict between good and evil in the world; it was a moral call to man to take the side of the teaching which was fighting for the victory of good over evil. The political implications of his teaching lay in the idea that the new religion represented the highest ideals of those religions which preceded it, and took over their rights.

Gifted with great imagination and an incorrigible love of writing, Mānī wrote many books in the Aramaic dialect of his native region. He wrote under the influence of the gnostic view of the world, the gnostic diction and fantastic style of those philosophers who used the melancholy mood of a decadent age to further their theosophical and mysterious speculations. He drew on the religious thought of his time, incorporating its ideas into his system, and creating a vast vision of the history of the universe and the conflict between the two opposing principles. He illustrated his own writings, thus earning the reputation of the greatest painter of his time. Judging from his later books, which evoked earlier Manichaean poetry, in his odes, hymns and sermons Mānī piled up metaphors. He used similes and paraboles to allow his moods and emotions to appear in his words; he alternated outpourings of melancholy with ardent enthusiasm, and indulged in lyrical descriptions of nature, based on very accurate observation. He represented the cosmopolitan trend of his time, regaling the intellectuals with science-fiction and the broader public with visions of a happier future.

Of his Aramaic works we know of his *Book of Secrets*, *Treasure of Life*, *Great Gospel*, *Book of Giants*, *Pragmateia* (On constant endeavour), a collection of psalms, a book of letters, and others; the works themselves, and their titles, have not survived. *Ardahang* was a description and illustration of the universe and its parts, like a fantastic atlas of struggles between angels, giants and demons, the tragedy of man in an abyss of darkness, and his subsequent liberation. It is a cosmic drama of symbols shrouding the process of salvation of the human soul and its liberation from the shadows. In his *Shāhpuhrakān* (For Shāhpuhr), written in Persian, Mānī used Biblical themes and quotations, often drawn from the *Apocrypha* and the *Book of Revelations*. His works were later destroyed, and are known today only from quotations found in hostile polemic writings and in translations and imitations in the remnants of Middle Persian and Coptic Manichaean literature.

Ch. H. Puech, *Le manichéisme, son fondateur, sa doctrine* (Paris 1949); G. Widengren, *Mani und der Manichäismus* (Stuttgart 1961); O. Klíma, *Mānīs Zeit und Leben* (Prague 1962). OK

Manūchehrī, Abu'n-Najm (d c1041), Persian poet from Damghan. He was first attached to the ruler of Tabaristan, Manūchehr b. Qābūs (d 1028), then joined the Ghaznavid court where he belonged to the circle of Maḥmūd's son Mas'ūd. For that reason he cannot be properly called a disciple of 'Onṣorī (qv), whose mastery he has praised in his famous 'Candle-*qaṣīda*'. Only 2758 distichs of his poetry have been preserved which, however, prove his skill in the use of similes and realistic detail-descriptions. His poetry shows many traces of classical Arabic tradition, so that some of his *qaṣīdas* follow almost exactly the pattern and imagery established by the Arabs. He also lays more stress on acoustic than on visual imagery. This, too, may be due partly to Arabic influence. Manūchehrī is credited with the invention of the *musammaṭ*, a strophic poem which became popular with later Persian poets. He is the only Ghaznavid court-poet whose *dīwān* (qv) was studied in full in the 19th century.

Trans.: A. de Biberstein-Kazimirski, *Menoutchehri, texte et traduction* (Paris 1886).
C.-H. de Fouchécour, *La Description de la nature dans la poésie lyrique persane du XIe siècle* (Paris 1969). AS

maqāma (pl. *maqāmāt*, 'Assembly'), a literary genre, introduced into Arabic literature late in the 10th century AD. Each *maqāma*, although it is an independent unit, is part of a cycle of *maqāmāt* and each one is connected with the others by a loose plot. Al-Hamadhānī is credited with inventing the *maqāma* and al-Ḥarīrī (qv) with perfecting it. Their main achievement was, however, in putting into polished writing a genre that was already being employed orally by story-tellers. The *maqāma* remained in rhymed prose but, unlike the stories told on street-corners or cafés, it ceased to be a popular form, due to its intentionally intricate language. Two characters appear regularly in the *maqāma*. One, frequently a merchant, travels around and has opportunities for meeting the second and telling his exploits. The second, who is the main character, reminds us of the Greek mime and the hero of the later picaresque novella. Prototype of the unscrupulous rogue, he wanders around and lives by outwitting the naive by his resourceful cunning, adroit disguises, literary talent, and linguistic eloquence. Completely amoral, the rogue is nonetheless likeable and must have had the sympathy of both author and audience. *Maqāmāt* were translated into Hebrew in the 13th century, by Yehūda al-Ḥarīzī, who also composed his own; the genre found its way into Persian, Turkish and Syriac. It was imitated in North Africa in the 17th century, and revived in Lebanon in the 19th century by Nāṣīf al-Yāzijī, and in Egypt in 1907, when Muḥammad Ibrāhīm al-Muwayliḥī (qv) published his *Ḥadīth 'Īsā b. Hishām* (The Story of 'Īsā b. Hishām).

Trans.: W. Prendergast, *The Maqāmāt* (Madras 1915); also translated into French and German. BGAL 100–2; 123–6; WILA 174–9. JML

al-Maqrīzī, Abu'l-'Abbās Taqīaddīn Aḥmad (b 1364 Cairo, d 1442 Cairo), Arab historian. After holding various offices in the state administration, in religious life, and as a teacher, he became a judge in Damascus and then in Cairo, where towards the end of his life he retired to devote himself to writing. He is one of the most important Egyptian historians both for the volume of his work and for his interest in the social and cultural aspects of history. The most important of his books, *al-Mawā'iz wa'l-i'tibār fī dhikr al-khitat wa'l-āthār* (Sermons and Learning Concerning Settlements and Monuments), deals with the ancient monuments, topography and history of Egypt, particularly of Cairo. He also wrote a history of the Fatimid dynasty and a detailed history of Egypt under the Mamluks, *Kitāb as-sulūk li-ma'rifat duwal al-mulūk* (Guidance to the Understanding of Royal Dynasties). Other works worthy of note are a monograph on plague epidemics and their consequences, a book on the history of Islam in Ethiopia, and a work on minting and coins in the world of Islam and particularly in Egypt.

BGAL II 38–41, S II 36–37. IH

maqta', final couplet of *ghazal* (qv), see **qit'a.**

Margiani, Revaz (b 1916 Svaneti), Georgian poet. His father was a village teacher and devoted to reviving the national feeling of the Svan people. Margiani learned Georgian both at home and in school, and went to the Languages Faculty of Tbilisi (Tiflis) University. In the last war he was on active service. His verse is mainly meditative and nature lyrics, and much of it echoes Svan folk poetry. His love for the mountains and their inhabitants enables him to understand the Abkhaz poets, whose work he translates.

Trans.: UAGP. VAČ

Marqeh, Samaritan poet and scholar in 4th century Neapolis (now Nāblus) in central Palestine, a Samaritan centre. He composed many hymns which were incorporated in the Samaritan liturgy. An extensive but clearly incomplete collection of commentaries on the *Pentateuch* (qv) and of thoughts on religious subjects has survived under the title *Memar Marqeh* (The Word of Marqeh). He is rightly regarded as the most important figure in Samaritan literature, which includes inter alia the Hebrew *Pentateuch* with many variants from the Jewish version, the Aramaic translation (*targum*, qv), and several chronicles.

J. Macdonald (ed.), *Memar Marqah: The Teaching of Marqah*, I, II (Berlin 1963); J. Macdonald, *The Theology of the Samaritans* (London-Philadelphia 1964). SZS

Marzbānnāme, Persian fable, see **Kalile o Demne.**

al-Mas'adi, Maḥmūd (b 1911), Tunisian writer and dramatist. He studied Arabic in Paris, taught in Tunis, and was for many years Minister of Education and active in politics. His chief work, the eight-act play *as-Sudd* (The Dam, 1955), a symbolical treatment of the conflict between dream and reality, was influenced by classical Greek tragedy. The central idea of the novel *Mawlid an-nisyān* (Birth of Oblivion, 1945) is oblivion in death; his short sketches in traditional Arabic manner are equally symbolical, with recurring themes of beauty, orgiastic dance and purification of the soul by fire. His great originality lies in the choice of themes and his delightful style in classical Arabic. SP

maṣnavī (Turkish: *mesnevi*), an Arabic word meaning 'double'. In Persian, and hence in Islamic treatises on poetry, the word is used for poems in which both lines (*misra'*) of each couplet (*bayt*) rhyme, so that the scheme is aa, bb, cc, etc. This form goes back to the Middle Persian period. The *maṣnavī* form was used for all the *dāstān* (qv) and the two terms are thus often used interchangeably. The *maṣnavī* may have an historical subject like *Shāhnāme* (see Ferdousī), or be romantic like *Shīrīn o Khosrou* (see Neẓāmī), didactic like *Būstān* (see Sa'dī), fairy-tale like *Haft peykar* (see Neẓāmī) or inspired by Sufi philosophy like the *Maṣnavī* of Jalāloddīn Rūmī (qv). In the seventeenth century a new type of *maṣnavī* emerged, the craft poem in praise of the different crafts and describing their tools and methods. Sayyidā Mīrābīd (qv) excelled in this form. JB

Mas'ūd, Dehātī Moḥammad (b c1905 Qomm, d 1948 Teheran), Persian novelist and journalist. He studied journalism in Brussels and on his return edited the weekly *Marde emrūz* (Man of Today); he voiced sharp criticism of social conditions, and this was undoubtedly the reason for his treacherous assassination. In the loosely-connected trilogy of his first three novels he analyzed the problems tormenting the young post-war generation in Iran, their attitudes and their inability to find their place in society. His own scepticism showed itself in a marked tendency to naturalism and to coarseness of expression, but the formal structure of his first novels, and the unfinished project of a many-volume novel, prove his undoubted literary talent. His work: novels *Tafrīḥāte shab* (Night Diversions, 1932); *Dar talāshe ma'āsh* (In Quest of Living, 1932), *Ashrafe makhlūqāt* (The Noblest of Creatures, 1934), *Golhā'ī ke dar jahannam mīrūyand* (Flowers that Grow in Hell, 1942), *Bahāre 'omr* (Spring of Life, 1942).

RHIL 408–9; KMPPL 66–8; AGEMPL 152–4. KB

Mas'ūd b. Sa'de Salmān (b 1046, d 1121), Persian poet who grew up and lived most of his life in Lahore as a feudal lord. His Persian panegyrics interfered with the political struggles of the Ghaznavids. For ten years, till the death of Ibrāhīm b. Mas'ūd, he was in prison. Then he became governor of Jhallandar, but was imprisoned once more for three years. His *ḥabsiyāt*, 'prison poems', are impressive and show his feelings and his nostalgia for Lahore. He also composed *robā'īs* (qv) and poems of the *shahrāshūb* type (describing sweethearts of different professions) which became common in Persian, Turkish and Urdu poetry. His *qaṣīdas* (qv) are not overburdened with rhetorical subtleties but written in clear everyday language, and his poetry often reaches practical wisdom. Mas'ūd used, for the first time in Persian, the Indian form of 'Poems on the Months' which served him for panegyric poetry. He is also credited with some Hindi and Arabic verses. AS

al-Mas'ūdi, Abu'l-Ḥasan 'Alī b. al-Ḥusayn (b Baghdad, d 956 Fustat, Egypt), Arab historian and geographer. The extensive education he acquired in youth was supplemented by wide travels both in the Muslim world and beyond (India, Ceylon, east Africa, and perhaps China). Later he lived in Baghdad and in various towns in Syria and Egypt. A follower of the Mu'tazilite school, he was deeply influenced by the progress then being made in the exact and the natural sciences. From his great knowledge, the experience gained during his travels and contacts with other peoples and religions, he was able to compile a thirty-volume encyclopaedia, *Akhbār az-zamān* (History of the Times), covering the history, geography, anthropology and philosophy of the Muslims, their neighbours and their predecessors. All that has survived of this work is an extract entitled *Murūj adh-dhabab wa-ma'ādin al-jawāhir* (Pastures of Gold and Mines of Jewels), which presents, in lively manner and with many keen observations, the scientific knowledge current in the Muslim world in the 10th century. Shortly before his death al-Masʻūdī summarized his philosophy of history and of nature, and the current philosophical ideas on the gradation between minerals, plants and animals, in *Kitāb at-tanbīh wa'l-ishrāf* (Book of Advice and Revision). He has often been called the 'Herodotus of the Arabs', on account of his wide experience and his new approach to history (not that of a chronicler), and for his ethnographical and geographical interests.

Trans.: Barbier de Meynard, *Les Prairies d'or*, 9 vols. (Paris 1861–77); Carra de Vaux, *Le Livre de l'advertissement et de la revision* (Paris 1897).
BGAL I 143–5, S I 220–1. IH

Matagdān I Yavisht I Friyān (Book of Yavisht the Friyān), a Middle Persian tale by a Zoroastrian priest, of unknown date, about the young saint, Yavisht of Friyān; the tale is also known from the *Avesta* (qv). The wicked magician Akht wants to destroy a town and all its inhabitants, unless someone can solve his riddles. With the help of heavenly creatures Yavisht solves his riddles and puts his own to the magician. Even with the help of Satan the evil creature cannot solve them, and Yavisht kills him. The tale is a collection of riddles in the framework of a simple story.

RHIL 43; TMSLZ. OK

Mathevosian, Hrant (b 1935), Soviet Armenian novelist. He came from the country, and studied education and film-making. In stories and novellas written during the last decade (*Ögostos*, August, 1967; partly filmed in Yerevan, 1968; *Xumhaz*, 1970, etc) he expresses in a highly individual manner the impact of town and country ways of living, the fragmentary nature of inter-personal relations and of the different strata within the same person, the tormented search for ways of achieving once again the lost unity and integrity of personality. In his use of non-conventional methods and in the wealth of information on the inner changes taking place in contemporary Soviet society, his work marks the peak of modern Armenian prose, extending beyond the boundaries of the national literature in its significance.

Trans.: B. du Crest, in OeO. LM

Mawlawi, Persian poet, see **Rūmi,** Jalālod-dīn.

Mazdak (fl c500, birthplace unknown), religious and social reformer under the Persian king, Kavād I (488–531). In the course of the great political and economic crisis which overcame Persia c494 he put forward a religious system which he thought would improve the miserable lot of the lower classes. He gained influence at court and initiated violent action against the wealthy. About the year 524 he was killed, with his principal followers. He called for a just distribution of property and of women and for support for village communities. His *Dīsnād* (or *Dabistān*, Teachings) was destroyed. His name soon appeared in a persiflage entitled *Mazdak-nāmak* (Book about Mazdak), intended to show him up as a charlatan; later writers drew on this source, the contents of which are preserved in later writings.

O. Klíma, *Mazdak. Geschichte einer Sozialen Bewegung im Sassanidischen Persien* (Prague 1957). OK

al-Māzini, Ibrāhīm ʻAbdalqādir (1889–1949), Egyptian novelist, poet, critic and essayist. He was educated at Teachers' Training College where he specialized in English language and literature and began to study classical Arabic literature. After many years as a teacher he turned to journalism and writing. In the first phase he wrote poetry based on personal feelings and experience, pessimistic in mood. In 1921, with al-ʻAqqād (qv), he published a volume of important critical studies attacking arid traditionalism in modern Arabic poetry and prose. In 1924–35 al-Māzinī published many articles and essays, later edited in book form, dealing mainly with European and Arabic literature, literary criticism, and contemporary social questions. In the 30's he turned more to short story and novel writing. His chief work is the novel, *Ibrāhīm al-kātib* (Ibrahim the Writer, 1931), with some autobiographical features, portraying the thoughts of a man unable to reconcile his absolute ideals with reality, failing to stand up to the decisive moments in his emotional life and giving way to resignation. Although the book ends on a pessimistic note, it is warmly humanist and filled with longing for a full life. The critics were highly appreciative of the wealth and high mastery of his language. JO

Mehmed Emin, Turkish poet, see **Yurdakul,** Mehmed Emin.

Melikh-Hakobian, Hakob, Armenian writer, see **Raffi.**

Memmi, Albert (b 1920), Tunisian writer and critic, writing in French. He studied philosophy at the Sorbonne and worked in Tunis. From 1956 in Paris he lectured on sociology at the Sorbonne and directed research into North African culture. His autobiographical novel *La statue de sel* (Pillar of Salt, 1953), tending to naturalism, describes Jewish life in Tunis. The psychological novel *Agar* (1955) deals with the conflict situation of an educated young Tunisian in tradition-bound society, and his wrecked marriage; it indicts colonialism. Several volumes of essays deal with sociological aspects (the colonizer and the colonized, Jewish problems, manipulated human beings). Memmi also edited anthologies and studies of North African literature, and wrote the novel *Le scorpion* (1969).

Trans.: E. Roditi, *The Pillar of Salt* (New York 1955 and 1963, London 1956).
DLMEF 269–99. SP

Merchuli, Georgi (early 10th century), Georgian hagiographer. A priest, he wrote the historical-biographical *Shromay da moghuatsebay ghirsad-tskhorebisay tsmidisa da netarisa mamisa chuenisa Grigolisi* (Deeds and Endeavours of the Meritorious, Holy and Blessed Father, Gregory), known for short as *Grigol Khandztelis tskhovreba* (Life of Gregory of Khandzta); this describes the foundation of the monastery of Khandzta and the spiritual and political work of the founder, Gregory. Written with an eye for detail and psychological observation, many of the episodes are revealing for the student of 9th and 10th century Georgian social and political conditions. There are signs of a latent novelist's talent. The language is rich and polished, embellished with metaphors and allusions to the Bible (qv) and to Greek works.

Trans.: P. Peeters (Latin) *Analecta Bollandiana*, XXXVI–XXXVII (1923); extracts in D. M. Lang, *Lives and Legends of the Georgian Saints* (London 1956). VAČ

Merjani, Shihabuddin (1815–1889), Tatar theologian and historian. He studied first at his father's *madrasa*, but in 1831 he left for Bukhara and later for Samarkand, where he also studied history. Returning in 1849 he was made *imam* of the great mosque in Kazan, and taught in the *madrasa*. He collaborated with the Historical, Archaeological and Ethnographical Society attached to Kazan University. In 1880 he went on a pilgrimage to Mecca, through Istanbul, where he met Ottoman writers and historians. Merjani was not a writer in the usual sense, but a personality influencing the cultural life of his day. He wrote a number of theological and historical works in Arabic, including the history of the Tatars, *Mustefad al-abhar fi ahvali Kazan ve Bulghar* (Information Gained about Events in Kazan and Bulgar, 1887).

PHTF II 766–7. NZ

Mert, Özkan, Turkish poet, see **Özel,** Ismet.

mesnevi, poetic form, see **mas̱navi.**

Mesrop or **Mashtop,** Armenian scholar, see **Koriun.**

Metsarents, Misak (real name Metsaturian, b 1886 Benka near Kharberd, d 1908 Istanbul), Armenian poet. He published from 1903 (also under pseudonym Shawasp Tsiatsan). He wrote nature and love lyrics (*Ciacan*, The Rainbow, 1906; *Nor taǵer*, New Songs, 1907), and prose essays. The main inspiration for his style was mediaeval Armenian lyric poetry, especially Grigor Narekatsi (qv); in form and subject he drew on Durian (qv). Metsarents was one of the West Armenian poets whose work reflected the extremely difficult conditions of life of the Armenian minority in Turkey of that time (see P. Sevak). The hopelessness of their situation drove them to seek an escape in nature and in abstract ideal love. Metsarents' poetry was subject to European influences, including French and English mysticism, pervasive in the local ambiance; it passed these influences on to East Armenian poets (seen, for example, in the early work of Tcharents and Terian, qqv).

HO 7; THLA. LM

Mevlana, Persian poet, see **Rūmi,** Jalālod-dīn.

Micah (end of 8th century BC), Israelite prophet. He declared that the wealthy and powerful who oppressed the people would be punished at the hand of their enemies; in the end God would grant victory and peace to His people. See also Bible, and Twelve Lesser Prophets. SZS

Midhat, Ahmed (b 1841, d 1912 Istanbul), Turkish writer and journalist of the late *Tanzimat* period. The son of a small trader, he led a stormy life, and for his contributions to various papers was exiled to the island of Rhodes. In 1876 he returned to Istanbul and began publishing the papers *Takvimi Vakayi* and *Tercümani Hakikat.* After the Young Turk revolution he lectured at the University of Istanbul, on secular and church history. He was an unusually fertile writer, publishing over 200 works on philosophy, history and religion, and original works and articles. His popular short stories often recall the art of the folk narrator, *meddah,* and use the language of the broad masses. His *Letaifi Rivayat* (Entertaining Tales) in 28 volumes have been analyzed by Horn. He was the creator of the national novel (*milli*), contributing about forty works to the genre. His style is careless and his stories read easily. He also wrote twelve plays of which *Eyvah* (Woe) was translated into German by Doris Reeck in 1913. Ahmed Midhat's books are not so much literary works as entertaining stories with historical and social attitudes and themes.

HGTM 12–30; PHTF II 494–500; KLL II 2617. OS

Midrash (Aramaic and Hebrew: explanation, study), one type of Biblical commentary, expanding details of an account in a literary manner. The earliest is an Aramaic *midrash* on *Genesis,* found in the Qumran caves; later *midrashim* are usually written in Hebrew.

H. L. Strack, *Introduction to the Talmud and Midrash* (New York 1969). SZS

Mir Jälal, Pashaoghly (b 1908 Iranian Azerbaijan), Azerbaijan writer and literary critic. Born in a peasant family, he spent his childhood in Gendje (now Kirovabad), where he later became a teacher. He then studied literature in Kazan and Baku, where he became Professor of Literature at the university. In addition to a number of studies of Azerbaijan literature, he has written many stories and novels describing the changes brought about in village life under Soviet rule, including the rise of a new intelligentsia and the heroism of the people during World War II. Some of his works are satires on bureaucracy and other shortcomings of contemporary life in his country. He also gives literary form to the results of his research in historical novels. Volumes of stories: *Ayaz* (The Great Cold, 1936); *Vätän* (My Country, 1944); *Ilk väsigä* (The First Lead, 1949); *Insanlyg fälsäfäsi* (Philosophy of Humanity, 1961); *Khatirä hekayäläri* (Tales of Memoires, 1962); novels, *Dirilän adam* (Man Awakes, 1935); *Bir gyandjin manifesti* (The Manifesto of a Young Man, 1939); *Yashydlar* (Contemporaries, 1964).

PHTF II 690. ZV

Mirshakar, Mirsaid (b 1912 Sindev, Badakhshan), Tajik poet and playwright. His father was a peasant, but he was orphaned early and grew up in a Khorugh boarding school. In 1930 he graduated from the Communist Party school in Dushanbe and took charge of the young people's paper, *Komsomoli Tojikiston.* Here he published his first verses, in the style of Badakhshan folk poetry. From 1940 for many years he was secretary of the Tajik Writers' Union. Much of his work is devoted to his native Pamir, written in simple language and on subjects that appeal to his Tajik readers; dialogue is introduced into his poems. Mirshakar was influenced by Mayakovskiy, whom he met in 1929.

The pre-war period of his writing culminated in the well-known poem, *Qishloqi tilloī* (The Golden Qishlaq, 1942), based on an old Pamir legend of a Happy Land which generations of the oppressed have longed for; the poet accompanies a band of pilgrims searching for the Happy Land, and finds it at home, where the Soviet had won through. During the war he wrote in praise of the heroism of the Soviet people at the front and in the rear; he did not condemn the German people, but blamed only the Nazis and Nazism. The satirical cycle of poems *Az daftari Afandī* (From Afandi's Notebook) belongs to this group. The best known poem of this period is *Odamoni az bomi jahon* (People from the Top of the World, 1943), about the heroism of civilians working for victory. The poem *Kalidi bakht* (Key to Happiness, 1947) returns to the half fairy-tale theme of hidden treasure in the Pamir, which only Soviet society is capable of discovering. After the war he also wrote *Panji noorom* (The Wild Panj, 1949), his longest poem, on the life of the peasants and frontier guards on the river Panj. The long poem *Dashti laband* (The Lazy Steppe, 1961), is about the campaign to reclaim land in the Central Asian steppes, and faces up to some of the negative aspects of the Stalin cult period. The *dostons* (see *dāstān*) *Ishqi dukhtari kūhsor* (Love of a Mountain Girl, 1966), *Dukhtari chūpon* (The Shepherd's Daughter, 1966) and others are on love themes. Mirshakar has written several plays, for example *Fojiai Usmonov* (The Tragedy of Usmonov, 1951) drawn from contemporary life. His verse for children has made him well known to young readers, and he was one of the first theoretical writers on the subject of children's books.

RHIL 580–1. JB

Mirtskhulava, Alio (real name Mashashvili, b 1903 Khorga), Georgian poet. Of a peasant family, he studied philology at Tbilisi (Tiflis) University and from 1923 at the Bryusov Literary Institute and the Journalists' College, Moscow. His work consist mainly of meditative lyrics, but also includes other lyric genres, epics (*Enguri* 1937) and plays (*Gangashi,* Trouble 1931). He translates classical Russian poetry into Georgian, as well as the work of Mickiewicz and others.

Trans.: UAGP. VAČ

Mishnah (Hebrew: repetition, teaching), a collection of commentaries on the Jewish Law (see *Pentateuch*) arranged in a legal system, consisting of 66 treatises in six series. After similar earlier attempts, Yehuda ha-nāsī ('Prince') edited them at the end of the 2nd century AD, in North Palestine. Containing for the most part legal instructions and comments, the treatise on the Fathers (*Abot*) is of greater interest, being a collection of utterances by Jewish scholars.

Trans.: H. Danby, *The Mishna* (London 1950). SZS

misra', a line of the poem, see **masnavi.**

Mkhithar Gosh (b 1130–40 Gandzak, d 1213 Getik), Armenian writer of fables. An erudite monk, he studied in Cilicia and founded the Nor (New) Getik monastery, a centre of learning. About 200 fables are attributed to him, adapted from Greek and other sources, and from Armenian folk lore; they are written in a laconic style, with lively imagery (*Arakk' Mxit'aray Goshi,* Venice 1842, etc). He introduced into Armenian literature the form of the fable, in which he had many followers (Vardan Aygektsi, qv). His authorship of the *Datastanagirk' Hayoc* (Armenian Code, German trans. by J. Karst, Strassburg

1905) modelled on Theodosius and Justinian, the first of its kind in Armenia, is doubtful.

Trans.: F. Macler, *Choix de Fables Arméniennes* (Paris 1902).
HO 7. LM

Mkhithar Sebastatsi (real name Manuk, b 1676 Sivas, d 1749 Venice), Armenian author. He took the name Mkhitar (the Consoler) at consecration and attained the theological rank of Vardapet (Magister). He is the author of a grammar of classical and modern Armenian, an Armenian dictionary (1745–69), textbook of rhetoric, religious verse, commentaries to religious texts, etc. In 1701 he founded a monastic congregation in Istanbul, recognized by Rome as belonging to the Benedictine order. In 1717 he secured the Island of St Lazarus in Venice for the congregation called Mkhitarist; in the monastery precincts a library, seminary, printing press and hospital were set up. A group of members of the congregation founded a similar order in Trieste, which soon transferred to Vienna. Venice and Vienna became centres of European Armenian studies, important in the history of Armenian culture especially in the early 19th century, when the contact with Europe furthered the modernization of Armenian literature, education and scholarship, and provided for the continuity of Armenian literature during a period of stagnation at home.

The Mkhitarist journals played an important role: the literary *Bazmavep* in Venice (from 1843 up to today), the philological and historical Viennese *Handes Amsōreay* (from 1887 up to today) and others. Among the scholars active in Venice were the historians T. Tchamtchian (1738–1823), author of *Hayoc patmut'iwn* (History of Armenia I–III, 1784–86, trans. by L. Avdall, Calcutta 1827), Gh. Intchitchian (1758–1833; *Hnaxosut'iwn Hayastaneayc Ašxarhi*, Armenian Antiquities, 1852), Gh. Alishan (qv), the Armenian classicist A. Bagratuni (1790–1866; *Hayk dyucazn*, Hayk the Hero, 1858), the philologist G. Avetikian (1751–1827), the literary historian S. Somal (1776–1846); in Vienna the philologist A. Aytenean (1825–1902; *K'nnakan k'erakanut'iwn*, Critical Grammar, 1866), H. Tashian (1866–1933, also a historian), and others. The Mkhitarists published hundreds of classical and mediaeval Armenian texts, catalogues of manuscripts, dictionaries, translations, scientific, historical, philological, philosophical, literary–historical, ethnographic and other studies. Most of the European Armenologists studied in their schools (F. Neumann, J. H. Petermann, etc) and thousands of Armenian children in exile acquired their education in their mother tongue there. The libraries of both centres are famous.

A. Goode, *A Brief Account of the Mekhitaristen Society* (Venice 1835); Le Vaillant de Florival, *Mekhitaristes de Saint Lazare* (Venice 1841 and 1856); G. Kalemkiar, *Eine Skizze der literarisch-typographischen Thätigkeit der Mechitaristen-Congregation in Wien* (Vienna 1898); HO 7; THLA. LM

Mkrtitch Naghash (the Painter, b c1393 Por near Bitlis, d c1470), Armenian poet, miniature painter and priest active in public life. He drew on his own experiences for his poetry, which is less ornamental and less musical than that of nature and love lyrics. His philosophy draws on Grigor Narekatsi and Nerses Shnorhali (qqv), while the emphasis on the social aspect comes from Frik (qv). Mkrtitch introduced to Armenian poetry the theme of the pilgrim seeking protection and subsistence in foreign lands, and usually dying in homesick poverty. These lyrics (*antuni*, qv) were a traditional Armenian folk form.

Trans.: TTA
HO 7. LM

moʻammā, muʻammā, literary verbal rebus, see **Vāṣefī**, Zeynoddīn.

Moʻezzī, Amīr ʻAbdollāh Moḥammad (b about 1048/9 Khorasan, d 1125/7), Persian poet. Son of the poet, Borhānī, he was introduced to the Seljuq court by Malek Shāh's son-in-law, ʻAlī b. Farāmorz Kākūye; he became poet laureate of Malek Shāh and Sanjar. Moʻezzī's *takhalloṣ* (pseudonym) is derived from the honorific name of Malek Shāh Moʻezzoddīn. Between the reigns of Malek Shāh and Sanjar the poet wandered to Herat, Nishapur and Isfahan as panegyrist to Arslan Arghū, Abū Shojā' Ḥabashī and others. Finally he was reinstated by Sanjar. Hit by a stray arrow from the bow of Sanjar, Mo'ezzī died after a prolonged illness caused by the wound. Life as a court poet was kinder to him than to his rival, Anvarī (qv) and brought him success and wealth. His fame rests mostly on his *qaṣīdas* (qv). In form he is inclined more to the early masters of the Khorasani school like Farrokhī and 'Onṣorī than to the artificial poets of the Azerbaijan school like Khāqānī (qqv). Fluency of language, appropriateness of similes and easiness of style characterize his poetry, which is not loaded with obscure allusions, rare words and deliberate difficulty like that of Anvarī and Khāqānī. In fluency, beauty of language and unforced metaphors this work is comparable to that of Sa'dī (qv) or Ẓahīroddīn Fāryābī. It shows the metaphorical language of Persian poetry at its best, with peculiar similes and metaphors, before it became the repetitive stock-in-trade of later poets. Mo'ezzī's love poetry is already tinged with Sufi concepts (see Sufism) and the *sūz o godāz*' of later periods, in contrast to the simple straightforward attitude of the Khorasanian period. He avoided satire and obscenities in his works. His *dīwān* (qv) consists in the main of *qaṣīdas*, while there are a few *ghazals*, *robā'īs* (qqv), *tarkībbands* and *moqaṭṭa'āt*.

RHIL 195; BLHP II 327–30. WH

Mordinov, N. Y., Yakut writer, see **Achchygyya,** Amma.

Moses (Hebrew Moshe), according to ancient tradition, the law-giver of Israel, active in the 13th century BC. For the books attributed to him, see *Pentateuch*. See also Bible. SZS

Moulavi, Persian poet, see **Rūmi,** Jalāloddīn.

Movses Khorenatsi (ie of Khoren), classical Armenian historian, author of the first systematic history of Armenia from the mythological beginnings to the fall of the Arsacid dynasty (428 AD), *Patmut'iwn Hayoc* (History of Armenia; Latin trans. by W. and G. Whiston, London 1736, and by Brenner, Stockholm 1733, French in LCHAMA II). Movses records folk myths, legends, ballads and fragments of old epics (literal or in paraphrase), criticizes his sources (especially Labubna, Mar Abas Katina, ie Maraba, etc) and tries to reach the truth; he rejects miracles and attempts to preserve the chronological order of events. This work, written in polished, grammatically and logically correct, lively and imaginative language and the style, variable but always passionately committed (praising ancestors, the building of cities, battles, Armenian landscapes, or condemning bad rulers and priests and despairing over the ruin of his country), is a valuable source of information about Armenian society of the time; it has influenced Armenian historians from then on. The polemic concerning the date of this work and the author has not yet been concluded; the majority of Western European scholars (including Inglisian, in HO) hold Movses to be later (9th century?), though many Armenian writers still consider him to belong to the fifth century.

A. Carrière, *Nouvelles sources de Moise de Khoren* (Vienna 1893, suppl. 1894); N. Akinian, *Moses Khorenatsi* (Vienna 1930); F. Macler, *Moise de Khoren et les travaux d'Auguste Carrière* (Paris 1902); AAPJ; HO 7; LACC. LM

Movsisian, Alekhsandr, Armenian writer, see **Shirvanzade.**

mu'allaqāt (sg *mu'allaqa*), the most renowned of pre-Islamic Arab odes, numbering from six to ten in different collections. The term means 'suspended'. A legend has it that these odes were so highly regarded that they were written in gold on fine cloth and suspended in the Ka'bah; however, the earliest mention of this is in the 10th century. The word may also be taken to mean 'retained' or, by the same analogy with the stringing of pearls that we find in *manẓūm* (versified), 'composed'. Of the poets to whom these masterpieces are ascribed the oldest is Imru' al-Qays (qv), said to have lived early in the 6th century; the latest may have been contemporary with the appearance of Islam. Doubts have been expressed about the attribution and age of all reputedly pre-Islamic poetry; at all events, the metrical forms, conventional structures, pastoral images and overtones shared by the *mu'allaqāt* leave no doubt that they are the choice products of a well-established poetic tradition.

A. J. Arberry, *The Seven Odes* (London 1957); Mary C. Bateson, *Structural Continuity in Poetry* (Paris 1970). PC

Mukanov, Säbit (b 1900 Kustanay region), Kazakh poet and writer. He was the son of a poor nomad, and was illiterate at seventeen years of age. After the Revolution he went to the university, and worked in various papers and journals. He began writing poetry during the Revolution; his version of the folk legend, *Sulushash* (1928), presented as a verse novel, brought him fame. After the Revolution he came under the influence of Mayakovskiy; his epic talent developed later, culminating in the major success of his novel *Botagöz* (1940). He also engaged in literary research, especially the analysis of folk-lore. It was due to him that Abay Kunanbayev (qv) was established as a progressive historical figure in Kazakh literature. Mukanov was chairman of the Union of Kazakh Writers for many years. LH

Muqimi (in contemporary Uzbek, Amin Mirzo Khūzha Muqimiy, b 1851 Kokand, d 1903), Uzbek poet who lived simply all his life. His lyrics were popular, particularly among folk ballad singers. He was also an excellent satirist, striking at the social inequalities of his time. Among his more important satirical works were *Hezhvi Viktor* (Satire on Victor), *Tanabchilar* (The Surveyors), *Tūy* (Feast), *Choyfurush* (The Tea Vendor), *Avliyo* (The Holy Men), many quotations from which have become proverbial. He gave a new social content to the traditional classical lyric.

PHTF II 398–400, 703–4. NZ

Muratsan (real name Grigor Ter-Hovhannisian, b 1854 Shushi, d 1908 Tbilisi), Armenian novelist. Of a craftsman family, he became teacher, clerk and journalist. From the 1880's he published mainly in journals, such as *Nor-Dar*, *Phordz*, *Luma*, *Meghu Hayastani.* He wrote historical novels of the Arab and Mongol invasions, in the manner of Raffi (qv); plays which are now forgotten; and novels on the decline of the patriarchal way of life and the unhealthy influence of bourgeois ways on the traditional Armenian family. Works: *Gevorg Marzpetuni* (1896); *Andreas erec* (Andreas the Priest, 1898).

HO 7; THLA. LM

Musa Jalil, Tatar poet, see **Jalil,** Musa.

Mūsā, Salāma (1887/8–1958) a Copt, one of Egypt's foremost essayists in Arabic. Born in Zaqāzīq, he graduated from primary school and entered high school, but never completed it. He then travelled to France (1907) and England (1909). His four-year residence in England was crucial to his intellectual formation, particularly his meetings with H. G. Wells, G. B. Shaw, and the Fabians. From his first article in 1909, Mūsā pleaded with his Arab readers to leave Asia and join Europe. For him, Europe meant England, and British civilization became his life-long love. His numerous articles in Arabic tried to synthesize Western thought and popularize it by concrete examples. His book *al-Yawm wa'l-ghad* (Today and Tomorrow), a collection of articles written from 1925 to 1927, caused a furore, because he, a Copt, had dared to lecture the Muslims on modernization. In another book, *Tarbiyat Salāma Mūsā* (The Education of Salāma Mūsā, 1947), he sums up many of his experiences and thoughts. Several of his works develop the idea of 'Pharaonism', promoted by some Egyptian Copts, viz, that ancient Egypt was the glorious period of history, and that the Copts are the true descendants of the Ancient Egyptians.

Trans.: L. O. Schuman, *The Education of Salāma Mūsā* (Leiden 1961).
BGAL S III, 213–5; GLAIG 219–21. JML

musammaṭ, Persian type of poem, see **Manūchehri,** Abu'n-Najm.

Mushfiqī, 'Abdurraḥmān (b 1538 Bukhara, d 1587/8 ibid.), Central Asian Persian-Tajik poet. He lived in Bukhara and Samarkand and twice tried his luck at the court of the Mughal Akbar in Delhi. He is regarded as the last good panegyrist in Central Asia; he wrote a number of *qaṣīdas* (qv) addressed mostly to the Amir of Bukhara, 'Abdullāh Khān. His *Shikāyat az ẓulm* (Complaint of Oppression) is an outstanding poem, describing the sufferings of the broad masses of the people and blaming them on an unscrupulous vizier who had not the good of the people at heart. Mushfiqī also wrote many quatrains (see *robā'ī*), and love *ghazals* (qv) in which the subject is more vitally expressed than was the rule. An important place in his work goes to his *Dīvāni mutā'ibāt*, or *Dīvāni hajvīyāt* (Collection of Satires) dated 1557/8 and comprising some 1,500 *bayts* (see *maṣnavī*); easily understood in both subject matter and language, it became popular reading and the poet passed into folk legend as a mischievous practical joker on the lines of the Turkish Nasreddin Hoca (qv). In addition to the traditional *dīwāns*, Mushfiqī wrote three *maṣnavī* (qv), the greatest of which is the *Gulzāri Iram* (The Garden of Iram, 1571/2).

RHIL 502–4. JB

Müsrepov, Ghabit (b 1902 Kustanay region), Kazakh writer. He was trained at the Agricultural Institute, but broke off his studies to become a writer. He earned his living on newspapers. His main theme was the social and economic changes taking place in Kazakhstan after the Revolution, and the part played by Kazakhs in World War II. His most successful novel was *Kazak soldaty* (Soldier from Kazakhstan, 1948). In his plays he adapts folk legends and presents revolutionary moments in the history of the Kazakhs. The influence of Gorki is clearly marked. LH

Mustafin, Ghabiden (b 1902, Karaganda region), Kazakh writer. He worked very hard to educate himself, and became a journalist. His principal theme is man and the way he changes in his working environment. He also devoted some of his writing to agriculture (eg the novel *Millioner*, 1948) and industry (the novel *Karaghandy*, 1952). In recent years he has turned again to the description of the first period of Soviet rule in Kazakhstan. LH

Mustaghni, 'Abdul'alī (b 1876 Kabul, d 1934), Afghan poet writing in both Pashto and Dari. A believer in enlightenment, he belonged to the Young Afghan movement. He was in charge of the literary page of Ṭarzī's (qv) paper, *Sirāj ul-akhbār*, and then edited *Ṭulūi afghān* in Kandahar. He wrote *ghazals* and *qaṣīdas* (qqv), giving his verse the new themes of the importance of education, work and perseverance. He did not agree with the contemporary movement closer to folk poetry. Most of his poems are scattered throughout periodicals; a collection of didactic *qaṣīdas* is contained in *Gulzāri Mustaghnī* (Mustaghnī's Bed of Flowers). JB

müstezad, Turkish poetic form, see **Haşim,** Ahmet.

al-Mutanabbi (b 915 Kūfa, d 965), 'He who claims to be a prophet'; an Arab poet, representing the climax of Abbasid panegyric and classical verse. Of low origin, he studied in Kufa and then lived for a time among the Bedouins to learn their language (because of its purity); on his return he decided to follow a literary career when he came among Shiites and agnostics. He went to Baghdad (928) and later to Syria, as a wandering singer. An unsuccessful attempt to found a religious-political movement was punished by prison (933), after which he devoted himself exclusively to poetry as a panegyrist in Damascus (937) and later in Aleppo (948) at the court of Sayf ad-Dawla; the nine years he spent there mark the height of his poetic achievement. Intrigues drove him away first to Egypt, back to Baghdad, and further east to Persia, but he was no longer successful, and on his return journey, he was killed in an attack by Bedouins. Arabic literary historians have engaged in great controversies round the personality of al-Mutanabbī, but he is generally agreed to be one of the greatest, if not the greatest, of all Arab poets. He began his poetic career as an admirer of the great panegyrists of the past (Abū Tammām, al-Buḥturī, qqv), on whom he modelled his verse. The time he spent in Syria introduced him to the neo-classical *qaṣīda* (qv), and in the years following his imprisonment his original talent could be seen while he was writing at the court of Sayf ad-Dawla. He was outstanding as a panegyrist, and his eulogies could serve as the prototype for later Arabic poetry of this genre; he was also a master, however, of philosophical verse, love poetry and other traditional genres, with a very subtle style sometimes verging on precocity.

BGAL I 86–8, S I 138–42; EI² 844 ff; NLHA 304–13; GSLA 167–70; A. J. Arberry, *Poems of al-Mutanabbi* (Cambridge 1967). KP

Mu'tazilite school, see **kalām**; **al-Mas'ūdī.**

Muṭrān, Khalīl (? 1872 Baalbek, d 1949 in Egypt), poet and journalist, of a well-known Christian family. After studying in Beirut, he went to Paris in 1890 to avoid Ottoman persecution. In 1892 he followed other prominent Lebanese emigrés to Cairo where he spent the rest of his life. He distinguished himself as a poet and journalist, and worked for *al-Ahrām* (then Lebanese-owned) for almost ten years, in addition to editing for a short time a periodical and a daily paper of his own. The first volume of his *dīwān* (qv) was published in Cairo in 1908, and the complete *dīwān* appeared in four volumes in 1949. His poetry demonstrates great versatility and variety, ranging from traditional 'occasional' verse to narrative poetry and poems of lyrical subjectivity where he concentrates on the emotions and crises of his individual predicament.

RCO

al-Muwayliḥi, Ibrāhīm (b 1845, d 1905 in Egypt), Egyptian journalist. The father of Muḥammad al-Muwayliḥī, he came to Cairo from Istanbul in 1895, and in 1896 published a book entitled *Mā hunālika* (Life Over There); this was about political life in Constantinople, and contained bitter satirical attacks on the Sultan's court and its many intrigues. As a journalist he cooperated with al-Afghānī and 'Abduh (qqv) in Paris in the production of *al-'Urwa al-wuthqā* (The Firm Bond). He owned and edited the newspaper *Miṣbāḥ ash-sharq* (Lamp of the East) which was published in Cairo between 1898–1903: this paper covered many general topics, with news from Istanbul and items of local interest, and articles on literature, politics and religious reform. Both Ibrāhīm and his son Muḥammad used the *maqāma* (narrative in rhymed prose, qv) form as a medium for social and political comment, and Muḥammad's famous book *Ḥadīth 'Īsā b. Hishām* (The Tale of 'Īsā b. Hishām) first appeared in serial form in *Miṣbāḥ ash-sharq*. RCO

muwashshaḥ, a form of Arabic poetry arranged in strophes. It was particularly popular in Muslim Andalusia, whence it spread throughout Arabic literature. Originally a folk form, it was recognized and codified by literary critics. It employs metre and the literary language, and the structure (four, five or six-line strophes) is capable of great variation. The most frequent scheme is *aa bbbaa cccaa* etc, the final *aa* of the last strophe constituting a kind of envoy. The theme most often found is that of love. The *muwashshaḥ* was often set to music and is still a living musical form. In many forms it is still used in Arab folk poetry, and has also been used by modern poets. The *zajal* is a similar form, differing from the *muwashshaḥ* in its use of the vernacular and consequently it has no metrical system.

A. R. Nykl, *Hispano-Arabic Poetry and its Relations with Old Provençal Poetry* (1964); H. Pérès, *La Poésie andalouse en Arabe classique au XIe siècle* (1953); AAP 26; BGAL 109. KP

N

Nabî Yusuf (b 1642 Urfa, d 1712 Istanbul), Turkish poet, of a family of Muslim religious dignitaries, himself an official and later a scribe at the Sultan's court, in which function he took part in military campaigns and civil journeys through different provinces; he later attained higher office at court. From his ethnically mixed birthplace he brought a knowledge of Arabic and Persian, used in his works. Fond of proverbs, Nabî formed new ones in apt similes, which made his verse popular with the common reader, unusual in Turkish classical literature. Besides scholarly translations from Arabic Nabî wrote poems celebrating Ottoman victories (*Fetihnâmei Kameniçe*, The Conquest of Kamenec), travel notes on journeys to Mecca, and descriptions of court ceremonies in Istanbul and Edirna which present a great deal of interesting detail (*Sûrnâme*, Book of Ceremonies). His most important work was the *mesnevi* (see *maṣnavī*) *Hayriye*, advice to his son Abdülheyr, which gives a picture of the life of the time and the author's ideas on society. The poets later classified in the second (realist) school of Bakî (qv) drew on this work.

GHOP III 325 ff; PHTF II 442 ff. ZV

an-Nābigha adh-Dhubyānī (6th century), Ancient Arabian poet at the royal courts of Ḥīra (Lakhm) and in Syria (Ghassān). There is no doubt about his historical authenticity; he came from central Arabia and won popularity and favour at the Lakhmid court. Although he was later driven out, and established himself at the rival Ghassanid court, he returned to his earlier lords, whom he had never forgotten in his panegyrics. His poems are not attributed to him with complete certainty, however, particularly his long *mu'allaqa* (qv). Their elegance and spontaneity are alike appreciated. His *dīwān* (qv) has survived and was first printed in 1868–69 (H. Dérenbourgh).

BGAL I 22, S I 45; EI[1] III 859; NLHA 121; C. J. Lyall, *Translations of Ancient Arabian Poetry Chiefly Preislamic* (London 1885). KP

Naci, Muallim (b 1850 Ömer Hulusi, d1893 Istanbul), Turkish poet, writer and critic, a teacher and later government official. He travelled much in the Ottoman Empire, but devoted his later life to literature. To save Turkish literature from a threatened decline, and from European influences, he attacked new trends (see Kemal Namık), and attempted to revive the tradition of the *dīwān* (qv). Many of his *ghazals* (qv) became popular songs. Although officially accepted as a poet, his verse was not in the mainstream of Turkish literature; his short stories later earned him a lasting place in modern literature, especially *Ömerin Çocukluğu* (Omar's Childhood), childhood reminiscences far in advance of his time in their form, and in fact running counter to his own principles, since their language was near to the colloquial. Chief works: *Ertoğrul Gazi; Sünbüle* (The Ear); *Küçük bir müzhike* (A Little Comedy); *Ateşpare* (Sparks).

GHOP III and IV; PHTF II 435 ff. ZV

Nāderpūr, Nāder (b 1929 Teheran), Iranian poet. After private schooling at home he went to a French University, and now devotes himself to writing. He is a leading exponent of the new poetic school, in his own verse, in his theoretical studies and in his personal creed. He believes that the poet's ways of expression and the structure of his verse are conditioned by his view of the world, and by his wealth of mind and imagination. He himself is indeed a poet of a broad gamut of idiom; he has a sense for melody in verse, and for a variety of lyric moods, varying from happy experiences, love which is far from mystical, to thoughts of man's eternal companion through life, the presence of death, not a horrifying thought, but the natural culmination of the human lot. His work: volumes of poetry *Chashmhā va dasthā* (Eyes and Hands, 1954), *Dokhtare jām* (The Daughter of the Bowl, 1955), *Sormeye khorshīd* (The Marvels of the Sun, 1960).

RHIL 406; AGEMPL 228–9. KB

an-Nadīm, 'Abdalla (b 1844 in Egypt, d 1896 Constantinople), one of the early Egyptian nationalists who took part in the abortive 'Urābī Revolution in 1881–82. When that movement collapsed after the British occupation, he was obliged to go into hiding. Others of his collaborators such as 'Abduh, 'Urābī and al-Bārūdī were exiled, a fate which he shared when he reappeared briefly in 1891. Al-Nadīm was an orator and political journalist of great talent and reputation, and was one of the chief spokesmen of the 'Urābī movement. Most of his writing is strongly patriotic, and not devoid of xenophobic elements, probably due to his extreme interpretations of parts of the teaching of al-Afghānī. A selection of his writings was published in Cairo 1897–1901, entitled *Sulāfat an-Nadīm* (Anthology of al-Nadīm). RCO

Nafīsī, Sa'īd (b 1896 Teheran, d 1966 Teheran), Iranian writer and critic, from a family with a long literary tradition. He completed his education in France. He was a Professor at Teheran University and devoted himself both to the study and practice of literature. He was a leading figure in the cultural life of Iran. He began by writing short stories about life in Iran at a time when modernization had just begun. Later he turned to historical themes, and his stories were popular for their patriotic ardour. His first novel, romantic in tone, was followed by critical satirical novels on contemporary Iranian life. The author's intention to criticize conditions in his homeland often became more important than the literary criteria in these works. Stories are: *Setāregāne siyāh* (Black Stars, 1937), *Māhe nakhshab* (Moon from Nakhshab, 1955) and novels: *Farangīs* (1931), *Nīme rāhe behesht* (Half-way to Paradise, 1953), *Āteshhāye nehofte* (Hidden Fires, 1960).

RHIL 412–3; AGEMPL 168–70. KB

Nahapet Khutchak (the Youth, or Bard; b early 16th century, d 1592), Armenian poet-*ashugh* (see *âşik*). Learned and well-travelled, he lived in the vicinity of Akn, a centre for trade and the crafts, and was buried in Kharakonis near Lake Van. Many legends are connected with his life and until recently his tomb was a place of pilgrimage. He wrote, in Armenian and Turkish, hundreds of *hay(e)ren* (quatrains, see Frik) on nature, love, pilgrimage and philosophy. Some are close to Armenian folk songs, and may be versions or adaptations. Nahapet is sometimes compared to 'Omar Khayyam (qv). He had a great influence on the Armenian *ashughs* of the 16th–17th centuries (Hovasaph and Ghazar Sebastatsi, Hovhannes, Gharip Mkrtitch, etc) and on 19th century Armenian verse.

Trans.: H. Bethge, *Die Nachtigall Armeniens* (Berlin 1924); TRA II; TTA; TPAAM.
HO 7; THLA; FGAL. LM

Nahum (end of 7th century BC), Hebrew book of prophecies, celebrating the judgment of God which befell Nineveh, the the capital of the cruel Assyrian kingdom. See also Bible, and Twelve Lesser Prophets. SZS

Naimy, Lebanese critic and writer, see **Nu'ayma,** Mikhail.

Nāji, Ibrāhīm, Egyptian poet, see **'Abdal-mu'ti Ḥijāzi,** Aḥmad.

Nalbandian, Mikhayel (b 1829 or 1830 Nor Nakhidjevan, d 1866 Kamyshin), Armenian writer and critic. Son of a smith, he studied, but did not finish the courses in theology, philology, philosophy, medicine and the natural sciences (Moscow, St Petersburg); parting ways with religion he made contact with Russian revolutionary democrats in Moscow and London. For political reasons he travelled in Europe and India, helping to prepare the Armenian revolt in Istanbul. In 1858–60 he was one of the most active collaborators in the journal, *Hyusisaphayl* (Aurora Borealis, edited by S. Nazarian, Armenian counterpart to the Russian *Sovremennik*), proclaiming the acceptance of colloquial Armenian (*ashkharhabar*) as a modern literary language. For his relations with Hertsen, Bakunin and others, Nalbandian was arrested and condemned to exile, where he died. His principal contribution to Armenian literature lay in the analysis of the writing of the mid-nineteenth century, especially that of Abovian and Proshian (qqv). He was the first theoretical writer on modern Armenian literature, and the founder of modern literary criticism and aesthetics. He also wrote patriotic and political verse, satires and novels, and polemics on many aspects of Armenian cultural life and European philosophy. His correspondence and translations (Béranger, Heine, Pushkin, etc) are also of value.

HO 7; THLA. LM

Namık, Kemal (b 1840 Tekirdağ, d 1888,

island of Chios), Turkish poet and writer. Of an aristocratic family, his education was sporadic. His plays and novels became literary models; he revived Turkish literature and revolutionized national life. He was a member of the Young Ottoman Committee (*Yeni Osmanlılar Cemiyeti*), published their papers *Muhbir* and *Hurriyet*, enlivened them with his passionate temperament and was rewarded with prison and exile. Freedom and the homeland were his slogans. His plays, *Vatan Yahut Silistre* (1873) and *Akif Bey* (1874), reveal his ardent love for his country; *Zavallı Çocuk* (Pitiful Child) deals with the sociological problems of Islamic marriage; *Gülnihal* (1875) is a play of woman's revenge, and *Karabela* (Terrible Disaster, 1910) of the suppression of woman's honour. Besides his historical play, *Celalettin Harezemşah* (1885), he wrote historical novels like *Intibah* (Awakening, 1876) and *Cezmi* (1880). He was firmly entrenched in the Islamic faith and its culture, and wrote *Renan Müdafaanamesi* (published by Fuad Köprülü, 1962) in reply to Ernest Renan, who regarded Islam as hostile to culture.

HGTM 30–4; PHTF II 470–3, 486–9. OS

Nar-Dos (real name Mikhayel Hovhannisian, b 1867 Tiflis (Tbilisi), d 1933 Tiflis), Armenian novelist. Teacher, later an editor, from 1884 he wrote verse, plays, and stories for the journals *Nor-Dar*, *Taraz*, *Aghbyur* and *Surhandak*. Later he wrote predominantly in prose and his main contribution was in the psychological novel describing urban intellectuals at the turn of the 19th–20th centuries in the great cities of Transcaucasia. Influenced by the Russian critical realists, especially Goncharov and Turgenyev, he comments on the changes taking place in social and family relationships, and analyses the minds of his characters (*Anna Saroyan*, 1888; *Spanvac aghavni*, The Dead Pigeon, 1898, new version 1913; *Payk'ar*, The Struggle, 1911; *Mah*, Death, 1912). Nar-Dos had a talent for narrative and paid great attention to the language of his books, thus helping to refine the eastern variant of modern literary Armenian.

THLA. LM

Narts, a legendary race of heroes in the folk epics of the peoples of the northern and western Caucasus. In the Circassian and Abkhazian versions the mother of the Narts was the Lady Satanaya, who had from 9 to 100 sons, in different versions. The youngest son, Sosruqo (Soslan in the Ossetian), was born of stone and tempered by a divine smith. In other versions the Narts comprised several families, according to which the epic has several branches. The epics, apart from mythologized explanations of the origin of things (eg iron, millet, wine) and trades (that of the smith and the farmer), seem to be based on historical reminiscences of fighting and migration experienced by different peoples. Some of the motifs were taken over by the Greeks (eg Prometheus) showing that the epics may have been in existence at the beginning of the first millennium BC. Some parts have been handed down in verse (2–4 stress tonic lines using alliteration, similar to Old Germanic), others in prose. The most highly developed and best known versions are the Ossetian; the Circassian and Abkhazian differ from them in having more archaic features. The epics have been recorded and studied since the first half of the 19th century, and many editions have been published both in the original languages and in translation.

Trans.: Georges Dumézil. *Le Livre des Héros: Légendes sur les Nartes* (Paris 1965). VAČ

Nāṣere Khosrou, Abū Mo'īn al-Marvazī al-Qobādiyānī (b 1004 Qobādiyān, d 1072–77 Yomgān), Persian poet. Descendant of a family of government officials, after an apparently cursory education he followed suit in Balkh. Confusion of his life with Naṣīroddīn Ṭūsī (qv) and a later spurious autobiography soon shrouded him in a mist of legends, until it was even suggested that two persons had borne the same name. The poet himself tells of a vision forcing him to renounce his post as an official and undertake a pilgrimage to Mecca. After this he claims to have been converted to the Ismaili sect in Cairo and attained the rank of *ḥujjat*, the fourth in the spiritual hierarchy. He could not remain in Balkh on account of his convictions and teachings. The rest of his life was spent at Yomgān in the mountains of Badakhshan, devoted to literature. The identity of the amir whom he served before his pilgrimage is unknown, as is that of the amir who gave him refuge after he was banished from Balkh; there are reasons to assume that it was the amir of Badakhshan, himself an Ismaili, who in both cases influenced the life of the poet. This would mean that he adopted the Ismaili creed as the outcome of a slow inner development, and that the pilgrimage was a journey to the centre of the sect, the Fatimid Caliphate in Egypt.

In any case, his description of this journey, the *Safarnāma*, is one of the best specimens of early Persian prose before the overladen florid prose of later times, and is his most famous work. It gives a vivid description of his journey from 1045 to 1052, starting from Juzjānān and passing through Marv, Nishapur, Tabriz, Aleppo, and Jerusalem before finally reaching Mecca, visiting holy places en route and meeting famous poets of the time like Qaṭrān and Abū'l-'Alā al-Ma'arrī (qqv). His *Safarnāma* and other works have come down to us in versions largely purged of Ismaili tendencies by the copyists. Ismaili tendencies, rejection of Sunnite views and personalities and some approval of Sufi ideas are more freely expressed in his *dīwān* (qv). Originally claimed to have contained over 30,000 verses, it now contains about 11,000 and consists almost entirely of *qaṣīdas* (qv). The *dīwān* yields as much material for a biography of the poet as the *Safarnāma*, perhaps even more, as some of the data in the latter seem to be deliberate fiction. Ismaili terminology, using common words in a hidden sense familiar only to the initiated, leaves much open to further discoveries, apparently clear as it may seem on the surface. Free use of *ta'vīl*, the allegorical interpretation of the *Qur'ān* (qv), is typical for Nāṣer as for all Ismaili authors. Remarkable is his championing of man's free will against predestination. The concept of man as a microcosm, although not only an Ismaili idea, underlies many of his verses, which are written in more of a philosophical spirit than from genuine poetic urge. Tradition also ascribes to him sceptical and cynical verses in the manner of 'Omar Khayyām (qv).

Further works are two ethical and didactic *maṣnavīs* (qv) *Rūshanā'ināme* and *Sa'ādatnāme* (Book of Bliss), while the *Zād ol-mosāferīn* (Sustenance for Travellers) and the *Vajhe dīn* (The Countenance of Faith) are in prose. They cover theology, metaphysics and science and are especially interesting for the author's coinage of philosophical concepts in Persian, the language for such topics at that time being usually Arabic. The Ismaili standpoint is less outspoken here, but the reader is gradually led to accept its ideas via discussions in the atmosphere of the *Bāṭiniyya*, the doctrine of the hidden interpretation of revelation. The same method prevails in numerous other tracts, aided by quotations from Greek philosophers and supported by an encyclopaedical unfolding of the knowledge of the time; the *Jāmi'ol-ḥekmateyn* and the *Omm ol-ketāb* may be mentioned.

Trans.: G. le Strange, *Diary of a journey through Syria and Palestine* ... (London 1888).
W. Ivanov, *Problems in Nāṣir-i Khusraw's Biography* (Bombay 1956); BLHP II 218–46; RHIL 185–9; SPL 1138–41. WH

nasīb, the amatory prelude of the classical Arabic ode. The earliest odes extant deal with a succession of conventional themes, of which the first was reminiscences about a lost love. To spark these off, the commonest opening situation was that the poet came across the traces of an abandoned desert encampment, and recognized it as one in which he had met his lady love; in another, a vision of the beloved appeared to him during a lonely night. A sensual and often detailed description of her charms followed. Ibn Qutayba (d c889, qv) suggested that it was in order to establish rapport with his audience that the poet started with a subject of which nearly every man has some experience, 'lawful or unlawful'. Some kind of amatory introduction remained a feature of many odes down to the 20th century. The term is sometimes extended to all love-poetry, which developed as a separate theme mainly in Umayyad times, though it has clearly traceable pre-Islamic roots in the *mutayyamūn*, the 'love-possessed' poets.

A. Kh. Kinany, *The Development of Ğazal in Arab Literature* (Damascus 1951). PC

Nāṣif al-Yāziji, Lebanese writer, see **maqāma.**

Nāṣeroddin Shāh, Qājār (reigned 1848–96), Persian writer. He was the most important ruler of the Qajar dynasty and in 1873, 1878 and 1889 travelled through Europe to acquaint himself with western progress. A number of administrative, military and cultural reforms were instituted during his reign. He was assassinated by a fanatical pan-Islamist. Two diaries, *Safarnāmeye Farangestān*, contain the impression and knowledge he gathered during his travels in Europe; the first appeared in 1874. He used the same method to describe his journeys through Iran and what is now Iraq. The diaries, which are still edited in Iran, reveal him as a sensitive and accurate observer of people, social life, technical progress and nature. His narrative accounts, extremely interesting in their subject-matter, suffer somewhat from excessive descriptiveness. The importance of the diaries lies in their stylistic and linguistic innovations; the aim of factual communication led the author to write simply, without rhetorical ornament, and even not to avoid colloquial constructions. Unique in the literary context of the day, the diaries made a considerable contribution to the formation of contemporary written Persian. These travel journals were later imitated by Nāṣeroddīn's courtiers and by his son and successor, Moẓaffaroddīn Shāh. He also wrote poems in the traditional style, particularly that of the Khorasan poets, and compiled an anthology of classical verse.

RHIL 336; BLHP IV 155; AGEMPL 25. JOS

Nasiri, Kayum (1825–1902), Tatar writer and educationalist. He studied at the *madrasa* in Kazan, where he learned Russian; he became a teacher at the Russian Mission School and met Russian writers and intellectuals. His interests were wide and varied; he studied linguistic theory and applied it to Tatar, studying the spoken language and codifying a written form on the basis of its principle features, writing a grammar, rules for spelling, and a dictionary. He also published a study of *Tatar syntax* (1860). He collected and published folk literature, and between 1871 and 1897 published *Kazan kalendari*, an almanack of articles on subjects from mathematics to psychology and health education; he also collaborated with the Historical, Archaeological and Ethnographical Society attached to Kazan University.

PHTF II, 764–5. NZ

Naṣiroddin, Abū Ja'far Moḥammad b. Moḥammad Ṭūsī (b 1201 Tus, d 1274 Baghdad), Persian polyhistorian writing in Persian and Arabic. He served under the last two Grand Masters of the Assassins, the Isma'iliya terrorist sect. Finally he became court astrologer to the Ilkhan Hülägü (1256–1265) and Abaka (1265–1283). He translated mathematical and astronomical works from the Greek, and himself wrote on many subjects, particularly philosophy, dogma, mathematics and astronomy. His best known works are the Persian *Akhlāke Nāṣerī* (The Ethics of *Nāṣer*) for Nāṣeroddīn 'Abdorraḥīm b. Manṣūr, the Isma'ili governor of Kuhistan; the Arabic *Tajrid al-kalām* (The Revelation of Scholastics), an outline of Shi'ite philosophy; the Persian *Zīje īlkhānī* (Ilkhan Astronomical Tables) drawn up by order of the Ilkhan Hülägü; and the Persian *Bīst bāb dar astorlāb* (Twenty Chapters on the Astrolabe). FT

Nasreddin Hoca, the hero of Turkish folk humour. Often appearing to be stupid, this sly, wise and good-hearted man loves life in spite of its miseries, and pits his common-sense against the treacheries of fate. Nasreddin is no hero; often time-serving, he is never cowardly or mean. His stupidity often serves to show up the absurdities of society and the even greater stupidity of the powers-that-be; such are the tales of Nasreddin's encounters with Timur the conqueror. The character of Nasreddin is not clear-cut; in the course of time comic, sarcastic, wise and profound episodes have accrued to the legend, as well as low jokes. Over 500 themes had been noted by the end of the last century. The core, however, remains that of the sly, simple fellow who gets by. So much for the Turkish Nasreddin, but he has found his way into the Balkans and Russia, to the Caucasus, Iran, and Central Asia, with new accretions to the legend. In Azerbaijan Molla Nasreddin has turned his satire on backwardness even in recent times; in Central Asia he helped to ridicule his feudal masters under the name of Mushfiqī (qv) or Apandī.

There has been much argument as to whether a real Nasreddin ever lived; it is denied by some. The Turkish legends cover two main periods of time: in the reign of the Seljuq Sultan Alaeddin II there actually lived a Hoca Nasreddin, whose grave gives 1284/5 AD as the date of death. The poet Lamiî (qv) believed him to have been the original Nasreddin. The second set of tales brings together Nasreddin and the great conqueror Timur (d 1405). A hundred and fifty years before Timur's invasion of Asia Minor there were raiding Mongol forces against whom the warrior Nasreddin Mahmud fought famously, becoming the hero of folk songs as Nasreddin. The people of Anatolia comforted themselves under nomad raids not only with the warrior Nasreddin, but with a poor teacher-preacher (*hoja*) who defined the enemy. Later on the sly *hoja* pushed the legendary warrior right out of the picture, and from these two roots grew the folk hero of Turkish humour.

Trans.: Jean Dj. Bader, *Les bonnes histoires de Mollah* (Neuchâtel 1962); Charles Downing, *Tales of the Hodja* (London 1964); Eric Daenecke, *Tales of Mullah Nasir-ud-Din: Persian Wit, Wisdom and Folly* (New York 1960); P. Garnier, *Nasreddin Hodja et ses histoires turques* (Paris 1958). ZV

Navā'i, Alīshīr (in contemporary Uzbek, Alisher Navoiy; in Tajik, Navoī, b 1441 Herat, d 1501 Herat), classical author in Chaghatāy literature, sometimes called Old Uzbek literature, also claimed by Afghan and Tajik literatures. Educated at the best schools in Herat, Mashhad and Samarkand, vizier to his childhood friend, Ḥusyan Bāyqarā, who ruled in Herat from 1496, he did much for culture and economic life. After trouble at court he retired from public life to literature, and wrote many varied works. An early work was the linguistic tract *Muḥākamat ul-lughatayn* (Contest of Two Languages, 1449) showing that Turkish can produce descriptions as masterly as those in Persian or Arabic. In his study of prosody *Mīzān ul-auzān* (Scales of Poetic Metres, after 1492), in which he refers to the literary language of his time as Chaghatāy, he praises the charm of the syllabic folk verse as well as the literary metres. He also wrote on literary history, the *taẓkira* (qv) *Majālis un-nafā'is* (Gathering of the Greats, after 1491), an anthology which provides valuable information on Persian and Turkish poets. His volume of lyrics *Chār dīvān* (Four Divans, 1498–99), in which the very popular *ghazal* (qv) form predominates, contains fresh descriptions of nature as well as philosophical meditations. He is also the author of a *dīwān* (qv) composed in Persian (1496), under the *takhalluṣ* (ie the literary pseudonym) of Fānī. The nucleus of his work is the *khamsa*, or quintet of works, formed by romantic and philosophical-didactic poems. Navā'ī follows the Persian poets, especially Neẓāmī and Jāmī (qqv) to whose work and friendship he dedicated his *Khamsat ul-mutaḥayyirīn* (Quintet of Confusion, 1492). Other works: *khamse: Ḥayrat ul-abrār* (Travellers in Confusion, 1483); *Farhād va Shīrīn* (1484); *Laylī va Majnūn* (1484); *Sab'ayi sayyara* (Seven Planets, 1485); *Saddi Iskandar* (The Bastion of Alexander, 1485); *Lisān uṭ-ṭayr* (The Language of the Birds, 1499).

Trans.: Robert Devereux, *Muḥākamat al-lughatain* (Leyden 1966).
PHTF II 326–61. NZ

Nayır, Yaşar Nabi (b 1908 Üsküb, now Skopje in Yugoslavia), Turkish poet, author and publisher. From 1926 to 1936 Yaşar Nabi published poems in literary journals. He belonged to a group known as the Seven Firebrands (*Yedi Meşale*), the other members being the short-story writer Kenan Hulusi, and the poets Sabri Esat Siyavuşgil, Muammer Lütfi, Vasfi Mahir Kocatürk, Ziya Osman Saba and Cevdet Kudret. In 1928 the group brought out a joint selection of their works, *Yedi Meşale*, and for eight months produced a review, *Meşale*. With youthful enthusiasm they called for originality and sincerity in art, declaring that the range of themes must be extended, and reality transcended. However this rather artificially formed group soon broke up. Its place in the history of the development of modern Turkish literature has largely gone unrecognized, its claims being immediately superseded by such extremely individual talents as Cahit Sıtkı Tarancı (qv), etc.

Yaşar Nabi, who is an ardent Kemalist, has been greatly influenced by Western culture in general, French in particular. Throughout his many activities he has always fought for the supremacy of artistic values and the rejection of fanaticism of whatever kind. His review, *Varlık*, since it first came out in 1933, and particularly in the period from 1940 to 1956, was the rallying-point of the most gifted Turkish writers. In its pages he championed the cause of Turkish new literature, and in particular the New Poetry. Thus *Varlık* has played an important role in the cultural life of Turkey. The same may be said of the *Varlık* publishing house, founded in 1946, which produces carefully prepared, well-chosen editions of Turkish works and foreign translations at prices within the means of the broad reading public. Hence Yaşar Nabi must be looked upon as not only an author and anthologist, but as the man who has done most to promote an interest in the arts amongst the public in republican Turkey. In 1966 Nabi founded the review *Cep Dergisi*. Collections of poems: *Kahramanlar* (Heroes, 1929), *Onar Misra* (Ten Lines Each, 1932); essays: *Balkanlar ve Türklük* (The Balkans and Turkishness, 1936), *Edebiyatımızın Bugünkü Meseleleri* (Problems of Our Literature Today, 1937), *Nereye Gidiyoruz?* (Where Are We Going?, 1948), *Yıllar Boyunca* (As the Years Pass, 1959). DL

Nāzik al-Malā'ika (b 1923), Iraqi poet, leading writer and theorist of free verse. After studying Arabic literature in Baghdad she went to the USA, where she studied English literature. Her field of study, as well as home influences (her mother was also a poet) led her to poetry, and she became one of the modernizers alongside Badr Shākir as-Sayyāb and al-Bayātī (qqv). She differs from them, however, in her rational approach to modern free verse and its theoretical aspects (introduction to *Shaẓāyā wa rimād*, Splinters and Ashes, 1949; *Qaḍāyā ash-shi'r al-mu'āṣir*, Problems of Contemporary Poetry, 1962, 1967[2]). Her theoretical studies served as the foundation for the free verse movement. Her chief justification for revolt against the old forms was the need for new forms which are related to the demands of modern life, modern poetry being the continuation of the old poetic tradition, in these new forms. Her poetry rises mainly from the conflict between a woman's inherent longing for life and the bounds which life enforces. She reflects the influence of English and American poetry (surrealism, the romanticism of Keats, the atmosphere of Poe, the influence of T. S. Eliot). The constant mood of her verse is melancholy, sorrow and pain, the world of the spirit; the external world is not to the fore. On the other hand she writes essays on modern problems and on political subjects. Although her poetry is popular, probably for its romantic mood, she is not the poet of the wide public. Of her many volumes of verse: *'Āshiqat al-layl* (Mistress of the Night, 1948); *Shaẓāyā wa rimād* (Splinters and Ashes, 1949); *Qarārat al-mawja* (In the Trough of the Wave, 1957).

CAR 217; NTALAC 176. KP

Nazim Hikmet, Turkish poet and dramatist, see **Hikmet Ran**, Nazım.

Necatî Bey (d 1509 Istanbul), Turkish poet, of unknown origin. He may have been a slave brought up by a wealthy woman in Kastamonu and sent to the university. He soon became famous as a poet. He was a scribe in the Sultan's chancellery, later an official of the Ottoman princes in various provinces, and finally court poet. Author of the Turkish versions of several well-known Arabic and Persian works; the excellent *mesnevi* (see *maṣnavī*) *Münâzerei Gül ü Hüsrev* (Husrev Arguing with a Rose), which has not survived, is attributed to him. His lyric *dīwān* (qv) is his most important work; the language is not difficult, and makes use of Kastamonu dialect words; the musical verse offers beautiful descriptions of nature (*Kasidei şitâîye*, Poem on Winter; *Behârîye*, on Spring). His poems dedicated to dead friends were taken as models by later 16th century poets, at the peak of Turkish classical poetry.

GHOP II, 193 ff; PHTF II 662. ZV

Nedim, Ahmed (d 1730 Istanbul), Turkish poet. Of a judge's family, he himself was a teacher who got to court as a friend of the Great Vizier and was put in charge of the library. When the Sultan's palace was partly demolished during a popular rising, Nedim lost his life. Nedim was the greatest poet of the reign of Ahmed III (1703–30), known as the 'tulip period' and marked by the extravagant magnificence of court festivities, costly buildings, and lavish dissipation of the country's revenues. Nedim was a scholar, and translated learned Arabic texts, but his poetry is fresh and simple, recalling folk songs; most of the poems were actually sung; he used folk poetry forms. Rejoicing in wine, love, beauty, his imagery is lively and original; his verse is still read today, although some of it has to be translated into the modern idiom.

GHOP IV 29 ff; PHTF II 442 ff. ZV

Nef'î (d 1634/35 Istanbul), Turkish poet, from a village near Erzurum. He went to Istanbul as an accountant, and lived at court after he secured the Sultan's favour. His satires made many enemies among the great, and he was executed for insulting one of the viziers. Nef'î's poetry is the height of involved and highly polished style. His *kasides* (see *qaṣīda*) took a curious form in later years at court; besides panegyric *fahrîye* he wrote *kasides* of flowery insult instead of praise. While singing of murmuring fountains he relentlessly showed up the faults of the courtiers, mullahs, lawyers, wandering dervishes and even of popular saints; the language he used to expose their sinful lives was harsh indeed. The colourful style and formal perfection of his verse was highly esteemed, but systematic analysis has still to be made to determine whether his stinging attacks on the upper classes of his day can be regarded as deliberate social criticism. Although it was not the custom to put a title to a *dīwān* (qv), Nef'î called his *Sihâmi Kadâ* (Arrows of Fate).

GHOP III 252 ff; PHTF II 442 ff. ZV

Nehemiah, governor of Jerusalem for the Persian king c400 BC, who took great care in the rebuilding of the ruined city. The chronicle of these operations, one of the few autobiographical writings of the ancient East, is thus of greater literary value than its style would warrant. It is part of the Old Testament (qv) *Book of Nehemiah*. See also Bible.

J. M. Myers, *Ezra, Nehemiah* (Garden City 1965). SZS

Nerses Shnorhali (the Gracious) or Klayetsi (of Romkla, 1102–73), Armenian poet. Of the Pahlavuni family, he was a consecrated priest at eighteen; he achieved high dignity in the Church and became a leading figure in Armenian cultural and political life, an enthralling orator. He wrote encyclicals, epistles and sermons in prose, didactic poems, popular riddles, a history according to Movses Khorenatsi (qv), a poem on Old and New Testament (qqv) themes and religious lyrics. He set some of his own verses to music and therefore paid great attention to their rhyme, rhythm and other formal aspects. Extremely popular was his 1,067-line poem *Oǵb Edesioy* (Lamentation of Edessa, 1146, French excerpts in E. Dulaurier, *Recueil des historiens des Croissades*, Vol. I, Paris 1869). In it, the city Edessa, personified as a woman, laments the destruction of thousands of Christians under the Seljuq attack of 1144. One of the earliest Armenian lyric epics, this poem gave rise to the tradition of long poems celebrating great moments of national history.

Trans.: *Jesus, Son, Only-Begotten of the Father* (New York 1947); *The Profession of Faith of the Armenian Church* (Boston 1941); *Preces Sancti Nersetis Clajensis Armenorum Patriarchae triginta sex linguis editae* (Venice 1882; a remarkable multi-lingual edition of the prayers of Nerses); TTA; TRA II.
G. Alishan (qv); E. Dulaurier, *Histoire, rites, dogmes et liturgie de l'Eglise Arménienne* (Paris 1855); THLA; FGAL; NACL; HO 7. LM

Nesimi, Sayyid Imadeddin (b after 1370 Shirvan, d 1417 Halab), Azerbaijan poet. A dervish who supported the popular heretical *hurufi* movement, he propagated these teachings in his verses, which were later collected in two *dīwāns* (qv), one Persian and one Azerbaijan-Turkish. It is assumed that he also wrote in Arabic but none of these verses has survived. His poetry aroused feelings of aversion to social inequality and to the existing social order, nor was it in accordance with the official Muslim ideology from the theological point of view. On account of his verses he was accused of heresy and cruelly executed. After his death Nesimi's work spread rapidly throughout the Ottoman Empire, where it exercised an influence on the dervish orders for many centuries. It also became well known in Azerbaijan, where it helped to establish the language as a means of literary expression. Many later poets drew on his work, including Füzuli (qv).

GHOP I, 343 ff. ZV

Nesin, Aziz (real name Mehmet Nusret Nesin; b 1915 Istanbul), Turkish writer and playwright. Nesin grew up in a poor milieu as a half-orphan and later became a boarding student in military schools. After various assignments in the army he resigned in 1944. Work for oppositional newspapers resulted in a total of five years of prison sentences for critical articles. After his release he held various jobs until he returned to writing in 1953. Since then he has made a living as a freelance writer and publisher. Nesin is a gifted writer of many facets. He owes his fame to his satirical short stories of which he produces about a hundred every year. They are witty and pointed, presented in an unpretentious, lively style, and often convey sharp social or political criticism. For them he was twice awarded high international prizes ('Golden Palmtree', Bordinghera, Italy, 1956 and 1957). His novels are similar in character, but his plays show him from a serious, artistically ambitious side: in them he creates a world of phantasy and presents basic human problems in symbolic language. Short stories collected in 35 volumes, 1948–68. Six novels, especially: *Gol Kıralı* (The Goal King, 1957), *Zübük* (1961). Autobiographical writings: *Bir Sürgünün Hatıraları* (Reminiscences of Exile, 1957); *Böyle Gelmiş, Boyle Gitmez* (As it Came, it Can't Go on, 1966). Five plays, especially: *Biraz Gelir Misiniz* (If you Come a Little, 1958); *Bir Şey Yap, Met* (Do Something, oh Tide, 1959). Remarkable as a successful modernization of a folk tradition are his shadow plays (*Üç Karagöz Oyunu*, 1968). AT

New Testament, a collection of 27 books dating from 50–130 AD and put together from the 2nd to the 4th centuries. All are in Greek, but the majority draw on Semitic material. The *Gospel* according to St Mark is based on Aramaic traditions. Those of Matthew and John are also based on Semitic traditions, probably in a Hebrew language. The *Revelation* (*Apocalypse*) of St John the Divine is written in Greek which is often unintelligible without knowledge of the Semitic expressions it is based on. The *Epistle* of James keeps closely to the Jewish tradition of the Twelve Patriarchs. From the literary point of view the language of the *Gospels* represents the Aramaic way of teaching in rhythmic sentences, using many illustrations from nature and from the work of man. The narrative in the *Gospels* reflects the style of folk story-tellers. In all the books of the New Testament themes, expressions and direct quotations from the Old Testament (qv) play an important part, while the *Revelation* also makes use of earlier apocalyptic visionary material from the ancient East. Scholars have devoted much study to the linguistic and literary aspects of the Semitic basis of the New Testament books

(J. Wellhausen, C. C. Torrey, M. Black), but their results have not yet been applied as they deserve in the literary translation of the texts. See also Bible.

Trans.: HBKJV; HBRSV; NEB.
Good News for Modern Man (New York 1966); V. Marxsen, *Introduction to the New Testament* (Oxford 1968, Philadelphia 1970). SZS

Neẓāmi 'Arūẓi of Samarkand, Aḥmad b. 'Omar b. 'Alī (12th century), Persian writer. He seems to have spent most of his life at the Ghorid court as official poet. Apart from doubtful autobiographical references nothing is known of his life, nor has any of his courtly poetry survived, including the epic *masnavī* (qv) which he is said to have written. Only one of his works has come down to us, the *Chahār maqāle* (Four Discourses), an original treatment of the popular 'Fürstenspiegel'. *Four Discourses* deals with the four professions he believed most important for the prosperity of the realm, and the qualities to be desired in their ideal representatives; the official (*dabīr*), the poet (*shā'er*), the astrologer (*monajjem*) and the doctor (*tabīb*). E. G. Browne considers it one of the most remarkable prose works in Persian literature and an important source of reliable information on life at Iranian courts in the 12th century. He translated it in 1899; a Russian translation by S. I. Bayevskiy and Z. N. Vorozheykina, with a comprehensive introduction by A. N. Boldyrev, appeared in Moscow, 1963.

Trans.: E. G. Browne, *Revised Translation of the Chahār Maqāla*, the *Four Discourses*... (London 1921, repr. 1955); I. de Gaslines, *Les quatre discours* (Paris 1968).
BLHP II 336–40; A. J. Arberry, *Classical Persian Literature* (London 1958), 100–2; RHIL 221–2. VK

Neẓāmi, Neẓāmoddīn Ilyās b. Yūsof b. Zakī Mo'yyad (b 1141/43 probably near Ganje, d 1209 Ganje), Persian poet. The *tazkeres* (see tazkira) tell us almost nothing about the poet. Although the *Lobāb ol-albāb* by 'Oufī (qv) was written between 1203 and 1227, only a few years after the death of Neẓāmī, it gives only general eulogies and no concrete information. Doulatshāh, the next *tazkere*-writer at the end of the 15th century, who tells much more, is regarded as notoriously unreliable, while all later works of this kind give secondhand data and repeat Doulatshāh's material. We thus have to rely on the information to be deduced from the work of Neẓāmī himself.

Persian literature produced its greatest epic poets in the first centuries, Ferdousī (qv) and Neẓāmī. While the former gave expression only to topics of Iranian origin, Neẓāmī drew both on Arabic and Persian tradition. In both cases it is of course mythical tradition, legends and saga cycles of different origin melted together, the past as willed as when Alexander the Great was described as being of Persian descent. The first culmination point of the epic, in the work of Ferdousī, may be characterized as static, the second, Neẓāmī's work, as dynamic. This can be seen in the social background as well as in the literary style of their work: feudalism and agrarian society in the case of Ferdousī, the growing city and the artisan milieu with its *Futuvva* affiliations in Neẓāmī. Archaic language, static description of the characters without development, underlined by the same *epitheta ornantia*, and chronological order limiting free treatment of the topic; all this (without diminishing Ferdousī's greatness) shows the opportunities open to an epic poet after him.

The world into which Neẓāmī was born and which formed him was characterized by Seljuq domination. The old strife between Sunna and Shī'a had been partly solved by the Seljuqs after they overran the Shiite Buyids, but the Fatimids and the sect of the Assassins survived. Azerbaijan and the northern provinces of Arran and Shirvan were being thoroughly Turkisized. Naturally Neẓāmī had to be a convinced Sunnite when he turned to the fanatical Seljuq rulers and the Atabegs of Azerbaijan, the Ildegisids, his sovereigns and patrons. He could hardly be a herald of Persian national sentiments, when the Turkish rulers and the cosmopolitan environment of the commercial centre of Ganje were indifferent to it. While the wave of Iranian revival bore Ferdousī to his height, the rising spiritual wave in Neẓāmī's time was that of Sufism (qv), Islamic mysticism.

Neẓāmī was a poet of *qaṣīdas* and *ghazals* (qqv) as well as of epics. His lyrical work, published by Vaḥīde Dastgerdī, and called *Ganjīneye Ganjevī*, is allegedly what remains of 20,000 verses. It consists of five *qaṣidas*, 56 *ghazals*, nine *robā'īs* and two *qiṭ'as* (qqv). Safavid times knew of a more copious *dīwān* (qv), and even this material was reported to have been lost, although three manuscripts were in European libraries. The credit for discovering Neẓāmī's lyric belongs to M. Houtsma and J. Rypka.

The fame of Neẓāmī rests on his five epics, later called *Khamse* (Five Poems), which was imitated in form and topic by generations of poets with lesser talent. The first of the five epics in order of origin bears the title *Makhzan ol-asrār* (Treasury of Mysteries, 1176). It belongs to the genre of ethical-didactic works so popular in Persian literature and was inspired by the *Ḥadīqat al-ḥaqīqat va sharī'at aṭ-ṭarīqa* by Ṣanā'ī (qv). Written in the *sarī'* metre (see *arūḍ*) and dedicated to Fakhroddīn Bahrām Shāh, this epic was mainly responsible for the idea of Neẓāmī as a mystic, but it lacks real mystical spirit. In twenty chapters, illustrated by stories and parables, the poet talks of the creation and ultimate destination of the world and counsels all kinds of virtues like honesty, justice, content etc; he addresses this advice mainly to rulers. Neẓāmī is revealed as a moralist, but not as a Sufi. Although later writers tend to interpret every poet as a mystic as far as possible, there may be a grain of truth in Doulatshāh's suggestion that a certain Akhī Farrokhe Zanjānī (variant Faraje Reyḥānī) was the spiritual teacher (*pīr*) of Neẓāmī. A moralist with principles, not blindly extolling the *bazm o razm* mentality of the ruling circles, Neẓāmī must have aroused their antipathy, as they cared only for dedication and eulogies. In fact Neẓāmī dedicated all his epics to different rulers and never seems to have attained riches by them. Complaints about poverty permeate his works, and he calls himself a prisoner of Ganje, the city which he never seems to have left for long. He may have been born here too, but we do not know this for certain, nor on what he actually lived.

His name and that of his father and grandfather are revealed in a passage in the epic, *Leylī o Majnūn*, and the name of his mother was Ra'īse (but Ra'īseye Kord may also point only to her descent from a ruling Kurdish family without mentioning her name). The mention of a maternal uncle as Khvāje 'Omar may hint at the social position of the poet's relatives, the title *khvāje* having sunk low enough at the time to designate an artisan or merchant. There is no foundation for the oft-repeated statement that Neẓāmī became an orphan in his early youth; it rests on an error in translation when W. Bacher, the first biographer of the poet, quoted a passage in *Leylī o Majnūn*, in which the poet mourns the death of his father. Neẓāmī was nearly fifty when he wrote this epic. Besides, the passage mentions the death as being no cruelty and due to the course of time, so that old age can be assumed. We may also assume from his work that in his youth Neẓāmī had the time and the means to acquire a good education, and to learn Arabic. His view of the world is rationalist in many respects, suggesting that he acquired his knowledge when it was still able to form his mind, and not later on. In the *Haft peykar* (Seven Portraits, 1197) he boasts of having mastered all the subtleties of astronomy and the other sciences, but the references to reading Jewish, Christian and Pahlavi historical works, in the *Sharafnāma* (Book of Honour), are of course oriental exaggeration. From the *Haft peykar* we learn that he early started to write poems and had success with them. Later on in three passages of the *Khamse* he turns to his son and admonishes him to become a physician or lawyer, but not a poet.

Neẓāmī married a slave girl named Āfāq and she bore him his only child, the boy Moḥammad. After about eight years of happy marriage she died in 1180. Verses about his love for her also occur in the latest works, and her memory must have cast its shadow on the two later marriages of the poet, both ending in the death of his wife. He had only one wife at a time and there are verses in which he defends monogamy, characteristically in the inrests of the children and their development, not in that of the wife. The picture of Āfāq was dominant in Neẓāmī's mind when he composed his second epic, *Khosrou o Shīrīn* (1177–81). Written in the *hazaj* metre (see *'arūḍ*), it has a traditional Iranian theme, the love of the Sassanid ruler Khosrou II for the Armenian princess, Shīrīn, which finds fulfilment only after many adventures. The most famous episode is that of Farhād, architect and rival of the ruler in his love for Shīrīn, which casts a tragic shadow on their life. As Farhād seems to be succeeding in the task of cutting a road through mountain rocks, the condition for gaining the princess, the Shāh resorts to the stratagem of spreading a false report of Shīrīn's death. Thereupon Farhād commits suicide, a theme treated in innumerable miniatures. Neẓāmī dedicated the epic to three rulers at a time and seems to have put great hopes in it. Eventually it was Qizil Arslan who invited the poet to his court, and we have a charming description of this audience. Neẓāmī received the little village, Hamdūniyān, but apparently it did not yield him much profit, as he puts taunting verses about the gift in the mouth of an enemy, being unable to speak in his own name.

As the poet's fame spread, the Shirvanshah Akhsetan invited him to treat the subject of the tragic love between Leylī (Lailā) and Majnūn (qv), taken from Arab lore. On the advice of his son, Neẓāmī accepted the theme for his third epic, *Leylī o Majnūn* (1188). Qays, a young Bedouin and the Romeo of Arab lore, falls in love with the girl Lailā, the Arab Juliet. The rejection of his suit drives him to wander through the desert as a maniac (*majnūn*), while Lailā is married to another. After her husband's death she returns to Majnūn, but dies, upon which he too seeks death. In spite of his original reluctance Neẓāmī created a masterpiece, the best treatment of the theme before Jāmī (qv), who gave it a mystical interpretation.

Neẓāmī was beginning to feel the burden of his years, when a new commission from Alā'oddīn Korp Arslan reached him, leaving him to choose the subject of his fourth epic himself. This became his greatest and most mature work, the *Haft peykar* (Seven Portraits, 1197) in the *khafīf* metre. Again the poet reached into Iranian history for the figure of the idealistically glorified Sassanid ruler, Bahrām Gōr. It gave him ample opportunity to present his ideas of an ideal ruler, just as the pictures of the Shah's seven princesses, one for each climate of the earth, permitted deeper thoughts about the meaning of love. From these the title is derived, and each princess tells a story describing the development of Bahrām Gōr's character and his deeds. The work is understood in Europe as description of natural and fantastic events, while Persian tradition sees in it the symbolic representation of the way of the mystic to God in purifying his soul.

The *Eskandarnāme* (Book of Alexander, 1200) is the last of the epics. It is written in the *motaqāreb* metre (see *'arūḍ*), traditionally used for heroic themes; it is divided into two parts called *Sharafnāme* (Books of Honour) and *Eqbālnāme* (Book of Happiness). Neẓāmī was the first poet in Persian literature to collect all the legends about Alexander the Great and his campaigns. He relied mainly on oriental material, both based on the pseudo-Kallisthenes and independent of it. The first part of the work depicts Alexander as the great general, while the second shows him as a seeker of knowledge in deep discussions with philosophers and wise men, embracing the knowledge of his time.

Neẓāmī has often been imitated, but never equalled. His poetic intention and his contribution to the development of the epic can be summed up as 'dynamization'. With verse based on the growing cities, and with a deep interest in the developing sciences, the poet has an open ear for the dynamism of all being and existence. His characters develop with their changing situations. He stands for the diminution of fantastic elements in the epic and the increasing importance of rational motivation. The dynamic principle is visible in his vocabulary too, deserting the fixed and limited language of the chivalrous epic and introducing common, everyday language. Even colloquial idioms are found occasionally, together with proverbs and sayings. Neẓāmī's conception of poetic idiom is much deeper than hitherto; the word is not merely the vehicle of meaning, but acquires deeper significance. His metaphors and similes are creative and original to a degree rare before the Indian style (qv). He follows the basic tendency of Persian poetry, not merely to describe natural scenes but to transpose the elements into a completely new network of relations, sharing a creativeness and better taste than many after him. This prevented Neẓāmī from stereotype concepts on the one hand and from extremes on the other. In the works of the Caucasian (Azerbaijan) School, especially Neẓāmī's, tendencies appeared, which later became stronger and led to the creation of the Safavid or Indian style, although they were not taken up in the next three centuries. However, they show that this style was a genuine Iranian phenomenon.

Trans.: G. H. Darab, *The Treasury of Mysteries* (London 1945); J. Atkinson, *Lailí and Majnún* (London 1836); C. E. Wilson, *The Haft Paikar* (*The Seven Beauties*), 2 vols. (London 1924); H. Wilberforce Clarke, *The Sikander Náma E Bará, or Book of Alexander the Great* (London 1881).
BLHP II 400–11; RHIL 210–13. WH

Neẓāmolmolk, Ḥasan b. 'Alī Ṭūsī (b 1018 Tus, d 1092 Sahna), Persian statesman, vizier to the greatest of the Seljuq sultans, Malekshāh. A competition arranged by Malekshāh gave Neẓāmolmolk the impulse to write the *Siyāsatnāme* (Book of Government), dealing with the art of governing a great empire. It contains much historical information, many anecdotes, and descriptions of events from his own life. After Neẓāmolmolk was murdered the work was completed by Moḥammad Maghrebī, a scribe in the court chancellery, who added several chapters giving details of political and religious conditions at the time.

Trans.: Hubert Darke, *The Book of Government, or Rules for Kings: the Siyāsatnāma or Siyar al-Mulūk* (London, New Haven 1960). FT

Niger (real name Ivan Dzanayty; Russian Ivan Vasilyevich Janayev; b 1896 Sindzisar, d 1947 Ordzhonikidze), Ossetian poet and literary scholar. He was educated at Ardon theological college and Saratov University. After the October Revolution he was a teacher, a Red partisan and a worker in public institutions. From 1927 to 1930 he studied literature at the North Ossetian Pedagogical Institute. From 1935 until the end of his life he was head of the Department of Literature at the North Ossetian Research Institute in Ordzhonikidze. From 1913 he wrote poems under the pseudonyms 'Sau laeppu' (Black Boy, ie 'lower class') and 'Niger'. After describing in his earlier works the hard life of poor Ossetians he later used revolutionary, optimistic themes. An example of his call to fight against the rich (the Aldars) is his poem, *Razmae!* (Go ahead!). In his longer poems, such as *Gycci*, *Terchy tsur* (On the Terek), he compares the sorrowful past with the new Soviet time. By introducing free verse and new rhythms he enriched Ossetian literary forms. ML

Niyazî Misrî, Turkish writer, see **Yunus Emre.**

Niyozī, Foteh (b 1914 Samarkand), Tajik writer. Of a clerk's family, he studied in Samarkand, where his first verses were printed, and then became a journalist. During the war he was active as a correspondent for several army papers. After writing Uzbek poetry he turned exclusively to Tajik prose. His first book was *Intiqomi tojik* (Tajik Vengeance, 1947), a collection of tales and reportages from the front which are also valuable historical material. His most important work is the novel, *Vafo* (Loyalty, 1954), describing the fate of a Tajik teacher at the front and after he returns home; it is an epic canvas of the part played by the Tajiks in World War II, stressing the friendship between soldiers of all nationalities in the Soviet army. The short stories on war themes collected in *Hissa az qissahoi jang* (War Tales, 1962) deal with a similar subject. The novella *Dukhtari hamsoya* (Girl from Our District, 1955) deals with a contemporary subject and exposes the danger of deep-rooted prejudice and superstition. He wrote sketches on his travels in Burma, India and Indonesia and in 1938 collaborated with S. Ghanī in a play on the life of the frontier guards. In 1965 he returned to drama with another war play, *Mash'alai jovid* (Eternal Torch).

RHIL 576–7. JB

Niyoziy (Niyazī), Hamza Hakimzoda (b 1889 Kokand, d 1929), Uzbek poet, dramatist and composer. He is considered the founder of Soviet Uzbek literature especially in the fields of poetry and the drama. He went to Muslim schools, and found for himself the works of Muqimī and Furqat (qqv), who influenced his early work. He was a pioneer of new teaching methods, which he applied in his reform schools in Kokand and later in Margelan. He was conscious of the need for broader education and prepared a number of Uzbek textbooks. He greeted the October Revolution in enthusiastic poems which soon spread throughout Central Asia. He was not only a talented poet who left the traditional manner behind, but also a skilful dramatist and composer. In his musical compositions, as in his poetry, he drew on the wealth of folk art. His main activities, at least in the last years of his life, were in the field of drama, and the organization of Uzbek theatrical life. His best and most successful plays include *The Bey and the Labourer*, attacking inequality and the system based on inequality. Main works: *Pedagogik asarlar* (Pedagogical Studies, 1911–1915); plays, *Zaharli hayot yokhud ishq kurbonlari* (A Ruined Life, or Sacrifices to Love, 1916), *Boy ve khizmatchi* (The Bey and the Labourer, 1918), *Burungi kozilar yokhud Maysaraning ishi* (Former Judges, or Maysara's Tricks, 1926), *Paranzhi sirlaridan bir levha yokhi yallachilar ishi* (Picture of the Secrets of the Veil, or Dancers' Amusements, 1927).

PHTF II 705 and 710. NZ

Nu'ayma, Mikhail (b 1889), Lebanese critic, poet, writer and dramatist. He went to Russian Palestinian Society schools in Syria and Palestine, then in Poltava, and studied law at Washington University. After fighting in Europe with the US army in World War I, he studied in France. He devoted himself to literature on his return to the USA, helped to edit Arabic publications and became secretary of the literary society led by Jibrān (qv). Ny'ayme wrote mainly in Arabic, later some works in English. His first published work was a play, *al-Ābā' wa'l-banūn* (Fathers and Sons, 1918). He was influenced by classical Russian writers. In 1959 he published an autobiographical trilogy and has also written many short stories and poems. Other works: novel, *al-Ghirbāl* (The Sieve, 1923); study, *Jibrān Khalīl Jibrān* (1934).

SP

O

Obadiah (exact date unknown, 8th–5th century BC), Hebrew book of prophecies directed against the people of Edom. See also Bible, and Twelve Lesser Prophets.

SZS

'Obeyd Zākāni (b 1270 or 1280 Zākān near Qazvin, d 1370/71), important Persian poet. His family was of Arab origin and his father held high office. He studied in Baghdad and then lived in Shiraz. A talented lyric poet, he is sometimes considered as good as Ḥāfeẓ (qv), his contemporary in Shiraz. His *'Oshshāqnāme* (Book of Lovers, 1350) is a love poem of 700 *bayts* (see *maṣnavī*) about seeking, loving and parting. His place in the history of literature, however, is ensured by his satires, in which he cleverly criticized the morals of his time. His best known works are the prose social satire, *Akhlāq ol-ashrāf* (Morals of the High-born, 1340), which compares the traditional moral ideal with contemporary behaviour using witty sarcasm; the prose *Sad pand* (100 Counsels, 1349) in which much of the advice is meant satirically. *Dah faṣl* (Ten Chapters), also called *Ta'rīfāt*, is a satirical encyclopaedia which is a caricature of old Arabic and Persian dictionaries. *Resāleye delkoshā* (Amusing

Treatise) is a prose collection of anecdotes inveighing against the evils of the time. *Rīshnāme* (Book of Whiskers) is a mischievous satire in prose and verse, about the whiskers which grow and mar the beauty of young ephebes. Best known of all is probably *Mūsh o gorbe* (The Mouse and the Cat), a *dāstān* (qv) in *qaṣīda* (qv) form less than 100 *bayts* in length, exposing the horrors of the never-ending wars between the feudal rulers of the time, and the meanness of religious leaders and 'holy' men.

Trans.: Mas'ūd Farzād, *Rats against Cats* (London 1946). R. Levy, *An Introduction to Persian Literature* (New York, London 1969), 150–5; RHIL 272–3. JB

Odes of Solomon, a collection of 42 songs almost completely preserved in a Syrian version of the early 2nd century AD, but probably written a century earlier in Hebrew. One song is attested in Greek and five in Coptic (*Pistis Sofia*). The odes combine themes and the form of the Biblical psalms with early Gnostic Christian motifs and bold metaphors and imagery.

J. Rendel Harris and A. Mingana, *The Odes and Psalms of Solomon I–II* (Manchester 1916, 1920). SZS

Oghuznâme, Turkish legend, see **Dede Korkut.**

Old Testament, a collection of ancient Hebrew literary works with a few shorter passages in Aramaic. The individual parts date from the 12th to the 2nd centuries BC, the whole being canonized towards the end of the 1st century BC, according to Jewish tradition. The earliest parts of the Old Testament are those which preserve the traditions of pastoral life and of fights of the tribes of Israel. Some of the historical accounts in prose and some sets of laws date from the time of the Judges (qv 12th–11th centuries BC) but most material of this type dates from the time of the kings of Israel and of Judah (10th–6th centuries BC). The utterances of the prophets and the accounts of their deeds are preserved from the 9th–2nd centuries BC.

According to tradition, the wisdom books were the work of King Solomon (qv, late 10th century), but they were given their present form several centuries later. Similarly the basis of the *Book of Psalms* (qv) came from King David (qv, mid 10th century), although the canon includes both earlier and later psalms. The Israelites made use of Egyptian, Babylonian and Canaanite models for some literary genres, adapting the themes to their own monotheism and their own view of history. Sets of traditions were preserved in oral, and partly in written, form at the royal court, round the shrines and in individual tribes. Traditions both oral and written were collected during the troubled times when the kingdoms of Israel and Judah were falling (722 and 586 BC).

It appears that the considerable collection of historical and legal material of earlier and later origin known as the *Law* (Hebrew *Torah*, the *Five Books of Moses*, *Pentateuch*, qv), was put together after the return from captivity in Babylon, 5th century BC. By the beginning of the Christian era the canon included also the prophetic books, including historical accounts (*Joshua*, *Judges*, *Kings*, qqv), the prophetic books proper (*Isaiah*, *Jeremiah*, *Ezekiel*, the *Twelve Lesser Prophets*, qqv) and the writings (*Hagiographa*) which include the great poetic books (*Psalms*, *Job*, *Proverbs*, qqv) and most of the books of a later type (such as the *Song of Songs*, *Ruth*, *Daniel*, qqv). The shorter poems and prose pieces were first put together into larger units, which were then arranged in books according to certain criteria; the various sources were combined to produce a work at least outwardly unified. When the canon of the Hebrew Bible (qv) was finally determined at the end of the 1st century BC (according to Jewish tradition), 24 books were included. The text was already established by that time, but it was not until the 9th century AD that scholars in Tiberias (the *masoretes*) added the current vowel marks and vocal symbols for the traditional liturgical chanting to the consonantal text.

Ancient Hebrew literature, preserved in a deliberate selection made for primarily religious purposes, is the only literature of antiquity in the Near East to preserve an unbroken tradition. The Old Testament has handed down to recent times the heritage of other peoples whose language and literature died out in ancient times; the Egyptians, Babylonians, and Canaanites. This synthesis of material and literary types taken over from elsewhere and clothed in a distinct style, which was conditioned both by the semi-nomad life of the Israelite tribes and by the impact of the sedentary Canaanite civilization, was formed by the Israelites into a literature which not only maintained its significant content and distinctive literary style, but was adopted by other peoples as well. Most of the Old Testament was created by men whose names were not recorded (eg the story of the forefathers in *Genesis*, the reigns of Saul and David, many of the Psalms). Fairly early, however, individual authors made their appearance; these were the prophets whose own utterances, for the most part in poetic form, were preserved along with the account of their deeds. They too respected the verse forms which enhanced the effectiveness of the text by symmetrical repetition and complementing ('parallelismus membrorum'). The rhythmic types of ancient Hebrew poetry developed with the evolution of the language. The prose, too, was meant to be recited, and so it too was rhythmical. Various utterances which are not in definite poetic form but nevertheless do not have a prose structure are on the borderline between the two. For later treatment of ancient Hebrew themes and material, see Bible.

Trans.: HBKJV; HBRSV; NEB; *The Holy Scriptures* (Philadelphia 1917 and reprints).
O. Eissfeldt, *The Old Testament: An Introduction* (New York 1965); A Robert and A. Feuillet, *Introduction to the Old Testament I–II* (Garden City 1970). SZS

Olimjon (Alimjan), Hamid (b 1904 Jizzakh near Samarkand, d 1944), Uzbek poet, journalist, critic and translator. He studied in Jizzakh and then in Samarkand, at the Uzbek Teachers' Seminary. Later he worked on the staff of several papers in Tashkent. In 1939 he became chairman of the Union of Uzbek Writers. His poetry is in the tradition of Mayakovskiy and Gorki. At first marked by his journalistic writing, his verse attained a more mature level during the war years. As a critic he aimed at furthering clear principles in Uzbek literature. He is also known for his translations of Russian classical literature (Pushkin and Lermontov). Chief works: verse, *Bahor* (Spring, 1929), *Daryo kechasi* (Night on the River, 1936), *Oygul bilan Bakhtiyor* (1930), *Semurgh* (The Phoenix, 1930); criticism, *Sotsialistik realizmi egallash yŭlida* (On the Road to Socialist Realism, 1932).

PHTF II 712. NZ

Olongkho, Yakut heroic epic poem, see **Oyuunuskay,** Bylatyan Ölöksüöyebis.

'Omar Khayyām, real name Ghiyāṣoddīn Abū al-Fatḥ 'Omar b. Ebrāhīm al-Khayyāmī (b first half of the 11th century Nishapur, d 1122/3 ibid.), Persian poet who is probably the best known in Europe and America, thanks to Edward Fitzgerald's translation *Rubāiyāt of Omar Khayyām* (first published 1859). The extraordinary success of the work caused interest in Khayyām's poetry to arise in his own country, Iran, where he had always been honoured as an all-round scholar, a mathematician, astronomer, physicist, philosopher and doctor. His contribution to mathematics and astronomy was an original one; from 1074 to 1079 he collaborated on a reform of the calendar for the Sultan Malekshāh. As a philosopher he is considered an epigone of Ibn Sīnā (qv). His scientific work is reliably documented (works written in Arabic) and so are some events in his life: his studies, life at court, pilgrimage to Mecca, and the course of the last day in his life. The question of the poetical legacy, however, is more involved.

The *robā'ī* (qv), the form in which he wrote, is a complete unit both in form and content, and has no continuation. It is found at the onset of Persian poetry in the new era, with a great variability in metre and a wide range of themes. No information has come down to us from his own day that he collected his quatrains. Apart from isolated examples in Persian chronicles, *robāʿīs* assignated to him do not appear until the following centuries. The ardent admiration of the poet, as he appears in Fitzgerald's version, was followed from the late 19th century onwards by detailed research aimed particularly at the analysis of the existing manuscripts and those since discovered. It would be desirable to arrive at a definite conclusion as to the original form of his work, in order to assess his personality as a poet properly. Some features of his personality can be outlined from the themes most stressed in his quatrains, ie profound pessimism, echoes from the lines on human helplessness in the hands of Fate, culminating in a protest to God. Complaints of the ephemeral nature of this life and doubts about the next, lead to the only reliable philosophy of life, '*carpe diem*'. This was the impression created by Fitzgerald's version in the minds of Western readers. In Iran interpretation of Khayyām's verse has followed the line of Islamic mysticism (Sufism, qv), and some Western scholars also share this approach. Research into 'Omar Khayyām's work, and translations of it, are continuing. Thanks to the translators his *Robāʿiyāt* have found their place among the treasures of world literature.

Trans.: A. J. Arberry, *Omar Khayyam: a New Version Based Upon Recent Discoveries* (London, New Haven 1952); R. Graves and Omar Ali-Shah, *The Rubaiyyat of Omar Khayyam* (London, Toronto 1967); M. Foulandvand, *Khayyam, Omar. Les quatrains* (Paris 1960).
A. G. Potter, *A Bibliography of Omar Khayyām* (London 1929); C. Rempis, *Beiträge zur Hayyam-Forschung* (Leipzig 1937). HS

Ömer, A., Turkish folk singer, see **âşık.**

ʿOmrānī (ʿImrānī, 16th century), Jewish-Persian poet from Shiraz, author of the *Fatḥnāme* (The Book of Victory, 1523) and the *Ganjnāme* (The Book of Treasures, 1536). Like Shāhīn (qv) he is called Moulānā in the manuscripts. Influenced by Saʿdī (qv) he takes up the tradition of Shāhīn, his ideal, and takes post-Mosaic texts of the Bible for epic treatment in his first book. The second is a poetical paraphrase of the *Mishnah* (qv) tractate *Pirqe Aboth*, with moral exhortations and much good advice added. Unlike Shāhīn he reveals few Islamic elements. His works have not yet been edited, but Wilhelm Bacher (see Shāhīn) has given a survey with occasional translations. JPA

ʿOnṣorī, Abolqāsem Ḥasan b. Aḥmad (b 10th century Balkh, d 1039/40 Ghazna), Persian poet. Allegedly the son of a merchant, he was introduced to the Ghaznavid court by the ruling Sultan's brother, Amīr Naṣr. ʿOnṣorī became the poet laureate of Sultan Maḥmūd in Ghazna. He was the most important of the allegedly 400 court poets, and the political propagandist of the ruler. ʿOnṣorī set the standards for the Persian *qaṣīda* (qv), which were then maintained. He favoured the form of the Arabic *qaṣīda*, with an introductory part (*nasīb*) and the laudatory part, linked by an appropriate line (*gorīzgāh*). Arabic influence is still strong in the content. Rhetoric and arabization of the vocabulary grow in importance. ʿOnṣorī wrote epics like *Vāmeq o ʿAẕrā*, *Khing bot o sorkh bot* (White Idol and Red Idol), *Shād bahr o ʿeyno' l-ḥayāt* (Happy One and Life Source) ao, of which only some verses have been preserved. The greater part of his *dīwān* (qv) is also lost; about 2,500 out of some 30,000 distichs are extant, consisting mainly of *qaṣīdas* and some *ghazals* and *robāʿīs* (qqv). In the epic ʿOnṣorī is the connecting link between Ferdousī and Neẕāmī (qqv), between the feudal chivalrous epic and the romantic epic.

BLHP II 120–3; RHIL 174–5. WH

Orbeliani, Grigol (b 1800 Tbilisi (Tiflis), d 1883 ibid.), Georgian poet and politician, descended from a famous princely family. He graduated from the artillery school in Tbilisi and served as an officer in the Perso-Russian and Turco-Russian wars, and in the Daghestan campaign. In 1832 he was exiled for three years for his part in the Georgian aristocratic plot against the Russian government, but pardoned on his return. He then made a great military career, in 1860 was appointed governor of

Tbilisi, and became a reliable servant of Russian policy in the Caucasus. In his poetry Orbeliani represents that trend of Georgian Romanticism which delighted in nostalgic longing for the past glory of the kingdom; he sang the praises of banquets and drinking parties, and liked to imitate folk singers, whose company he enjoyed. In some of his poems he even refers to himself as 'the little tailor Bezhana'. In spite of frequent criticism of his political attitudes (especially from I. Chavchavadze, qv), Orbeliani's poetry is still very popular.

Trans.: UAGP. VAČ

'Orfī (1456–1591), Persian poet born in Shiraz. He was disfigured by smallpox and migrated to India, reaching Ahmadnagar in 1585. He then proceeded to Fathpur Sikri where he attached himself to Khān-khānān 'Abdorraḥīm. Through him 'Orfī gained access to Akbar, whom he accompanied to Kashmir in 1588. He died suddenly in 1591 and was buried in Lahore, but was transferred later to Najaf. 'Orfī has composed two *maṣnavīs* (qv), the *Majma 'ol-abkār* and the incomplete *Farhād o Shīrīn.* But his fame rests upon his *qaṣīdas* (qv) which contain 'new and original combinations of words, fine metaphors and a congruity of topics' (Shiblī). When his first *dīwān* (qv) was lost he boasted of being able to easily produce many more. His *ghazals* (qv) are powerful in feeling. 'Orfī has often been blamed for the intolerable arrogance he displays in his poems; but exaggerated self-praise in *qaṣīdas* is traditional. His *qaṣīdas* are among the most enrapturing poems in Persian poetry and, in spite of their difficulty, often convey a feeling of his innate fire in both passion and despair. Many commentaries have been written on them (ie by Munīr Lahōrī). He influenced Turkish poetry, where five commentaries exist. His powerful imagery inspired the poets of the Indian style (qv) up to Ghālib (see Vol. II). AS

'Oufī, Moḥammad (d after 1228), Persian poet. He spent his youth in Bukhara and Khorasan, then took refuge from the Mongols in Sind, where he dedicated his biography of poets, *Lobāb ol-albāb* (Quintessence of the Hearts), to the vizier of the ruler, Sultan Qubācha. After Qubācha's defeat, 'Oufī joined the court of Iltutmish in Delhi where he completed the *Javāmi' ol-hekāyāt va lavāmi' or-revāyāt* (Necklaces of Anecdotes and Splendours of Tales), a collection, in four parts composing 25 chapters each, of more than 2,000 anecdotes and stories which offer precious literary and folklore material. The *Lobāb ol-albāb* contains the biographies of all Persian poets known to 'Oufī and thus constitutes a primary source for the study of early Persian poetry, although the author never mentions any dates, usually only the dynasty, and the place where the writer in question flourished. Nevertheless he preserved a large number of otherwise unknown names, verses and facts.

M. Nizamuddin, *Introduction to the Jawāmi'u 'l-Ḥikāyāt wa Lawāmi'u'r-Riwāyāt, a Critical Study of its Scope, Sources and Value* (London 1929). AS

Oybek (Muso Toshmuhammadov, b 1905 Tashkent, d 1968), Uzbek writer, poet and critic. In 1925 he graduated from Tashkent Teachers' Training College and became a teacher; later he studied economics and worked in the Institute for Cultural Progress, the Institute of Language and Literature, and the educational publishing house. His first volume was *Tuyghular* (Emotions, 1926). At first influenced by the Russian Symbolists, later his verse was realist in spirit and marked by formal perfection. His greatest work is in prose: the historical novel *Kutlugh kon* (Sacred Blood, 1940) depicts Uzbek society before the Revolution, while *Navoiy* (1944) is a historical biography of the great poet (see Navā'ī) showing him not only as a poet and thinker, but also as an enlightened humanist. The post-war ballad *Hamza* is devoted to the last years in the life of *Hamza Hakimzoda Niyoziy* (qv), the poet, dramatist and composer. Other works: *Mash'ala* (Torch, 1932); *Tamirchi Zhŭra* (Zhyra the Smith, 1933); *Kahramon kiz* (Heroine, 1937); novels, *Oltin vodiydan shabadalar* (The Winds of the Golden Valley, 1950);

sketches, *Pakiston taassurotlari* (Impressions from Pakistan, 1950).

PHTF II 711, 713. NZ

Oyuunuskay (Oyunskiy), Bylatyan Ölöksüöyebis (real name Platon Alekseyevich Sleptsov; b 1893 Zhekhsogonsk, Alexeyev district, d 1939), Yakut poet. His childhood was one of great poverty, but he managed to attend school. In 1911 he joined in establishing one of the many secret literary groups then being organized. In 1917 he graduated from the Yakut Teachers Seminary and the Teachers' Training College in Tomsk, where he also began writing. He returned to Yakutsk in March 1918, to fight the Whites, but was arrested and sent into exile for his revolutionary activity. In 1935 he gained his PhD, in philology. He wrote revolutionary songs and political lyrics, his verse being close in spirit and content to that of Demyan Bedniy and A. I. Bezymenskiy. He was the first great Yakut poet and writer. From his childhood he had been fascinated by the Yakut folk songs and the heroic epic *olongkho*, and echoes of his folk tradition can be found throughout his work. His most important poetic work was *Kyhyl oyuun* (Red Shaman, 1917–25). He also wrote original prose, taking folk motifs for inspiration: *Iirbit Nyukuus* (Mad Nyukus, 1923), *Jeberetten takhsyy* (Digging out of the Mud, 1936). He translated the *Internationale* into Yakut. During the Stalinist cult he was arrested on false political charges, and died in 1939 of tuberculosis. His reputation was later restored. MM

Özel, Ismet (b 1944 Kayseri), Turkish poet. He abandoned his studies at the Faculty of Political Science of Ankara University. Özel is the owner and one of the founders of *Halkın Dostları* (Friends of the People), a monthly journal for progressive art and culture, which has come out regularly since March 1970 in Istanbul, and has become the focus for the youngest generation of Turkish writers and Marxist critics. Özel is an ardent and sincere revolutionary poet. He sees his poetry as an integral part of the revolution, and as a weapon in the struggle to which he has dedicated himself in the interests of his country and his culture. A successor to the tradition of progressive Turkish poetry of the republican period (Nazım Hikmet, qv, and Ahmed Arif), Özel is an outstanding member of the younger generation of revolutionary poets who have made their names since 1960. This group, which includes Özkan Mert and Ataol Behramoğlu (qv), is characterized by its modern sensibility and its rejection of the vapid aestheticism of the school of *İkinci Yeni*. Collections of poems: Geceleyin *Bir Koşu* (The Night Race, 1966); *Evet, Isyan* (Yes, Rebellion, 1969, 1970). DL

P

pand-nāmak, see **handarz.**

Paralipomenon, see **Chronicles.**

Paronian, Hakob (b 1843 Adrianopolis, d 1891 Istanbul), Armenian writer and dramatist. With incomplete education, he worked as a teacher, journalist, and later office worker, mostly in Istanbul, publishing work in democratic journals such as *Meghu*, *Thatron*, *Masis*, *Luys* and *Khikar*. Paronian was the creator of realistic satire in prose and plays in modern West Armenian literature, and a leading figure in the cultural and public life of Armenians in Turkey. He wrote comedies, satirical portraits, critical and political articles. His comedies are in the stock repertoire of the Armenian theatre, and some of his characters have become the personification of certain comical human traits. Works: *Paǵdasar aǵbar* (1886; English trans., *Uncle Balthazar*, Boston 1930; French, *Maître Balthasar*, Paris 1913); *Azgayin ǰoǰer* (Great National Figures, 1879–80); *Mecapativ murackanner* (Highly Placed Beggars, 1880).

Trans.: *Gentlemen Beggars* (Boston 1933).
F. Feydit, *Comédie et la satire en Arménie* (Venice 1961); HO 7. LM

Parvin E'teṣāmi, (b 1916/17 Tabriz, d 1941 Teheran), Iranian poetess in the traditional manner. She was the daughter of a well-known man of letters, Yūsof E'teṣāmī, called E'teṣām ol-molk (Support of the Realm). Her father taught her Arabic and Persian, while she went to the American High School for girls in Teheran to acquire a modern education. After graduation she spent some time as a teacher there. She was married in 1934 but soon divorced. She began publishing verse early, in the journal *Bahār* (Spring) edited by her father, and later mainly in the traditional monthly *Armaghān*. She was not one of the women poets committed to the feminist struggle, nor was she actively concerned in any aspect of the modernization of the country going on all through her life. Her poetry is firmly rooted in the classical tradition, which in the context of Iranian life is undoubtedly one of the reasons for her popularity and for the praise bestowed on her by important critics (Bahār, Nafīsī, qqv). She preferred the *qaṣīda* and *maṣnavī* (qqv), and frequently wrote *monāẓere* (the Persian counterpart of the tenzon of the troubadours) and fables. Her philosophical and social themes were treated in the classical manner, in the abstract (life and death, wealth and poverty, generosity and egoism, etc) and the didactic point was always brought home. Her *dīwān* (qv) appeared during her lifetime (1935) and several new editions were issued.

M. Ishaque, *Four Eminent Poetesses of Iran* (Calcutta 1950); MLIC 59–65, with examples in English or French translation. VK

Patkanian, Raphayel (1830–1892 Nor Nakhidjevan), Armenian writer. He studied in Moscow, Dorpat (Tartu) and St Petersburg, and worked as a teacher. With K. Kananian and M. Timurian he founded the literary group *Kamar-Katipa*, later using the name of the group as his own pseudonym. Patkanian published from 1855 onwards, some of his works in the dialect of his birthplace and some in the manner of Abovian (qv). One of the first authors of lyrics on social and patriotic themes in modern East Armenian, he was an important influence on several generations of readers and poets. He also wrote patriotic epics, satirical lyrics, prose and poetry for children, realistic stories of provincial life, readers and textbooks, articles on education. He translated Robinson Crusoe, the fables of Aesop, etc, and collected and edited folk tales and legends. Works: *Banasteģcut'yunner* (Poems, 1864); *Azat erger* (Free Songs, 1878); *Nor Naxiǰevani k'nar* (The Lyre of New Nakhidjevan, 1879).

Trans.: TPAAM; AB I, II.
HO 7; THLA; LLS. LM

Pazhwāk, 'Abdurraḥmān (b 1919 Ghazni), Afghan poet and writer, in both Pashto and Dari. Of a judge's family, he studied in Kabul and became a journalist, later taking up a diplomatic career and becoming the United Nations representative for Afghanistan. His plays and stories are in the tradition of folk literature and stress local colour. Many of his themes come from history, for example that of the Dari play in verse, *Waẓīfa* (Duty), about an Afghan girl-patriot in the period of Aḥmad Shāh Durānī, and the Pashto verse play *Kalimadāra rūpey* (Inherited Money) of the British-Afghan war in the 19th century. *Gulhāi andīsha* (Blossoms of Thought, 1965) is a volume of verse. JB

Payrav (real name Otajon Sulaymonī; b 1899 Bukhara, d 1933), Tajik poet. He came from a well-to-do merchant family, and attended the Russian school in Kogon (near Bukhara) secretly; he thus learned something of European literature. He welcomed the fall of the Emir of Bukhara in 1920 with an enthusiastic poem. For a time he was secretary of the Bukhara Republic in Kabul, and then lived briefly in Iran. On his return home he devoted himself to poetry; his work shows his profound knowledge of classical forms, to which he gave a new content, primarily influenced by Mayakovskiy. He wrote of the spread of education and the emancipation of women in poems like *Qalam* (The Pen), *Ba sharafi qiyomi zanoni Sharq* (In Honour of the Emancipation of the Women of the East, 1927). He was one of the first poets to write of international events, eg in *Hinduston* (India, 1928), *Ba militaristoni Yapaniya* (To the Japanese Militarists, 1932), etc. His poetry was at its best in the early thirties. The *maṣnavī* (qv) *Takhti khunin* (The Bloody Throne, 1931) exposed the cruelty and decadence of the Emir's regime, as did an unfinished poem, *Manorai marg* (The Tower of Death). Payrav also wrote love poetry and many satirical poems repelling the attacks of those who did not understand him, and ridiculing formalism and the empty posing of left-wing writers who did not recognize the value of classical poetry. He translated plays and poems from or through Russian: Chekhov, Langston Hughes, Furmanov. A volume of his verse appeared: *Shukūfai adabiyot* (The Flourishing of Poetry, 1931) and his *Majmūai she'rho* (Collected Poems) was repeatedly republished.

RHIL 567–8. JB

Pentateuch (the Five Books of Moses), Hebrew *Torah* (Law), the oldest of the sets of books comprising the Hebrew Bible (qv), the Old Testament (qv). Ancient tradition gives Moses as the author; some of the subject matter goes back to the time of Moses, some took form after the tribes of Israel had settled in Palestine. These ancient traditions were apparently collected in four 'streams' or 'sources'; passing through various combinations and revisions, of which the last editing probably took place in the 5th century BC, these brought the *Pentateuch* into being, comprising narrative, legal and ritual matters. The first book (*Genesis*) describes the creation of man (Adam and Eve), the Flood (Noah), the forefathers of Israel (Abraham, Isaac, Jacob and his twelve sons, including Joseph who later lived in Egypt). The second book (*Exodus*) is devoted to the delivery of the Israelites from Egypt, and also includes the laws given to them on Mount Sinai through the medium of Moses (the Ten Commandments, the Covenant). The following three books (*Leviticus*, *Numbers*, *Deuteronomy*) are mainly concerned with rules for fulfilling the Law, but the history of Israel is also traced from Sinai to the borders of the promised land (Palestine), ending with the death of Moses. Of literary worth are especially the ancient poems inserted between narrative accounts (the *Song of Lamech*, the blessing given by Moses and Jacob, the song after passing through the Red Sea), and the accounts of the lives of the forefathers of Israel, and of the life of Joseph in Egypt.

The Torah: The Five Books of Moses (Philadelphia 1962 and reprints); E. A. Speiser, *Genesis* (Garden City 1964); G. von Rad, *Genesis* (London, Philadelphia 1961); M. Noth, *Exodus* (London, Philadelphia 1962); *Leviticus* (London, Philadelphia 1965); *Numeri* (London, Philadelphia 1968); G. von Rad, *Deuteronomy* (London, Philadelphia 1966). SZS

Pīramērd, Tawfīq (b 1867 Sulaimaniya, d 1950 ibid.), Kurdish poet, writer, journalist and educationalist. Acquiring a basic education in Sulaimaniya and Bāne, he worked in the civil service in Istanbul from 1898 and studied law at the same time. He became a member of the Mejlis in 1899. After graduation he became a lawyer, and in 1909 was appointed *kāimakām* in Julamerk, in 1918 *mutasarrif* in Amasīya. Both as a student and in government service he kept active contact with patriotic Kurdish circles in Istanbul and in 1907 joined the *Kurdistān* organization. In the early 20's he left government service and returned to his home town to devote himself to literature and journalism. In 1926 he took over the weekly *Zhiān* (1920–26 *Zhianawa*) and became its owner in 1934; he edited it until the paper was closed down in 1938, and in 1939 began editing the newspaper *Zhīn*. Apart from his own original contribution to Kurdish writing he played an exceptionally important part in saving and popularizing many Kurdish folk works. His paper served to educate his people through their literature; he wrote, translated and published classical Kurdish poetry and opened his paper to young poets. His care for the purity of the language helped to establish the Sulaimānī dialect as a literary language. He collected six and a half thousand Kurdish proverbs and versified them, publishing one in every number of *Zhīn*. His own verse is romantic in mood; its subjects are love, patriotism and later also political themes. His translations of poetry from the Gorānī to the Sulaimānī dialect were important. Unfortunately his poems have not yet been collected and published in their entirety. In his prose he turned mainly to the glorious past of the Kurds. In 1935 he published a literary version of the folk legend *Duvānza suvārai Marīvān* (Twelve Riders from Marivan) and a prose version of the folk legend *Mam u Zīn*. In 1942 he published the story *Mahmūdāghā Shēvakal*. The poet A. Gōrān (qv) paid tribute to him in *Yādi Pīramērd* (1951).

FKH XLV–XLVI. AK

Phaphazian, Vrthanes (b 1866 Van, d 1920 Yerevan), Armenian novelist. Of a poor but well-read family, he studied in Baku, Istanbul, Moscow, and the Social Science Faculty at Geneva; he later worked as a village teacher in Armenian communities in Turkey and Iran. He collaborated with the liberal democratic journals *Ardzagank* and *Mshak*. His novels, stories, sketches, plays, allegorical tales and occasional verse give a realistic picture of Armenian life in Iran and Turkey, especially in the countryside. Protest against social inequality, the call for individual freedom, and sympathy with suffering, are his constant themes. The tragedy of the Armenian minority in Turkey shocked him, and his heroes began to reflect a mood of hopeless despair. Phaphazian also wrote a short history of Armenian literature (Tbilisi 1910) and ethnographical studies of the Armenian gipsies (in French trans., *Santho, Scènes de la vie des Bochas, Bohémiens d'Arménie*, Paris 1920). Further works: *Patkerner t'urk'ahayeri kyank'ic* (Scenes from the Life of the Armenians in Turkey, 1891); drama, *Žayŕ* (The Rock, 1904). LM

Phavstos Buzand[atsi] (Faustus of Byzantium, later 5th century), Armenian historian, who may have been from Greece. According to some his name shows that he lived or studied in Byzantium. Following Koriun and Agathangeghos (qqv) he wrote a history of Armenia under the Arsacid dynasty, during the 4th century AD: *Buzandaran patmut'yunk'* (Legends of Buzand; French trans. in LCHAMA I; German trans. by M. Lauer, *Geschichte Armeniens*, Cologne 1879). A supporter of the unionist and anti-Persian movement, he nevertheless criticized social conditions in the early feudal state. He quoted myths, legends, and fragments of folk epics. His treatment of history is dramatic, not without hyperbole, and his language simple and lively, probably close to the colloquial language of the day. His work is a succession of episodes with historical grounding in fact, rather than a strictly historical account; he does not respect either inner relationships or chronological ties.

J. Markwart, *Untersuchungen zur Geschichte von Eran* (Göttingen 1897); C. Toumanoff, *Studies in Christian Caucasian History* (Georgetown 1963); HO 7; THLA; AAPJ. LM

Prophets, authors of the utterances and collected sayings preserved in the Hebrew Old Testament (qv, and see Isaiah, Jeremiah, Ezekiel, Twelve Lesser Prophets, Daniel), active 11th–2nd centuries BC. The earlier prophets are usually the subject of a narrative recounting the remarkable deeds they performed (as they themselves insist) on direct orders from God (Nathan, Elijah, Elisha). From the 8th century BC the prophetic books began to contain both the prophets' utterances and the events in which they took part, and took on poetic form (Amos, Hosea, Isaiah, qqv). These and some other prophetic books (Nahum, Habakkuk, Zephaniah, qqv) are also directed against alien peoples, reproaching them and prophesying punishment for their oppression of Israel. The prophets foretold the destruction of the states of Israel and Judah as a punishment for disregard of God's commandments and for social injustice (ie Jeremiah). In the Babylonian captivity, on the other hand, the prophets encouraged the people in faith and hope (Ezekiel, qv, the 'Second Isaiah'), and later in the effort to restore the religious community in Jerusalem (*Haggai*, qv, Zacharias). Prophecy then became apocalyptical (Daniel, qv). The moving appeals of the prophets, with their abundant imagery and metaphor, were and still are effective weapons in the effort for live, non-formal religion and for social justice. See also Bible.

J. Lindblom, *Prophecy in Ancient Israel* (Oxford 1962 and reprints); A. J. Heschel, *The Prophets* (Philadelphia 1962 and reprints). SZS

Proshian, Pertch (real name Hovhannes Ter-Arakhelian; b 1837 Ashtarak, d 1907 Baku), Armenian writer. Of a craftsman's family, with incomplete education, he became a teacher and contributor to the journals *Pordz*, *Murtch*, and others. Proshian is a successor to Kh. Abovian (qv), one of the earliest modern prose writers, creator of the rural social novel. His first book, *Sos ev Vardit'er* (1860), was praised for its contemporary theme, the social motivation of the conflicts, the realistic treatment of the characters and especially for the lively colloquial style drawing freely on the popular speech. M. Nalbandian (qv) criticized the book for inconsistency in the treatment of social conflict and for the fatalistic conclusion. His later books gave a realistic picture of Armenian village life in the later nineteenth century, using emotional and family relationships to illustrate the process of social differentiation taking place. Proshian translated L. N. Tolstoy and Ch. Dickens. Further works: *Haci xndir* (The Problem of Bread, 1880, German trans. by J. Lalayan, Leipzig 1886); *Cecer* (The Moths, 1889).

HO 7. LM

Proverbs, one of the Hebrew books of the Bible (qv); after an introductory section praising wisdom it presents a wide collection of proverbs and wise sayings, based on the example of King Solomon (qv). Besides Israelite proverbs it includes adages of Egyptian and Edomite origin.

R. B. Y. Scott, *Proverbs, Ecclesiastes* (Garden City 1965). SZS

psalm (Greek *psalmos*), originally a song accompanied on a stringed instrument. The *Book of Psalms* (150) is part of the Old Testament (qv), many of them attributed (a few probably correctly) to King David (qv, 10th century BC). In this collection from the post-exilic period (probably 4th–3rd centuries BC) there are old psalms (Psalm 29 is close to ancient Ugarit poetry; Psalm 104 is inspired by Akhnaton's *Hymn to the Sun*, Egypt), as well as later psalms like 137, from the captivity in Babylon. Most of the psalms were used in the rites and liturgy of the Temple in Jerusalem, and later sung in synagogues and in Christian churches. The depth of feeling with which the psalms purvey the sufferings and hopes of man, and his deliverance, has given them a lasting significance and popularity far beyond the bounds of their original function. Other collections were formed on the model of the Old Testament *Book of Psalms*: the Essene *Songs of Praise* (*Hodayot*, qv), the Pharisean *Psalms of Solomon*, the *Odes of Solomon* (qv) preserved in a Syrian version, and the Greek Christian odes. Their appearance in the Catholic liturgy and the hymnbooks of the Reformation churches made them an important influence on European poetry. See also Bible.

Trans.: *The Psalms for Modern Man* (New York 1970).
M. Dahood, *Psalms I–III* (Garden City 1966, 1968, 1971). SZS

pseudepigrapha (Greek: [writings] wrongly entitled), works of Jewish origin which have been preserved, some in Hebrew, others in Greek, Latin, Syrian, Ethiopian and Old Slavonic; they are related to Biblical narratives. Many of these works are apocalyptic, giving a detailed description of events to come, especially of the imminent end of this world. They are usually linked with the names of early Biblical figures (Enoch, Moses, Isaiah, Baruch, the *Fourth Book of Esdras*). They tend to be long-winded and fantastic in style, and are more important as evidence of their time (3rd century BC—1st century AD) than as literature.

R. H. Charles (ed.), *The Apocrypha and Pseudepigrapha of the Old Testament* (Oxford 1913 and reprints); O. Eissfeldt, *The Old Testament: An Introduction* (New York 1965). SZS

Ptahhotep, vizier to Egyptian King Djedkarē Isesi (c2380 BC), assumed author of a set of maxims preserved complete in a hieratic papyrus in the Bibliothèque Nationale, Paris. Dating from the XIIth dynasty (after 1786 BC), the work was still being read and copied in schools of the New Kingdom (after 1570 BC) not only for its rules of conduct of a successful life, but as an example of elegant style.

Trans.: Z. Žába, *Les Maximes de Ptahhotep* (Prague 1956). JČ

Q

Qā'āni, Ḥabībollāh Fārsī (b 1808 Shiraz, d 1854 Teheran), Persian poet. Considered the last of the classical poets, he was the son of the Shiraz poet Golshan, whose premature death left his family in poverty. Nevertheless Qā'ānī was able to study, thanks to a scholarship given him by the governor of Fars in reward for his verse eulogies. Studies in Shiraz were followed by Isfahan, where he was interested in the exact sciences as well as in philology, and wrote commentaries on the famous panegyrics of Khāqānī and Anvarī (qqv). His exceptional talent and learning earned him the favour of the new governor, and employment in his service. He accompanied the governor on travels through Iran, and collected old manuscripts. Later he moved to Teheran, where he led the carefree life of a court panegyrist; he indulged too much in debauchery and drugs, however, which undoubtedly hastened his death at a relatively early age. He was primarily a panegyrist, with a ready and facile pen to produce eulogic *qaṣīdas* (qv) on every court occasion and for any personage, and the ability to adopt whatever views were officially held at any given moment. His poetic talent and his classical learning made him a ready and witty improviser, sure of the admiration of his public. He also wrote many *ghazals* (qv) on philosophical, erotic-mystical and nature themes, didactic verses and passionate elegies on the sufferings of the Shiite saints. Just as much of his poetry is an echo of the classics, so his *Ketābe parīshān* (Book of Confusions, or Book of the Confused One) is a parallel to Sa'dī's (qv) *Golestān*. In this work Qā'ānī's wit is often on the gross side, and he does not avoid perverse and lascivious scenes. In the fashion of the day he learned French and perhaps English as well, but this in no way detracted from the absolute traditionalism of his verse. His *dīwān* (qv) has appeared in numerous lithograph editions.

V. Kubíčková, *Qā'ānī, le poète persan du XIX siècle* (Prague 1956); RHIL 329–31; BLHP 326–35. VK

Qabbānī, Nizār (b 1923 Damascus), Syrian poet. The son of a well-to-do merchant, he qualified in law in 1945, then entered the diplomatic service and held positions in various capitals, rising to ambassadorial rank. He now runs a publishing concern in Beirut. His early work was mostly on sensual love; but since the 1967 clash with Israel the dominant note has been one of distress and anger, especially at ineffectual leadership. His diction is that of everyday speech, sometimes dipping into the colloquial and drawing on folklore. His stronger effects are achieved by vehemence and startling juxtapositions rather than elaborate imagery. The combination of elemental themes and lucid statements has made him by far the most popular of contemporary Arab poets. Main works: *Qālat li's-samrā'* (The Brunette Said to me, 1944); *Qaṣā'id* (Poems, 1956); *ash-Shi'r qindīl akhḍar* (Poetry is a Green Lamp, mostly in prose, 1964); *al-Mumaththilūn, al-istijwāb* (The Actors, Interrogation, 1969).

Trans.: Pedro Martínez Montávez, in: *Poemas Amorosos Arabes* (Madrid 1965). PC

Qādiri, Abdullo, Uzbek writer, see **Qodiriy,** Abdullo.

Qahhor (Qahhar), Abdullo (b 1907 Kokand), Uzbek writer, dramatist and translator. He studied at the Teachers' Training College and later at the University of Central Asia, and joined the editorial office of the paper *Qizil Ŭzbekiston.* His first poems and stories appeared in the periodical *Mushtum yangi Farghona*. He has attained a high degree of skill in the composition and style of his short stories and novellas. His pre-war novels are represented by *The Mirage*, which puts paid to pan-Turkish illusions. After the war he published another novel, *The Fire of Koshchinar*, devoted to the development of kolkhoz farming, as well as feuilletons, a series of short stories, and a number of plays. His name is also well-known in connection with translations from the Russian, including Chekhov, Pushkin, Gogol, Tolstoy and Gorki; their influence on his work is evident. Main works: novels, *Sarob* (Mirage, 1937), *Koshchinor chiroghlari* (The Fire of Koshchinor, 1947); stories and novellas, *Olam yasharadir* (The World grows Younger, 1932), *Tŭy* (Festival, 1939), *San'atkor* (The Artist, 1936), *Qizlar* (Girls, 1941), *Asrorbobo* (1943), *Oltin yŭlduz* (Golden Star, 1944); plays, *Oghrik tishlar* (Toothache, 1954).

PHTF II 713. NZ

Qamberdiaty, Mysost (Russian Kamberdiyev, Misost Bimbolatovich; 1909–1931), Ossetian poet. Son of a peasant family, he joined the Komsomol in 1923 and was active in this Youth Organization all his life. In 1927 he started studying literature at Workers Faculty in Moscow, but fell ill with tuberculosis in 1929. Nevertheless he continued working in Ossetia on the editorial staff of the newspaper *Vlast' truda*, especially as a poet. His first Ossetian poems, published in 1926 in Ossetian newspapers, describe the building of a new world, the role of communist education and the struggle against an antiquated order. His early death did not allow him to fulfil his promise. A short time before his death his anthology *Tsin* (Delight) was published. Qamberdiaty composed some long poems (*kadaeg*): *Ard* (The Oath), *Kurdzuan* (Going to the Sun), *Jyzaeldon* (The Jyzaeldon-River), poems in Russian (the cycle *Flag na minarete*) and some prose tales. He was one of the innovators of Ossetian poetry for his fine expressiveness as well as for his talent in form. ML

Qārizāda, Ziyā (b 1921 Kabul), Afghan poet writing in Dari. He studied in Kabul and then became a journalist, spending some time in Turkey. In his verse he is an innovator, taking new themes and deserting the traditional system of images. His poetry is strikingly melodious; it is inspired by the ideas of patriotism, the longing for peace, respect for the work of simple people, criticism of backward ways, and the demand for equality for women. He has published many volumes of verse, including *Naynawāz* (The Piper), *Hadaf* (The Goal), *Payāmi bākhtar wa zabāni tabīyat* (The Message of the East and the Language of Nature, 1952), *Muntakhabi ash'ār* (Selected Poems, 1967) and others. JB

qaṣīda, a term used in Arabic, Persian (*qaṣīde*) and Turkish (*kaside*) poetics for a long poem with a regular rhyme-scheme of *aa ba ca* (etc) in the hemistichs, and a single metre based on quantity. In Ancient Arabian (pre-Islamic) poetry the term *qaṣīda* was connected with *qaṣīd*, meaning verse written as an art form with highly developed metrical schemes, compared with the more or less occasional and improvised verses in the *rajaz* metre (see *'arūḍ*). Later the *qaṣīda* developed into a form, or rather a certain gamut, characterised by its structure and subjects. The traditional features of the *qaṣīda* were maintained, however, and determined the metre and rhyme-scheme. The *qaṣīda* thus reveals two strata, the prosodic and the semantic-compositional.

The earliest *qaṣīdas* are short, but later theorists considered seven lines the minimum. The introduction of the form is traditionally ascribed to al-Muhalhil, or to Imru' al-Qays (qv) in the sixth century, but this is no more than legend. This was probably the century in which the *qaṣīda* acquired its rudimentary form, although not yet the mature form devised by later theorists. The basic themes of the *qaṣīda* were apparently (in this order): love (*nasīb*), description (*waṣf*) and panegyric (the praise of others, *madḥ*, or of oneself, *fakhr*). The poet took one theme which he developed to a greater or lesser degree; the themes were not causally linked, a feature deriving from Ancient Arabian verse;

this aimed at concision of thought, and rejected transition both in form and matter. The interest of the literary theorists petrified the form and in later traditionalist Arabic literature it was a set pattern which survived into the present century. The Bedouin motifs partially receded, however, and particularly in modern Persian poetry the *qaṣīda* has been primarily a court panegyric. In the course of time satirical elements entered, and still later the *qaṣīda* became didactic and religious, or acquired elements of social criticism or of philosophy. This was also the character of the *qaṣīda* in Turkish literature. Today the term is used simply to mean a poem, mainly of the traditional type.

BGAL S I 27–31, EI[1] IV, 401, II 843; NLHA 76 ff. KP

Qaṭrān, Abū Manṣūr 'Aẓodī (b c1009–14 Shādīābād, d after 1072 Ganje), Persian poet. Panegyrist of the local rulers of Azerbaijan, he was a simple person without high pretensions. In his youth he squandered his heritage, but was able to continue a simple life on his income as a panegyrist in Tabriz and Ganje. Nāṣere Khosrou (qv) mentions his acquaintance with Qaṭrān in the *Safarnāme*. A more prolific use of rhetoric figures renders him a connecting link between the Khorasanis like Rūdakī (qv) and the later school of Azerbaijan. His *dīwān* (qv) consists chiefly of *qaṣīdas* (qv), but he also wrote *moqaṭṭa'āt*, and *robā'ī* (qv). A Persian dictionary, which he is alleged to have composed, has not survived. Qaṭrān's poems show no trace of the lack of knowledge of (Eastern) Persian alleged by Nāṣere Khosrou.

BLHP II 271–2; RHIL 194. WH

Qazbegi, Aleksandre (b 1848 Qazbegi, d 1893 Tbilisi (Tiflis)), Georgian writer. Of an ancient aristocratic family, he was brought up in the Khevi region. He was a typical Georgian Neo-romantic. During his short life he followed many trades, being a shepherd and an actor amongst others before devoting the second half of his life to writing. He remained in his native mountains, drawing the inspiration for most of his hastily and feverishly written stories from mountain folklore. His best known novel, *Elguja*, appeared in serial form in a newspaper; based on a detailed knowledge of legends and folk customs, it is played out during the rebellion led by Prince Alexander against the Russians. The novella *Khevisberi Gocha* is really the account of a folk legend, about the tragic conflict between love for a woman and the obligations laid down by tradition. It is one of the most popular works in Georgian literature. VAČ

al-Qazwini, Abū Yahyā Zakariyā b. Muḥammad (b 1203 Qazwin, d 1283), Arab cosmographer and geographer. He fled from Iran before the Mongol invasion, first to Damascus and then to Wasit and Hilla in Iraq, where he was a judge. His cosmographical work *'Ajā'ib al-makhlūqāt* (Marvels of Creation) is a systematic survey of knowledge about the universe and nature, including man, but the first part also includes angelology and demonology. The second part deals with the four elements, physical geography and geology, mineralogy, botany, and zoology and anthropology. Side by side with much factual material there is a good deal of fantasy. His geographical work *Āthār al-bilād* (Monuments of the Countries) is divided into seven climates after the Greek manner adopted by Arab scholars. Under each climate, towns, countries, mountains, rivers, islands and seas are described in alphabetical order; notes on important men are given under their birthplace. It is not a work of much originality, but it has preserved much material from lost Arabic and Persian works. Both of his works enjoyed great popularity in the Islamic world and were early translated into Persian and Turkish. Al-Qazwīnī is considered the foremost Arab cosmographer.

Trans.: H. Éthé, *Kazwinis Kosmographie* (Leipzig 1868). IH

al-Qazwini ar-Rāzi, Arab philologist of Persian descent, see **Ibn Fāris,** Abu'l-Ḥuseyn.

qit'a, Persian *qet'e*, a type of poem in classical Arabic and Persian poetry, and in that of other nations whose poetry is based on Arabic and Persian models. In form it is close to the *ghazal* and *qaṣīda* (qqv), but the *matla*ʻ is not used, ie the rhyme scheme is ab, cb, db etc. The subject varies: humour, satire, autobiography, views on society. The term *qit*ʻ*a* is sometimes used for short poems in *ghazal* form, in which the *takhalluṣ* (author's poetic name) is not found in the *maqta*ʻ (final couplet). JB

Qodiriy (Qādirī), Abdullo (pseudonym Jolkunbay; b 1894 Tashkent, d 1939), Uzbek satirical writer. He introduced the novel form to modern Uzbek literature. He began writing while still at the religious schools and the Russian school; he learned Arabic, Persian and Russian, and read the classics of Russian and oriental literature. After the October Revolution he became a journalist and wrote for the satirical periodical *Mushtum*; he became one of its editors after graduating from the Journalists' College in Moscow. After a number of shorter prose pieces he wrote his first longer work, on an historical subject; it clearly showed that he had left behind the traditional style of legendary narrative. His first and most popular novel, *Past Days*, and the one which followed, *The Scorpion from the Altar*, were both historical novels, while his third major book, *Obid Ketmon*, written 1935–36, described Soviet life and the changes in Uzbek village life during the collectivization of the countryside. He is renowned for his colourful descriptions of the lives of his characters, and makes masterly use of the linguistic possibilities of Uzbek. He also collaborated in producing an Uzbek-Russian dictionary, and translated from Russian some of the works of Gogol and Chekhov. Chief works: *Ahvalimiz* (Our State, 1913); novels, *Ŭtgan kunlar* (Past Days, 1926), *Mehrabdan chayon* (The Scorpion from the Altar, 1929), *Obid Ketmon* (1936).

PHTF II 711. NZ

Qohelet, Hebrew book of the Bible, see **Ecclesiastes.**

al-Qur'ān, or **Koran** (lit. recitation, reading, text to be read or recited), the Holy Scripture of Islam. The *Qur'ān* is the verbal expression (later fixed in writing) of the Prophet Muhammad's (b 570 Mecca, d 622 Medina) religious experiences in the course of his twenty-two years of prophetic activity. The longer or shorter passages of rhythmic prose with rhyme or assonance (*saj*ʻ, qv), in which he expressed his belief and preached the fundamentals of the new religion of Islam to his fellow Arabs, were to him and his followers messages or a revelation sent by God (Allāh) through the intermediary of the Angel Gabriel. In the Islamic doctrine the *Qur'ān* is the eternal and uncreated Word of God.

The revelations were at first preserved mainly by memory by Muhammad and his followers, some of whom later wrote them down. After Muhammad's death collections of revelations appeared, originally for the use of private individuals; around the year 650, however, the Caliph ʻUthmān ordered all the fragmentary revelations to be collected and the whole text of the *Qur'ān* to be written down; thus the *Qur'ān* was given the form in which we know it today. This Uthmānic recension is accepted by the Sunnites as well as by the Shiites; the latter, however, consider it incomplete.

The *Qur'ān* was divided into 114 chapters (*sūras*) which were arranged not in chronological order, but according to the mechanical principle of length, so that the longest are found at the beginning (except for the introductory *Sūra al-fātiḥa* which is rather a prayer, similar to the Lord's Prayer), while the shortest are at the end, although they are in fact the oldest. The *sūras* are divided into verses (*āyas*) running from 3 to 286 in number; they are of different lengths, the oldest consisting of a few words, while later *āyas* run to ten or even fifteen lines. Altogether the *Qur'ān* has 6,236 verses, 77,934 words and 323,621 letters. Modern scholarship has established that the *Qur'ān* is an authentic book, and that nothing essential has been added or taken away, so that it does in fact represent Muhammad's complete teachings. According to style and the content the *Qur'ān* can be divided into four periods.

1 The first Meccan period comprises the old core of the *Qur'ān*, namely the very short *sūras* stressing the ideas of one God and his mercy, but also his power to chastise. They contain therefore numerous eschatological passages, especially descriptions of the Last Day, of Heaven and of Hell. Towards the end of this period already, arguments against the Meccan pagans already appear. These *sūras* belong to the most poetical parts of the whole *Qur'ān* and are characterized by short effective verses with rich rhymes and a wealth of admonition, rhetorical questions and visions; the Prophet's passionate conviction of his vocation breathes from every word.

2 In the second Meccan period the fundamental theme remains the unique character of one God, but at the same time attacks on Muhammad's opponents are more frequent. They are warned to heed the fate of ancient peoples who did not listen to their prophets and were accordingly punished. The examples are taken partly from the Old Testament stories (Moses, Abraham, Noah, etc), and partly from Arabian legendary history (the people of 'Ād, Thamūd, Midyan etc). The language is calmer, with less poetic imagery, and a shift in style can be observed towards larger units of composition, often in the form of three-part homilies. The rhymes are already more monotonous and nearer to assonance than to genuine rhymes.

3 The *sūras* of the third Meccan period offer little new material, but repeat, elaborate and emphasize the themes of the previous period. A few new stories were, however, added, among them the famous biblical tale of Joseph, and stories of Jesus and Mary. Some of the verses reveal Muhammad's grief over the hopeless struggle against the pagans, and express his conviction that God has predestined some for damnation and others for salvation. In style and language the *sūras* of this period are marked by a decrease in the number of poetical passages; rhetoric passages and didactic purposes are in the fore.

4 In the last, Medinan, period Muhammad became the organizer and head of a politico-religious community, and combined the functions of law-giver, judge and general. The *Qur'ān* is now a moral and legal code setting down the relations between believers, and between them and nonbelievers. In the polemics against the Jews (and later against the Christians) Islam achieved its final form as an independent religious system. From this period stem the prescriptions about the basic duties of those who practise the Islamic cult, the food and drink prohibitions, the rules regulating marriage, family life and inheritance, as well as the rudiments of criminal and commercial law. As literature these *sūras* are prose in which assonance is still used but without aesthetic effect, being no more than a punctuation element. Sentences are longer and lack the former poetical flavour, often repeating stereotyped admonitions, exhortations and commands. The long *sūras* are no longer compact units, but conglomerations of longer or shorter revelations with no unifying principle.

The *Qur'ān* is the earliest written monument of the Arabic literature, and exercised an immense influence on its later development; a large part of Arabic (as well as other Islamic) scholarly production is directly tied with the Qur'ānic studies (linguistics, lexicography, historiography and all branches of theology). The language of the *Qur'ān* has become the norm of literary Arabic and, being the vehicle of Islam, Arabic thus achieved the status of a world language. Since according to Muslim doctrine of *i'jāz* the style and contents of the *Qur'ān* are inimitable, they exercised little influence in the direction of style; on the other hand, however, its ideological influence is immeasurable and far-reaching. The orthodox Muslim recites the *Qur'ān* many times every day in the course of his prayers, fragments of *sūras* are used as amulets and the whole of Islamic literature is pervaded by quotations from the *Qur'ān*. It has been translated into almost all the languages of the world; the earliest European translations (into Latin) date back into the 12th century.

Trans.: J. M. Rodwell, *Translation of the Koran*

(London 1876); Richard Bell, *The Qur'an. Translated with a critical rearrangement of the Surahs*, 2 vols. (Edinburgh 1937–39).
A. J. Arberry, *The Koran Interpreted* (London 1955); Richard Bell, *Introduction to the Qur'an* (Edinburgh 1953); W. Montgomery Watt, *Companion to the Qur'an* (London 1967). IH

R

Rab'e, Iranian literary group, see **Hedāyat,** Ṣādeq.

Raffi (real name Hakob Melikh-Hakobian, b 1835 Payadjiuk in Iran, d 1888 Tbilisi (Tiflis)), Armenian writer. Son of a tradesman, without completing his education he worked as a teacher and journalist. He collaborated with the democratic journals *Hyusisaphayl* (see Nalbandian), *Mshak*, *Ardzagankh*. Raffi made his debut in 1858 with sketches of Armenian life, past and contemporary, in Iran and Turkey. He wrote romantic lyric poetry (*P'unj*, Bouquet, 1874), later mainly novels with historical and contemporary themes (*Davit'-bek*, 1881; *Samvel*, 1886, French trans. by Altian and H. Kibarian, Paris 1924; *Xent'*, 1880, trans. Boston 1950), following Abovian (qv). His novel *Kaycer* (The Sparks, 1883) is also a treasure-house of ethnographical information. Raffi's work contributed outstandingly to the formation of a national consciousness and the spread of the national liberation movement in the second half of the 19th century, as well as to the formation of the eastern variant of modern Armenian.

Further trans.: A. Hovhannisian, *Bilder aus Persien und Türkisch-Armenien*, in AB III; the same, *Contes Persanes* (Paris 1902).
Mrs. Raffi, *Raffi's Works* (London 1915); THLA; HO 7. LM

Raḥmān Bābā, Pashtun poet, see **'Abdurraḥmān Mōhmand.**

Rases, Razes, Razis, Persian physician, see **ar-Rāzi,** Abū Bakr Muḥammad.

Rashid Vaṭvāṭ, real name Rashīdoddīn Moḥammed al-'Omarī (b 1114/16 Balkh, d 1177/83), Persian panegyrist and epistolographer. He held high office at the court of the Khwarazmshah Atsyz (1127–1156) and was a collector of manuscripts and patron of poets. His *qaṣīdas* (qv) were written in praise of the ruler and his wars, and are overburdened with high-flown ornament. Of greater significance is his prose treatise on rhetoric, *Ḥaqāyeq os-seḥr* (The Gardens of Magic). He also wrote Arabic *qaṣīdas*.

RHIL 200. JB

Rashidoddin, Faẓlollāh Ṭabīb (b 1248 Hamadan, d 1318 Tabriz), Persian statesman, doctor and historian. A Jew, converted to Islam, he became the doctor of the *īlkhān* Abaka (1265–81), and then vizier to the *īlkhāns* Ghāzān (1295–1304) and Oljeytū (1304–16); under the latter's successor, Abū Sa'īd, he was executed. His writings, in Persian or Arabic and sometimes in both, deal with many subjects including history, theology, medicine, agriculture and horticulture. The most famous is his *Jāme'at-tavārīkh* (Summary of History), one of the greatest historical works of the East; written in Persian, it was translated into Arabic. It consists of three parts; the last, geographical, has been lost. The first part contains invaluable information about Mongol and Turkish tribes, the history of Chinggiskhan and his ancestors and descendants, particularly the Iranian *īlkhāns* down to the time of Ghāzān. The second part gives the history of the patriarchs and prophets, the history of ancient Iran, the life of Muḥammad, the history of the Caliphs and dynasties ruling Iran up to the time of the Mongol invasion, the history of the Chinese, the Franks and the Indians, with a long chapter on Buddha. Rashīdoddīn's letters have also come down to us. Different parts of his historical work were translated into several European languages.

SPL I No. 106 and P 1230–2; RHIL 439–40, 455. FT

ar-Rayḥāni, Amīn (1876–1940), Lebanese writer and poet. He emigrated to the USA as a young man and lived a bohemian life, writing for papers and appearing on the stage. At first he wrote in English, later in Arabic, his favourite forms being short essays on ethical, literary or political subjects, and prose poems in the manner of Whitman. Rayḥānī's great importance was in the introduction of the prose poem to modern Arabic literature in the fullest measure. His work reveals a love of nature which is almost pantheistic. His work is also remarkable for its critical approach to western innovations. He returned home after World War I, and after travelling in the Arab countries published a two-volume travel book, *Mulūk al-'Arab* (Arabian Kings). He also wrote a history of the Najd. Other works: novels, *Khārij al-ḥarīm* (Outside the Harem); *Zanbaqat al-jawr* (Lily from the Depths). SP

ar-Rāzi, Abū Bakr Muḥammad b. Zakariyā, also known as Razes or Rhazes (b c864, d c925, or later, Rayy), Persian physician, philosopher and scientist most active in Rayy and Baghdad, where he ran a hospital. His philosophy was based on the five eternal principles: the creator, the soul, matter, space and time, and contains marked gnostic elements: the myth of the soul imprisoned in matter and liberated by Reason through philosophical learning. In physics Rāzī was an atomist whose conception was nearer that of Democritus than that of the Islamic *kalām* (qv). His approach to religion and prophetism was critical, and he believed that liberation of the soul could only be achieved through philosophical learning; his attitude to traditional religion was one of negation. Rāzī's greatest contribution was in the field of medicine, where his *al-Hāwī* (Latin: Continens), *Liber Almansoris* and other writings influenced European, as well as Arabic thought. He also made a significant contribution to alchemy.

J. Ruska, *al-Bīrūnī als Quelle für das Leben und die Schriften al-Rāzī's* (Brussels 1922); J. Ruska, *Al-Rázī's Buch Geheimnis der Geheimnisse* (Berlin 1937); S. Pines, *Beiträge zur islamischem Atomenlehre* (Berlin 1936); CHPI 194–201, 354. vsd

Razikashvili, Luka, Georgian poet, see **Vazha-Pshavela.**

Rezāqoli Khān Hedāyat (b 1800, d 1871/72 Teheran), Iranian scholar well known at the Qajar court, sometimes known as Lālābāshī (teacher of princes). Of a family of high officials, he received a thorough traditional education and became court poet, first in Shiraz and then at the court of Faṭh 'Alī Shāh in Teheran, later holding other offices there. He was made director of the newly established modern Polytechnic school, Dār ol-fonūn (1851) and proved very successful. He wrote studies on history, diplomacy, Persian lexicography and literary history. His two literary *taẕkiras* (qv) are of particular value. The first, *Majma' ol-foṣaḥā* (Congregation of the Eloquent) is a collection of older *taẕkiras* supplemented by writers of his own time, while the second, *Riyāz ol-'ārefīn* (Garden of the Knowledgeable) deals exclusively with the mystic poets of *taṣavvof* (see Sufism). His poetic *dīwān* (qv) of over 50 thousand lines shows him a typical scholar whose classical education and brilliant learning enabled him to write in imitation of the old masters, and although without special talent, to produce verse quite in conformity with the rules of traditional poetics.

RHIL 339–40. VK

Rifat, Oktay (b 1914 Trabzon), Turkish poet, son of the poet Samih Rifat, who was governor of Trabzon. He studied law in Paris, and is now a lawyer. He formed a literary group with Orhan Veli Kanık and Melih Cevdet Anday (qqv). Up to 1956 (publication of the poem *Perçemli Sokak*, Street with a Mane) he wrote in the manner of the *Garip* group. His new trend, sometimes accused of abstract writing, attracted younger poets; together they formed the *İkinci Yeni* (Second New) group or trend, basing poetry on verbal association; unusual combinations of words are declared to be an active approach to reality. Rifat has also translated classical Greek and Latin poetry, and written plays. Other works: poetry, *Yaşaıp Ölmek, Aşk ve Avarelik Üstüne Şiirler* (Poems of Life, Death, Love and Wandering, 1945); *Güzelleme* (1945); *Aşağı Yukarı* (Roughly Speaking, 1952);

Karga ile Tilki (The Crow and the Fox, 1954); *Âşık Merdiveni* (Stairway of Singers, 1958); *Elleri Var Özgürlüğün* (Freedom has Hands, 1966); plays, *Kadınlar Arasında* (Between Women, 1948); *Oyun İçinde Oyun* (Play within a Play, 1949). LH

Rishtīn, Ṣadīqullāh (b 1919 Ghāzīābād), Afghan writer, philologist and literary historian, writing mainly in Pashto. Born in the family of an enlightened theologian, of anti-British opinions, he first studied theology in Kabul and then turned to writing and scholarship, and took an active part in public life. Since 1963 he has been director of the Pashto Academy (*Pashto ṭolena*). His travel sketches of Kataghan describe the life of simple people; they preach love of the native country and the mother tongue, attack imperialism and plead for a free Pashtunistan. His works are often didactic and philosophical, but include love poetry. He has made a significant contribution to Pashto philology and written a number of linguistic studies as part of the endeavour to raise the level of knowledge of Pashto in Afghanistan. Works: *Khwakhē qissē* (Pleasant Tales, 1943), *Pashtō qissē* (Pashto Tales, 1952), *De Hind safar* (Travels to India, 1953), *De pakhtō de adab tārīkh* (History of Pashto Literature, 1954). JB

robā'ī or **rubā'ī,** an Arabic word meaning 'fourfold', used for one of the most widespread lyric forms among the Islamic peoples of the East, especially in Persian poetry whence it passed to Turkish, Pashto and other literatures. It is a quatrain composed of two *bayts* (see *maṣnavī*), in which the first, second and fourth lines rhyme. In special cases all four lines rhyme, and the form is then known as *robā'īye tarāne* in Persian poetry. The *robā'ī* is one of the oldest forms in Persian poetry and its origin is connected with folk poetry. The subject may be love, social philosophy, or it may have a didactic aim. The greatest writer of *robā'ī* is usually held to be 'Omar Khayyām (qv). The form is still frequently used by Iranian, Tajik, Afghan and other poets. It is distinct from the *dobeytī*, a quatrain, rhyming *ab*, *cb*. JB

Rūdakī, Abū 'Abdollāh Ja'far b. Moḥammad b. Ḥakīm b. 'Abdorraḥmān b. Ādam (b 850/60 Rudak, d 940/41 ?Rudak). Central Asian poet who is often designated as the founder of Persian and Tajik poetry. Although several earlier poets had written Persian verse, Rūdakī was a personality who exercised a strong influence on the further development of poetry. He practised all the typical genres (*qaṣīda*, *ghazal*, *maṣnavī*, *robā'ī*, *qit'a*, qqv) and was probably the writer who settled the *'arūẓ* (qv) metric system firmly in Persian poetry. He probably lived at the court of the founder of the Samanid dynasty, Naṣr I b. Aḥmad, in Samarkand; the ruler himself wrote Arabic poetry. After the death of his patron in 892 Rūdakī moved from court to court and seems to have acquired considerable wealth. He was invited to Bukhara either by Esmā'īl Sāmānī or by his vizier Bal'amī (qv) but in 937 was driven out, possibly because he belonged to the Karmat religious sect. According to the literature he is generally agreed to have been blind in old age, some believing that he had been so from youth, others that he had had his eyes put out in Bukhara when he fell into disfavour. His work seems to have been wide and diverse. He drew on the riches of folk literature and on the earlier literary traditions of the Iranian peoples of Central Asia. Only fragments of his work remain, however, amounting to about 1,000 lines (*bayts*, see *maṣnavī*). The *qaṣīda* entitled *Shekāyat az pīrī* (Complaint in Old Age) and that entitled *Mādere mey* (Mother of Wine) have survived in entirety. Parts of the *maṣnavī Kalīle va Demne* (qv) and *Sendbādnāme* (Book of Sindbad), translated from Pahlavi, are preserved in various anthologies, as are also about 40 *robā'ī* (qv) and some of the *ghazals* and *qit'a*. Much light has been thrown on the personality and work of Rūdakī by Nafīsī (qv) and especially the Tajik scholars S. Aynī (qv), Mirzoev, Mirzozoda, Tagirjanov and others.

A. G. Arberry, *Classical Persian Literature* (London 1958), 32–5; RHIL 144–5. JB

Rūmī, Jalāloddīn (b 1207 Balkh, d 1273 Konya), called Moulavī or Mawlawī by the

Persians, Mevlana by the Turks; he was the son of Bahā'oddīn Valad, a mystic theologian who left the town with his family, wandered through the Middle East and eventually settled in the Seljuq capital, Konya, where he died in 1231. Rūmī succeeded him in teaching. In 1244 the wandering dervish, Shamsoddīn Tabrīzī, reached Konya, and he appeared to Jalāloddīn as the manifestation of the Eternal Beloved, as the 'chosen mouthpiece of the Deity'. Their relation grew so close that Jalāloddīn's disciples and family strongly disapproved of it, and Shams left Konya (1246) for a while. After a second short stay he again disappeared, probably killed by the jealous disciples. In the search for his beloved Rūmī became a poet. The lyrical poetry in which he sings his yearning and love bears the name of Shams, as a sign of his complete identification with whom he had, eventually, found himself. Later, mystical love with a goldsmith, Ṣalāḥoddīn Zarkūb, inspired some poems. His last love experience was with Ḥosāmoddīn Chelebī, who inspired the didactic *Maṣnavīye ma'navī*; he became Rūmī's first successor in the spiritual leadership of the Mevleviya order into which his disciples were organized, mainly by his son Solṭān Valad (to whom we owe interesting poetry describing his father's spiritual experiences). The Mavlavis are known as Whirling Dervishes, from their ritual whirling dance.

Rūmī is the most fertile and, at the same time, the most fascinating mystic poet writing in Persian. His *Dīvāne Shamse Tabrīz* comprises about 30,000 verses; his *Maṣnavī* (in the *ramal musaddas* metre) has more than 26,000 couplets. Besides these, he left a smaller collection of treatises, *Fīhi mā fīhi*, and a number of letters. His lyric poetry is, for the largest part, extremely rich and beautiful, full of images and music. In many cases the reader can almost follow the flow of inspiration during the enraptured state, and its ebbing away or breaking off. Rūmī has taken a large number of metaphors and images from everyday life so that one could almost reconstruct parts of the life of mediaeval Konya from his poems. Often written under inspiration, his verses do not lack the refinement of Persian rhetorics, yet they are not overburdened with rhetorical tricks. The *Maṣnavī*, exalted by Jāmī (qv) as 'the Qur'ān in Pahlavī' is a complete encyclopaedia of mystical thought and motifs which grow out of each other without logical order, leading to new tales; they even sometimes contradict each other, if one reads them in search of a common denominator for all his poetry.

Rūmī never attempted a coherent system of thought, although his work expresses the teachings of classical Sufism (qv). Every generation and every reader has therefore found what he sought in it, whether Ibn 'Arabī's monistic doctrines which had reached him through Ṣadroddīn Qōnavī, or a personal love-relation between man and God. The yearning for mystical union is expressed in every possible symbol. Stories from different sources are introduced and offer rich material for the student of folklore, and the linguist can study some Turkish and Greek expressions in Rūmī's verses. Small wonder that as early as the late 14th century the *Maṣnavī* was read in the Muslim world as far as Bengal, and inspired commentators in Turkey, Iran, and the Indian subcontinent. There are numerous translations into Islamic languages (Turkish, Sindhi, Pashto, Arabic). Western orientalists, attracted first by the ritual of the Whirling Dervishes in Turkey, began to be interested in Rūmī's work at the beginning of the 19th century. A large number of translations and adaptations into Western poetry (since Hammer-Purgstall 1818, Rückert 1821) have made Rūmī the best-known mystic of Islam in Europe. But a conclusive evaluation of his style and his mystical thought is still lacking although he has always been praised as the greatest and most ardent herald of mystical love.

Trans.: R. A. Nicholson, *Mathnawī-i ma'nawī* (London 1925–40); A. J. Arberry, *The Ruba'iyyat of Rumi. Selected translations* (London 1949), *More Tales from the Masnavi* (London, New York 1963).

A. J. Arberry, *Discourses of Rumi* (London 1961); R. A. Nicholson, *Rumi, Poet and Mystic* (London 1950). AS

Rustaveli, Shota (late 12th–early 13th

century), the greatest Georgian poet. The information given about his life is not reliable; according to the legend he was a high dignitary and court poet to Queen Tamar, who sent him on a mission to Palestine, perhaps in 1192. In the Georgian monastery of the Holy Cross in Jerusalem there is a portrait of one Shota Rustaveli, and manuscript in the monastery contains references to the Queen's treasurer, Shota; they are generally believed to be identical. Professor P. Ingoroqva identified (1970) the poet with the Bagratian Prince Shota III, Duke of Hereti, residing at Rustavi; he lived 1166–1250 and was a distant relative of Tamar.

Rustaveli was the author of the heroic-romantic epic *Vepkhistqaosani* (literally, Clad in a Tiger [or Leopard] Skin), the greatest literary monument of the 'golden age' of Georgian literature. In the introduction, which is clearly not the work of Rustaveli in entirety, we are told that the epic is a new version of a Persian tale, but no such model has yet been found to substantiate this. The story tells how an unknown mad knight, clad in a tiger-skin, found his way to the kingdom of the Arabian king Rostevan. Princess Tinatin sent her knight Avtandil to seek him out, and he discovered after much trouble that the unhappy man is an Indian prince, Tariel, who has gone mad after losing his beloved Nestan-Darejan. He offers to help Tariel, and searches the whole world for the Indian princess, finally discovering, with the help of the King of Mulgazanzar Pridon, that she is imprisoned in the demon country. All three knights set out to free her, and there follow two weddings in Arabia: Tariel marries Nestan-Darejan and Avtandil marries Tinatin, who in the meantime has inherited the throne of Arabia. None of the episodes is placed in Georgia, and none of the characters are Christian, and yet the whole epic breathes the spirit of Georgia.

Frequent allusions to Arab and classical authors show Rustaveli to have been a man of profound learning. The whole work of 1669 rhymed quatrains is written in the classical Georgian *shairi* metre of 16-syllable lines with a caesura after the 8th and frequent internal rhymes. The end rhymes are rich (as many as six rhyming phonemes) and the instrumentation and formal polish of the strophes is inimitable. The epic became the classical model for all Georgian poetry, and every poet set its perfection as his ultimate goal. The mediaeval church tried in vain to weaken the influence of this great work; from the earliest times it was to be found in every bride's dowry, while many people knew (and today many still know) whole passages by heart; sayings from the poem became proverbs, and thus part of the folklore heritage. There are many copies and variants of the poem, while the most famous poets wrote commentaries on it, as did kings; for example, Archil (qv) and Vakhtang VI, who prepared the first critical edition, printed in Tbilisi in 1712. Magnificent manuscript copies have been preserved, illuminated with Georgian and Persian miniatures. The piece inspired Georgian and foreign painters (eg the Hungarian painter Zichi) and has been translated into many languages. VAČ

Ruth, one of the Hebrew books of the Bible (qv), it is a story of folk origin describing family loyalty which overcomes difficulties and wins just reward. In ancient and more modern times, *Ruth* was popular reading, for the simple idyllic way the story is told. See also Five Festal Scrolls. SZS

Rytgev (Rytkheu), Yuriy Sergeyevich (b 1930 Uellen), Chukot writer. The son of a poor hunter, he was also a hunter and sailor. He attended elementary school in Uellen and in 1946 went to study in Anadyr, where he also began writing. His verse and sketches appeared in the paper *Soviet Chukotka* and were later translated into Russian for the volume *Druzya-tovarishchi* (Friends and Comrades, 1953). In 1953 he published a volume of stories *Pyn'yltelte*; in 1955 *O'ravetlen vagyrgyn* (People of our Coast). In 1955 he returned to his native parts and wrote tales and sketches, and the novel *Tite tylgyrkyn y'lýl* (When the Snows Melt, 1958). The volume of verse, *Bear Stew*, dates from 1965. He also translates classical and contemporary authors from Russian to Chukot. MM

Rza, Räsül (b 1910 Gyökchay), Azerbaijan poet, writer and dramatist. Graduating from the Transcaucasian Communist University and the Moscow Film Faculty, he has held several leading posts in the cultural life of Azerbaijan. Although he wrote effective sketches from the front during the last war, it is as a poet that Rza excels. His verse records the changing life of the villagers of Azerbaijan, the struggle of the peoples of the East against imperialism, and impressions from the front. He dedicated long poems to Lenin and to the life of the Azerbaijan poet, Füzuli (qv). He has translated several important poetical works of world literature, and his plays are also popular. Chief works: poetry, *Chapai* (Chapayev, 1932); *Güvvät bairamy* (Festival of Strength, 1933); *Lenin* (1950); *Illär vä sätirlär* (Years and Poems, 1961); *Duyghular . . . düshünjälär* (Feelings . . . Memoires, 1964); short stories, *Oghul gatili* (The Boy's Murderer); *Dilara; Hekayälär* (Stories, 1956); *Teatr hekayäläri* (Tales of the Theatre, 1964); dramas, *Aligulu evlänir* (Aligulu Marries, 1962); *Väfa* (Fidelity, 1943).

PHTF II 690. ZV

S

Sabir, Mirzä Aläkbär (b 1862 Shemakha, d 1911 Baku), Azerbaijan satirical poet. Of a merchant family, after secondary school he travelled in Iran and Central Asia. In 1909 he opened a school in Baku, and also edited a paper. He occasionally used the pseudonym Aghlar Gjüläjän (lit., he who laughs, cries). He left many excellent satirical poems, prompt reactions to various events, especially political events. His bold satires appeared mainly in the popular periodical *Molla Näsräddin*, widely read in Turkey and Iran as well as in Azerbaijan itself, in the revolutionary years 1905–08. His sharp attacks on the ills of contemporary society were collected after his death in *Hop-hop-namä* (The Hop-hop Book), still republished and popular today.

PHTF II 679 ff. ZV

Sabit, Alauddin (b c1650 Užice in Yugoslavia, d 1712/13 Istanbul), Turkish poet, of a Bosnian craftsman's family, speaking South Slavonic. He studied Muslim law and became a judge; he lived in Edirna, Çorlu, Sarajevo and elsewhere, often in poverty. Sabit wrote several *mesnevi* (see *maṣnavī*) and a number of short poems, full of an almost too drastic humour; some of his verse narratives are somewhat immoral (*Berbernâme*, Story of the Barber). He usually achieved humorous and even absurd effects by using well-known proverbs in the wrong place, giving rise to ambiguous interpretations. Sharp criticism appears especially in the *ghazals* (qv). The romantic mystical epic *Edhem ū Hümâ* is free of rough humour; it is the story of the dervish who longed for the beautiful Hüma, the symbol of good fortune. Sabit's lyrics show the influence of Bosnian folk poetry, both Slavonic and Turkish. See also Mahmud Bakî.

J. Rypka, *Beiträge zur Biographie, Charakteristik und Interpretation des türkischen Dichters Sābit* (Prague 1924); GHOP IV 14 ff. ZV

Ṣabri, Ismā'īl (b 1854 Cairo, d 1923), Egyptian poet. A member of one of the educational missions sent to France, he obtained a law degree from the University of Aix. On his return to Egypt he was appointed to various posts in the Ministry of Justice, and became Governor of Alexandria. On his retirement, he devoted himself exclusively to literature, and poetry in particular. Although very much of the generation of Shawqī and Ḥāfiẓ, he was much less prolific, and was not so preoccupied in his verse with social and political issues; he tended to concentrate more on quiet, personal themes. An edition of his works appeared in 1938.

RCO

Sa'dī, Sheykh Moṣleḥoddīn (b c1208 Shiraz, d 1292 Shiraz), the Persian author most widely known in the West since his

Golestān (Rose Garden) was the first Persian literary work to be translated into European languages. Sa'dī, whose pen-name is derived from the name of the family of Abū Bakre Sa'd b. Zangī who ruled Fars in the first half of the 13th century, was born probably around 1208 (not, as most earlier scholars held, c1182). For nearly 30 years he travelled extensively in the Islamic countries although we no longer take at their face value all the travel accounts and anecdotes in his books. When Sa'dī returned, in 1256, to his birthplace he found it in a peaceful state thanks to the political acumen of the Zangid ruler for whom he composed the *Būstān* (Garden, 1257), a didactic poem in ten chapters in the simple metre *motaqāreb*. In 1258, he completed the *Golestān*, which deals with different aspects of human life in eight chapters of mixed prose and poetry. Besides these works, Sa'dī composed charming *ghazals* (qv). He can even be considered the first Persian poet to use the *ghazal* for pure love-lyrics. He further wrote a number of *qaṣidas* (qv) in honour of the Zangid and other rulers. After 1260, political changes affected Shiraz, which eventually became a Mongol province. Sa'dī's *Ṣāḥebnāme*, dedicated to the influential minister Shamsoddīn Joveynī, contains lyrical and didactic poems. In accordance with the taste of his time, the poet also wrote some *hazaliyāt*, frivolous or obscene verses which some of his European admirers would have liked to expurgate. Sa'dī spent his last years in a dervish convent in Shiraz where he died; a beautiful structure has been erected over his tomb.

Sa'dī is regarded, with full right, as the perfect master of classical Persian style. His prose and verse in the *Golestān* are so flowing and graceful that the amount of artistry hidden behind their harmonious and refined language can be discovered only by repeated and careful study. He is perhaps the most human Persian author, the 'philosopher of common sense' who teaches correct and elegant behaviour in every possible situation. Yet his is not a superficial etiquette but a way of life, deeply tinged by the values of the inherited religious, and especially mystical, teachings of Islam which taught him to love all God's creation. His poetry often sings the joy and pain of love, using the traditional imagery with extreme skill. However there is scarcely a poem which does not teach a small moral lesson to the reader. Out of the rich treasure of Islamic and Iranian pre-Islamic legends and stories Sa'dī chose, with great taste, those topics which appealed most to his audience, and he offered them in such a form that they were easily memorized by everyone.

Small wonder that his *Golestān* has been, for centuries, the first reading-book in Persian classes throughout the Islamic world, and was likewise recommended as a text-book in Europe by Sir William Jones in 1787. Numerous commentaries in Persian, Turkish and Indian languages have been written to both the *Golestān* and the *Būstān*. Translations, to mention only a few examples, into Turkish go back to the 14th century, and into Pashto to the late 17th century. The first Urdu translation is dated 1802. A book wrongly attributed to Sa'dī, the *Pandnāme* (*Karīmā*) was very popular in India. Libraries in many countries can boast of beautifully illustrated and illuminated manuscripts of the *Golestān* and *Būstān* and also of Sa'dī's lyrical poetry, which is highly esteemed in the Persian speaking areas for its lucid style, though less popular in the West. Numerous imitations of the *Golestān* can be traced in the countries under Persian cultural influence, the most famous of them Jāmī's (qv) *Bahārestān*. The first complete translation of the *Golestān* into a Western language was the German one by A. Olearius, published 1654; then followed translations into French, Latin, Dutch and English. The *Būstān* was translated into English in the 17th century by Thomas Hyde.

Massé's bibliography shows the immense amount of material about Sa'dī and his influence on European thought during the last three centuries. His didactic approach as well as his chaste language appealed more to a Western audience than the high-soaring, ambiguous verses of many mystical poets of Iran, and Sa'dī's 'humanist' teachings are as relevant today as they were 700 years ago.

Trans.: E. Rehatsek, *Sa'di, The Gulistan, or Rose Garden* (London 1964); Omar Ali Shah, *Le jardin de roses—Gulistan* (Paris 1966); A. J. Arberry, *Kings and Beggars, the First Two Chapters of the Sa'di's Gulistan* (London 1945). Henri Massé, *Essai sur le poète Sa'di* (Paris 1919). AS

Ṣā'eb, Mīrzā Moḥammad 'Alī (b 1601/02 Tabrīz, d 1677/78 Isfahan), Persian poet. Son of a merchant from Tabriz, who had settled in Isfahan by order of 'Abbās II, he early met poets like Shefā'ī and Ḥakīm Roknā. Disappointed by lack of recognition for his poems, Ṣā'eb went to India like many of his compatriots. Finding a patron in Ẓafar Khān, Governor of Kabul, he stayed there and accompanied him to the Mughal court at the ascension of Shāhjahān who honoured the poet with rank and title. Afterwards he went to Kashmir with Ẓafar Khān, but longing for Persia, from where his father came to accompany him home, Ṣā'eb obtained permission to return after about six years' sojourn in India. Honoured by Shāh 'Abbās II, Ṣā'eb became his poet laureate; he retired from court life after Shāh Soleymān's ascension and died soon after.

Ṣā'eb's work consists of *maṡnavīs* (qv) of which the best known is his *Qandahārnāme*, an account of 'Abbās II campaign, and *qaṣīdas* (qv). *Ghazals* (qv) form the bulk of his poems, of which he allegedly composed more than 300,000 verses. Coming from Tabriz, he also wrote poems in Azerbaijan Turkish. Ṣā'eb was the unrivalled master of the first phase of the Indian (Isfahani) style (qv), restored to favour in Persian after two centuries of disgrace due to the reactionary *bāzgasht* movement (return to the clichés of the Iraqi style). In his poetry new metaphors and similes break through the traditional stereotype; words and topics of everyday life, together with a growing number of idioms, are to be found. Formal rhetoric is greatly reduced, while the conception and idea are developed and play a prime role. A certain similarity to the English metaphysical poetry of Marvell, Herbert and others is obvious. The structural principle of the *ghazal* is developed and extended to the structure of the single verse, which later often contained a number of topoi. Words of untraditional and often contradictory association are used to awaken the reader to a higher consciousness of language. Ṣā'eb's work is a link in the chain of development of Persian poetry between the Iraqi style (qv) and the later second phase of the Indian style (Bēdil, Nāṣer 'Alī, qqv) and shows that the latter was a purely Iranian phenomenon.

BLHP IV 265–76; RHIL 301–2. WH

Sait Faik Abasıyanık, Turkish writer, see **Abasıyanık,** Sait Faik.

saj'. In Arabic, Persian and Turkish poetics, it signifies a type of rhymed non-metrical prose, not formally arranged, but with a certain rhythm. It was originally used in Arabic literature for the pronouncements of seers (*kāhin*), and seems to have been the basis on which poetry (rhyme) developed. The tradition culminated in the *Qur'ān* (qv) written in *saj'*. Later the form was used in official correspondence, whence the scribes took it over; it reached its highest point of development in the *maqāmas* (qv, short episodic works in *saj'*). Passages written in *saj'* occurred both in popular literature and in learned writings, especially in prologues. It has entirely disappeared from modern Arabic literature. In Persian literature the style of Sa'dī's (qv) *Golestān* saw the height of the *saj'* form, while in Turkish literature this was reached in historical prose works.

EI; BHLA 188 ff. KP

Ṣāliḥ, aṭ-Ṭayyib (b 1929), Sudanese novelist and short-story writer, studied at London University and is now Head of Drama in the BBC Arabic Service. His first novel, *Mawsim al-hijra ilā 'sh-shimāl* (Time of Migration Northwards) aroused unusual interest in all Arab countries; the impact of two contrasting cultures and mentalities, African and European, is portrayed in the dramatic story of a European-educated young Sudanese finding himself in a tragic conflict situation in the alien environment of England; his true life only begins when he returns home. The novel *'Urs az-Zēn*

(Zēn's Wedding, 1964) describes village life and customs with insight into characteristic types. Ṣāliḥ uses the methods of modern European prose-writing to present purely Sudanese themes. Other works: novel, *Bandar Shāh* (1971), story, *Dawmat wad Hāmid* (1962). JO

Salmāne Sāvejī, Jamāloddīn (b c1300 Sāve, d 1376 Sāve), Persian poet. The son of a revenue officer, he was panegyrist of the Jalāirides in Tabriz and Baghdad. He entered the service of Sultan Ḥoseyn Bozorg as court poet and became the tutor of Prince Oveys, the Sultan's successor, whom he continued to serve till failing health forced him to retire to quiet country life. He wrote one of the earliest works of the *Sāqīnāme* genre (Book of the Cupbearer), the romance *Ferāqnāme* (Book of Separation), and the *maṣnavī* (qv) *Jamshīd o Khorshīd* on the lines of the traditional epic. His lyrical works include *qaṣīdas, ghazals* and *robā'īs* (qqv). His predilection for a difficult form of the *qaṣīda* overladen with rhetorical artifice and excessive use of amphiboly reflects the last phase of the Iraqi style. With Shiite tendencies in the *qaṣīda*, and emphasis on new metaphors, and similies, the subsequent Indian style (qv) already casts its shadow on Salmān's work. Although he kept in line with new trends in this period, his mediocre talent could not do justice to them.

BLHP III 260–71; RHIL 261–2. WH

Samuel, Books I and II, is a part of the Old Testament (qv), with the following main characters; the prophet Samuel, the first king of Israel, Saul (fighting against the Philistines, death), and the second king, David (qv, victorious battles, family worries). Events during the reign of David are described particularly vividly, suggesting that eye-witness accounts were used. See also Bible. SZS

Sanā'ī, Abu 'l-Majd Majdūd (d c1131, extreme dates 1124 and 1150), Persian panegyrist of the Ghaznavid rules and a master of *qaṣīdas* (qv) after the models of the leading poets. For some unknown reason he became a mystic and left the court, devoting himself to mystical poetry. The *monājāt* of 'Abdollāh Anṣārī of Herat (qv) with their interspersed verses had prepared the way for mystical Persian poetry on a larger scale which was elaborated by Sanā'ī in both lyric and epic form. His main work, *Ḥadīqat al-ḥaqīqat* (The Garden of Truth), dedicated to Bahrām Shāh of Ghazna, is the first great mystical *maṣnavī* (qv). It discusses, in about 11,000 verses and ten chapters, different topics of spiritual life, like reason, love, inner knowledge, prophetology and philosophy; it is more didactic than truly mystical. A number of parables and tales are inserted to illustrate the author's intentions. Thus the model for the later *maṣnavīs*, like those of 'Aṭṭār and Rūmī (qqv), is given; the illustration of mystical and didactic topics by stories without fixed order. Western readers find the *Ḥadīqat* a rather boring poem, but it was a pioneer work which opened the way for a development that lasted almost eight centuries. Sanā'ī wrote six more *maṣnavīs* of which the *Sayr ol-'ebād elā al-ma'ād* (The Journey of the Servants to the Place of Return) is interesting because of its subject, a description of the development of the human spirit through different stages. This idea of heavenly journey, probably influenced by the legend of Muhammad's ascension to heaven, besides Iranian ideas, was for the mystics an appropriate symbol for their spiritual experiences, and has been imitated even in our century (Iqbāl's *Jāvīdnāme*, see Vol. II). Sanā'ī's lyrical poetry is much more graceful and charming than his epics, and full of originality.

Trans.: J. Stephenson, *The First Book of the Ḥadīqat ul ḥaqīqa, or, The Enclosed Garden of the Truth* (Calcutta 1911).
R. A. Nicholson, *A Persian Forerunner of Dante* (Towyn-on-Sea 1944). AS

Ṣan'atīzāde Kermānī (b c1885 Kerman), Iranian novelist. He was one of the first to write historical novels, drawing on the ancient and more recent history of Iran. His literary models were the novels of the West, especially French literature. The composition of his novels is not well-balanced, as a result of his insistence on the dramatic development of the plot. In addition, his interpretation of historical facts is often decidedly individual. His novels include *Dāmgostarān yā enteqām-khāhāne Mazdak* (Trap-setters, or the Avengers of Mazdak, 1921), *Dāstāne Mānīye naqqāsh* (The Story of Mani the Painter, 1927) and *Selāḥshūr* (The Warrior, 1933).

RHIL 370–1; KMPPL 47–51; AGEMPL 120. KB

Sanchuniaton, Phoenician writer, according to sources from the Roman times living in 13th century BC in Beryt (now Beirut). His book presenting the ancient Phoenician myths has survived only in part (the origin of the world, the gods, and human civilization) in the very free Greek re-working given by Philo of Byblos (1st century AD). A follower of Euhemer, Philo presented the gods in such a rationalist light that the form and meaning of the original traditions has been overlaid. Nevertheless the excerpts from Philo preserved in the works of the Greek writers Porphyry, Eusebius, etc, are all that remains of the once famous and extensive literature of Phoenicia. The nature of this literature can be seen, too, from certain inscriptions, the epitaph of the king of Sidon, Eshmunazor (4th century BC), and historical inscriptions referring to the kings of the Phoenician cities and some of the neighbouring states.

C. Clemen, *Die phönikische Religion nach Philo von Byblos* (Leipzig 1939). SZS

Sasuntsi Davith (David of Sassoun), Armenian national epic, a saga of four generations of heroes of Sassoun: the twins Sanasar and Balthazar, founders of the fortress and town of Sassoun; Sanasar's son Great Meherr, the Lion-tearer; his son David; and David's son Meherr the Younger, who at the end goes down into the underworld to remain until justice and good prevail in the world above. The monumental work presents a hyperbolized, stylized world inhabited by noble and ignoble characters, moved by wars and the longing for peace, the struggle for subsistence and dreams of prosperity and happiness. The historical core reflects the period of Arab invasions of Armenia, but some elements are drawn from mythology, some from utopian ideas of the ideal future. Composed on a unified plan, with dynamic action presented compactly and in vivid imagery, the major part of the text (which was originally recited) is interspersed with lyrical passages meant to be sung. There are over 100 characters in the epic, the chief of them being the heroes and their wives, foster-parents and enemies. Like these main characters, the episodic figures are marked by one specific trait, of which love for the mother-country is valued the highest. After being handed down by word of mouth for 1,000 years, the epic was recorded in over 50 versions, appearing in as many editions and in many translations. It has provided inspiration for Armenian literature, music and the graphic arts, and in its aesthetic and ethical aspects is one of the greatest folk epics in world literature.

Trans.: L. Sourmelian, *Daredevils of Sassoun* (London 1966); A. K. Shalian, *David the Sassoun* (Ohio 1964); *The Saga of Sassoun* (in prose, London 1970); F. Feydit, *David de Sassoun* (Paris 1964). LM

Satylganov, Toktogul, Kirghiz poet, see **Togolok Moldo.**

Sayath-Nova (real name Arutin Sayadian, b 1722 Sanahin, d 1795 Tbilisi, or Tiflis), Armenian poet. A weaver by profession, he was famed as an *ashugh* (minstrel, see *âşık*). He travelled with the army in Iran and India, and then became court singer to King Irakli II of Georgia. According to tradition, he was exiled to the Haghpat monastery because of a love intrigue; later he fled and was probably murdered when Tbilisi was conquered by the Persian Army. In the *ashugh* tradition he wrote in the Tbilisi dialect of Armenian, in Georgian, in the Turkish of Azerbaijan, and even passages in Persian. He composed his own melodies or paraphrased those of others, and accompanied himself on the *saz* or *kemancha*. In the 225 poems so far identified the principal theme is love and the poet's mission as comforter, teacher, counsellor and prophet, and above all as the servant of his people (*khalkhi nokhar*). His verse is highly polished, with an intricate system of alliteration and assonance, ingenious use of synonyms and a wealth of imagery, and traces of Persian influence. It has no equal in Transcaucasia, and is the crowning achievement in *ashugh* poetry. He had many disciples and imitators (Ghunkiannos, S. Levonian-Djivani, Zargiar, Miskin-Birdji, Shirin, Shamtchi-Melkho, and others) and no celebration or feast in Armenia today is complete without his songs.

Trans. into English, French, German and Italian: *Sayat-Nova, His Life and some Poems* (Yerevan 1963); TTA.
HO 7; THLA. LM

as-Sayyāb, Badr Shākir (b 1929, d 1964), Iraqi poet, one of the first exponents of modern poetry and free verse in Iraq. Born in Basra, he studied Arabic and later English literature in Baghdad. He worked as a teacher and civil servant, and welcomed the revolution of 1958; later he objected to the Qāsim regime on principle, and retired from public life, a sick man, in a spiritual exile until his death, which was lamented throughout the Arab world. He ardently supported the movements which aimed at improving the social lot of the Arabs, at first Communism and later Arab nationalism; his own attitude finally broadening into a progressive humanism. His chief volumes of verse illustrate the development of his thought clearly: in *Azhār dhābila* (Faded Blossoms, 1947), he was a romantic pure and simple; in *Asāṭīr* (Myths, 1950) symbolism has been added, continuing in *Ḥuffār al-qubūr* (Gravediggers, 1952) and moving towards realism in *al-Mūnis al-'umyā'* (The Blind Prostitute, 1954). The remaining volumes are consistently realist: *al-Asliḥa wa'l-aṭfāl* (Weapons and Children, 1954), *Unshūdat al-maṭar* (Hymn to the Rain, 1960), *al-Ma'bad al-gharīq* (The Drowned Shrine, 1962), *Shanāshīl ibnat al-chelebi* (The Fancy Balcony of the Marquis's Daughter, 1965). He made considerable use of impulses from English poetry (T. S. Eliot, Edith Sitwell, Shelley and Keats) as well as from classical and modern traditional Arabic poetry (see al-Jawāhirī). His path towards poetic freedom both in form and content ran parallel to his progress in social questions: from his ivory tower to freedom for the whole of society. His style is complex, reflecting the complexity of the questions his poetry deals with.

CAR 223; NTALAC 191. KP

Sayyidā (Tajik: Sayyido), Mīrābīd (b early 17th century Nasaf, now Qarshi, d between 1707/11), Central Asian Persian poet. He was a typical writer of the best traditions of craft poetry, intended for a broad public among the craftsmen and small traders; this is seen especially in the *maṣnavī* (qv) *Shahrāshūb* (Throwing the City into Confusion, which is also the title of short humorous poems). His language is close to the vernacular and the poems often touch on social themes. The verse fable *Bahāriyāt* (Spring Songs), also called *Hayvānātnāma* (Book of Animals), shows the strength of the people, for instance the importance of unity in the ant world. He also wrote panegyric *qaṣidas* (qv) but instead of addressing them to the ruler or high dignitaries he praised the crafts and their master craftsmen (see *maṣnavī*). He complained of the disturbed and split state of the country and the hopeless economic situation which forced the best scholars and poets to live abroad.

RHIL 509–10. JB

Sevak, Paruir (real name Ghazaryan, b 1924 Tchanakhtchi, now Sovetashen, d tragically nearby, 1971), Soviet Armenian poet and literary historian. He came from the country, studied philology in Yerevan, worked in the Literary Institute and gained the degree of Doctor of Philology for a thesis on Sayath-Nova (qv). He was a member of the executive of the Union of Soviet Armenian Writers for many years, and on the editorial board of a number of literary journals. He collaborated in the revised translation of the Bible for the 1970 edition. The most important of his seven volumes of poetry is the long poem *Anlřeli zangakatun* (The Never Silent Bells, 1959).

Around the life story of the Armenian composer, Komitase, he presents his conception of Armenian national history, freed from the obligation to pay for centuries of suffering; he later turned towards the search for positive values and for a dignified form of existence for the whole nation both at the present time and in the future. This work aroused a profound and widespread mood of soul-searching throughout the nation, and thus was the most significant event in modern Armenian poetry from the social point of view. In his volumes of lyric poetry (particularly the last two: *Mard ap'mei*, Man in the Palm of the Hand, 1963; *Eģici luys*, Let There Be Light, 1969) Sevak went to the roots of the problems of modern life in the second half of our century, seeking the meaning of life. Intellectually his verse is difficult; his poetic invention is lively, developing the finest traditions of mediaeval Armenian verse (especially that of Grigor Narekatsi, Sayath-Nova, qqv) and of more modern writers like Tcharents (qv). At the same time his work is modern poetry in the true sense of the word, and the peak of post-war Armenian poetry.

Trans.: I. Zheleznova, in SLM; J. Gaucheron, in OeO. LM

Seyfettin, Ömer (b 1884 Gönen near Balikasir, d 1920 Istanbul), Turkish writer. Finishing at the military academy in 1903, he served as an officer in various places, until he resigned from the army in 1910 to become a regular contributor to a literary journal, *Genç Kalemler* (Young Pens) published by Ziya Gökalp (qv) in Salonica. Joining the army again in the Balkan War, he spent a year in captivity, returning to Istanbul in 1913, where he lived as a teacher of literature in a High School and freelance writer until his early death in 1920. Seyfettin was one of the great innovators of Turkish literature; he broke away from the over-refined traditional style and used in his writings the simple spoken language, first arousing sensation by a proclamatory article on the 'new language' (1911). A patriotic idealist, he fought for a radical modernization of the country without advocating a slavish imitation of the West. Consequently, he was connected with the 'Establishment' of 1909–18. In view of all this, he naturally had many critics and enemies, but his style of writing had a tremendous influence on the younger generation. His short stories, some of which took their subjects from the glorious past of the Turks, others from the contemporary scene, were well told though sometimes marred by a certain ironic tone and an all too obvious moral. Other works: 138 short stories, collected in ten volumes; a one-act play *Mahcupluk imtihani* (Test of Chastity).

PHTF II 576 ff. AT

Seyfullin, Säken (b 1894 Karaganda, d 1939), Kazakh poet and writer. He graduated from teachers' training college, took part in revolutionary activities and later worked as a journalist and held high political office. Unlike other Kazakh poets who developed folk forms, Seyfullin was inspired by Mayakovskiy and dealt with themes of revolution and modern civilization in a non-traditional form. Most of this prose depicts the revolutionary period. Works: poetry, *Ötken künder* (Days Gone by, 1914); *Ekspress* (1926); *Kökshetau* (1929); stories, *Zher kazghandar* (Those who Conquer the Earth, 1928); plays, *Bakyt zholyna* (Towards Happiness, 1917); *Kyzyl sunkarlar* (Red Hawks, 1920); novel, *Tar zhol, tayghak keshüv* (Difficult Road, Difficult Transition, 1927). LH

ash-Shābbī, Abu'l-Qāsim (b 1909 Shabbiya, d 1934 Tunis), Tunisian poet and critic. Like his father, who was a judge, he had a traditional Islamic education at the Zaytūna mosque (1920–28) before he qualified in law in 1930. He had heart disease, however, so he never practised any profession. Ash-Shābbī knew no foreign language, but he read translations of Lamartine and Goethe and imbibed modernistic influences through Jibrān and other Syro-Americans, and through Ṭāhā Ḥusayn, al-'Aqqād (qqv), and Aḥmad Ḥasan az-Zayyāt. His first poems were published in the Tunisian press in 1927; then, through the Apollo Group and its Journal, he reached a wider public. The only book he had published in his lifetime was the text of a long lecture on the poetic imagination of the Arabs (*al-Khayāl ash-shi'rī 'ind al-'Arab*, 1929). It created a stir as a bold and spirited trumpet-call for modernism; and despite its youthful overstatements and sweeping rejection of the past, it is not without critical penetration and aesthetic directness. It is as a Romantic poet, however, that ash-Shābbī has been most widely and justly acclaimed. So short a life could not be expected to produce a bulky contribution or a fully-formed style. He was somewhat uneven in temper and quality, often uneconomic in expression, and not particularly adventurous in metrical forms. But his zest for life under the shadow of death, his patriotic and reformist fervour, his love of nature and his meditativeness produce poems of ringing integrity; his imagery is simple and direct and always of a piece with his experience. Further books: collected poetry, *Aghānī al-ḥayāh* (Songs of Life, 1955); letters, *Rasā'il* (1966); fragment of diary, *Mudhakkirāt* (1966).

Trans.: in *Anthologie de la Littérature Arabe Contemporaine*, Vol. I, *Les Essais* by Anouar Abdel-Malek (Paris 1965); Vol. III, *La poésie*, by Luc Norin and Edouard Tarabay (Paris 1967). PC

Shabestarī, Maḥmūd (c1250–1320), Persian poet born in the neighbourhood of Tabriz, where he spent most of his life. His *Golshane rāz* (Rosegarden of Mystery), became the most widely read poetical introduction to Sufi thought as developed after Ibn 'Arabī (d 1240). Composed in 1311, it consists of fifteen poetical answers to questions about mystical terminology posed by a friend. It deals, in comparatively simple terms, with such problems as *vaḥdat al-vojūd* (Unity of Being), of 'Essence', mystical stations, and the descent and ascent of the Perfect Man. Shabestarī's work is not very elevated poetry, but that is, perhaps, one reason for its success. It should be studied with the excellent commentary written by Lāhejī in 1473. In our days, the style and form of *Golshane rāz* influenced Moḥammad Iqbāl (see Vol. II), who composed a *Golshane rāze jadīd* (The New Rosegarden of Mystery, 1927), a poem in which Iqbāl discusses his philosophy.

Trans.: E. H. Whinfield, *Rose-Garden of Mystery* (London 1880); Anonymous, *The Dialogue of the Gulshan-i Rāz* (London 1887); J. Paske, *The Secret Garden* (New York-London 1969). AS

Shādmān, Fakhroddīn (b 1910, d 1971), Iranian writer. The principal theme of his work is the question of cultural relations between Iran and the West. He saw the native Iranian culture gradually giving way and swallowed up by western ways, and he discusses this problem directly in his essays and indirectly in novels. His remedy would be for the Iranians to make a thorough study of both native and Western culture, and apply their deductions to their own cultural life. This would imply the formation of a discipline corresponding to the Oriental studies of the West, 'Occidental studies', in Persian: *farangshenāsī*. Essays: *Taskhīre tamaddone farangī* (The Capture of European Culture, 1948); novel: *Tārīkī va roushanā'ī* (Darkness and Light, 1950).

AGEMPL 224. KB

Shāhīn of Shiraz (13th–14th century), Jewish–Persian epic poet, writing chiefly during the rule of the Mongol, Il-khan Abū Sa'īd Bahādor (1316–35). No biographical details are available, but the honorary title of Moulānā (Our Master), and the literary material used suggest a man of learning and thus probably a rather high social level. Tradition has no authentic title of the works of Shāhīn, save for the informative and colourless *Tafsīr* and *Sharh* (Commentary). In manuscripts the titles *Ketābe Shāhīn* (The Book of Shāhīn), and *Dāstān* (Story), occur. The titles chosen by Wilhelm Bacher have been commonly adopted, viz. *Book of Genesis*, *Moses Book*, *Ardashīr Book*, and *Ezra Book*, all of which, in *hazaj* metre (see '*arūḍ*), are poetical paraphrases of Old Testament themes combined with genuine, otherwise partly unknown, Iranian material. Although Shāhīn was an epigone of classical Persian tradition (especially Ferdousī, qv), as far as form, structure, and language are concerned, this happy combination of material shows the literary originality that made him the greatest poet in Jewish-Persian literature. His mastery, in this biblical poetry, of the *Qur'ān* (qv) and Islamic tradition is remarkable. A fifth work is only known from one manuscript and is probably incomplete, but although of lesser value it is apparently authentic. This is the story of King Kishvar and his seven pieces of advice to his son Bahrām. It is significant because of its being purely Iranian, devoid of any specific biblical influence.

Wilhelm Bacher, *Zwei jüdisch-persische Dichter. Schahin und Imrani* (Strassburg 1908); Dorothea Blieske, *Šāhīn-e Šīrāzīs Ardašīr-Buch* (Tübingen 1966). JPA

shahrāshūb, Persian humorous poem, see **Mas'ūd** b. Sa'de Salmān and **Sayyidā,** Mīrābīd.

Shahriyār, Moḥammad Ḥoseyn (b 1905/06 Tabriz), Iranian poet in the traditional Khorasan style. The son of an important Tabriz lawyer, he went to school there and then to Teheran to study medicine. He did not graduate, however, preferring to turn to the writing of poetry while earning a modest living in an unimportant position. The success of his lyric *ghazals* (qv) which proclaim him a follower primarily of Ḥāfeẓ (qv), and earned him from some critics the designation 'the second Ḥāfeẓ', is a good illustration of the Iranian literary atmosphere with its ever-living sense of continuity in the classical tradition. His poetry has appeared in literary magazines, especially in the traditional *Armaghān*. His first volume of verse appeared in 1931 (*Dīvāne Shahriyār*) with enthusiastic introductory articles by three major figures of contemporary literary life, M. T. Bahār, S. Nafīsī (qqv) and Pazhmān Bakhtiyārī. The polythematic character of the *ghazal* (so characteristic of Ḥāfeẓ) is seen in Shahriyār's poetry in allusions to contemporary events, important people, etc. His first *dīwān* (qv) includes the lyrical *maṣnavī* (qv) *Rūhe Parvāne* (The Soul of Parvāne), a tragic elegy on the premature death of a talented poetess. Two more volumes of his selected verses were published in Teheran in 1949 and 1953; they contain new strophe forms (eg in the lyrical autobiography) and attempts at free verse which reveal an all-round poetic talent (eg the elegy on the poet's dead mother).

MLIC II, 111–19 with examples in French translation; AGEMPL 194–96. VK

shairi, Georgian metre, see **Rustaveli, Teimuraz I** and **Vazha-Pshavela.**

Shamo, Arabe (b 1897 Susuz), Kurdish writer, journalist, translator, ethnographer and educationalist. He comes from a poor village family and went into service as a child, learning Armenian, Turkish and Russian. 1914–1916 he served in the Russian army as an interpreter. He was imprisoned for his illegal work in Bolshevik organizations. He began to write in the 30's, and was one of the compilers of a system of Latin script for Kurdish (in USSR) which was in use until 1945, when it was replaced by Azbuka. Since 1968 he has been Senior Lecturer in Kurdish at the University of Yerevan. He writes in the Kurmānjī dialect. His first work was the novella *Amre Lenin* (Lenin's Life, 1930). In 1935 he published the autobiographical novel *Şivānē kurd* (A Kurdish Shepherd). Here, as in his later novels, *Barbāng* (Dawn, 1958) and *Zhīyīna bakhtawar* (A Happy Life, 1959), he makes successful use of his knowledge of Kurdish rural life and folk customs. His last novel, *Dimdim* (1968), goes back to Kurdish historical legends about the fortress of that name.

J. Blau, *Kurdish Kurmandji Modern Texts* (Wiesbaden 1968); FKH xi. AK

ash-Shanfarā, a 6th-century Arabian poet for whom there is no historical evidence. Arabian tradition attributes to him the famous *Lāmiyat al-'Arab* (Poem with Rhymes in L), an eloquent expression of the fundamental attitudes of Bedouin poetry. The European critical tradition tends to regard the poem as a forgery.

EI[1] IV 321; BGAL I 25, S I 52; NLHA 79; WILA 31. KP

ash-Sharqāwi, 'Abdarraḥmān (b 1920), Egyptian dramatist and journalist. He graduated in law at Cairo University, and spent one year in Paris and other places in Europe and Asia. He was active in political life, and as a member of the Egyptian Peace Committee. Ash-Sharqāwī is one of the most outspoken Egyptian critical realists of the post-war generation, characterized by marked social and political commitment. His stories draw on the popular struggle against pashas and landlords, and against foreign invaders (*Arḍ al-maʿraka*, Land of Struggle) and on the life of the village and urban poor (*Aḥlām ṣaghīra*, Little Dreams). In 1954 he published a novel, *al-Arḍ* (The Earth), a truthful and knowledgeable picture of Egyptian village life and the peasants' social problems in the 30's, stressing the need for revolutionary action against oppression and exploitation. His novel *al-Fallāḥ* (The Fellah, 1970) describes the conflict-ridden process of transformation of the Egyptian village today, and the growth of a new *fellah* mentality, seen in the clash between the remnants of feudal and reactionary forces and the *fellahin* determined to attain their revolutionary goal. In style and composition it reflects the author's mastery of modern European prose techniques. In 1965 his historical verse-drama *al-Fatā Mahrān* (The Boy Mahrān) was produced; set in the Mameluke period, it deals with the antagonistic ruler–serf relationship. Other works: novels, *Qulūb kāliya* (Empty Hearts, 1957); *ash-Shawāriʿ al-khalfīya* (Back Streets, 1958).

JO

Shatberashvili, Giorgi (b 1910 Tbilisi, d 1965), Georgian writer. He studied history at Tbilisi (Tiflis) University and on graduating in 1932 became a teacher. After the war he held various posts in the Union of Georgian Writers. His work is in the romantic Georgian tradition, but he tried, especially towards the end of his life, to find in Georgian history the answer to the eternal question, the purpose of human life. He wrote in almost all genres: stories, poetry, novels and plays, and several children's books.

Trans.: *A Mother's Guest*, in *Mindia, the Son of Hogay and Other Stories by Georgian Writers* (Moscow, 1961). VAČ

Shavteli, Ioane (late 12th century), Georgian poet. Nothing is known about the author, to whom the panegyric *Abdulmesiani* is also attributed (Georgianized Arabic style: *Slave of the Messiah*) as well as poetic eulogies on Queen Tamar (1184–1213) and David (probably either her great-grandfather, David the Builder, 1089–1125, or her husband, David Soslan). The work is written in *chakhrukhauli* lines (viz, the metre used by Chakhrukhadze, qv), in an involved archaic style, and is permeated with a clear Christian, anti-Muslim spirit.

Trans.: UAGP. VAČ

Shawqī, Aḥmad (1868–1932), one of Egypt's foremost poets and playwrights. Born into an Egyptian family of Turkish extraction, he was educated in Egypt. The two years he spent in France, and a period in Spain (to which he was exiled for anti-British propaganda during World War I), had little influence on him, except for broadening his knowledge. His main sources of inspiration remained Arabic and Islamic. Indeed, Shawqī's literary work appears to span the old and the new. The three historical novels he wrote (the first appeared in 1897) glorify ancient Egypt and Persia. His poetry, much of which was later collected in the three-volume *Shawqīyāt*, is, to a large degree, patterned on the aesthetic and artistic values of classical Arabic poetry. Frequently encomiastic, it obtained favours for Shawqī with Egypt's rulers and princes. It was distinguished by its wealth of images, incisive expression and rhythmic sensitivity. These qualities made it popular; Shawqī's poems were avidly read and quoted and many were set to music. Shawqī's long career as a poet ensured acceptance by the public of his plays too. He wrote a rather mediocre comedy and six historical plays on ancient Egypt, Mamlūk Egypt, pre-Islamic Arabic, Ancient Persia, and Muslim Spain. These are more distinguished for their felicity of rhyme and rhythm than for their feel for the stage. The plots are frequently over-involved, as in *Amīrat al-Andalus* (Princess of Andalusia), which is his only historical play in prose, and his longest. Most are, indeed, more like historical pageants or court-plays than the

Shakespearean dramas he seems to have wanted to write.

Trans.: A. J. Arberry, *Majnūn Laylā, a Poetical Drama* (1933).
BGAL S III, 21–48; LSATC 125–38; GLAIG 221–25. JML

Shaydā, pen-name of the Pashtun poet, **Kāẓim Khān Khaṯak.**

Shenute (4th–5th centuries AD), monk, founder of the national Egyptian (Coptic) church, and writer. Born (date unknown) in the village of Shenabolet near Akhmun in Upper Egypt, he became a monk c370 AD and c385 abbot of the White Monastery, west of Sohag. He accompanied Cyril, Patriarch of Alexandria, to the third ecclesiastical Synod in Ephesus in 431 and died in 451 at a great age. A passionate opponent of paganism, heresy and sinful living, he wrote many sermons and epistles to monks and nuns, in the Upper Egyptian (Said) Coptic dialect. They constitute the only original Coptic literature, the rest being translations from the Greek. Shenute's style was lively, but sometimes confused and even cryptic.

Trans.: Sir E. A. Wallis Budge, *The Book of the Dead* (London-New York 1928). JČ

Sheybānī, Abū Naṣr Fatḥollāh Khān (b 1830 Kashan, d 1891 Teheran), Persian poet. Of a wealthy military aristocratic family, he was reduced to poverty by arbitrary official action. He retired from public life and lived in the manner of a dervish while devoting himself to poetry. Written in the classical tradition, his poems often touch on contemporary social problems: the evils of absolute rule, corruption, the disastrous state of the army, etc. Not even his panegyrics are free of critical and warning tones. His style is simple and natural, the language subordinated to the ideas. Besides *qaṣīdas* (qv) he published a volume of stories and poems, *Dorje dorar* (Box of Pearls, 1882). A selection of his verse was published in Istanbul in 1890/91, entitled *Montakhab az majmū'eye bayānāte Sheybānī.*

RHIL 346; BLHP IV 344. JOS

al-Shidyāq, Aḥmad Fāris (b (?)1801 Ashqut, d 1887 Constantinople), a member of an old-established Lebanese family of Maronites, his chief contribution to the *nahḍa* (renaissance) was as one of the earliest representatives of the great Lebanese tradition of journalism in the later 19th century. He travelled widely, to Malta, England, France, and Tunisia, before finally moving to Constantinople at the invitation of the Sultan. While in Paris, he wrote his best known book *al-Sāq 'alā 'l-sāq fīmā huwa'l-fāryāq*, which is a combination of social comment and autobiographical detail. In 1861 he began publication from Constantinople of *al-Jawā'ib* (Current News), the first Arabic newspaper to claim wide circulation throughout the Arabic speaking world. RCO

Shiraz, Hovhannes (real name Karapetian; b 1914 Alexandropol, now Leninakan), Soviet Armenian poet. Of a peasant family, he became a textile worker, and then studied philology at Yerevan University. His first volume of spontaneous and sincere lyrics, *Garnanamut* (The Beginning of Spring, 1935), was followed by many more; his style is simple, with folk and colloquial elements (eg *K'nar Hayastani*, The Lyre of Armenia, I. 1958, II. 1964). Although his work reached its peak in 1944, in the philosophical poem *Bibliakan* (Biblical), he remains one of the most popular poets in contemporary Soviet Armenia.

Trans.: D. Rottenberg, in SLM; J. Gaucheron, in ELA; L. Mardirossian, in OeO. LM

Shirvanzade (real name Alekhsandr Movsisian; b 1858 Shirvan, d 1935 Kislovodsk), Armenian writer and dramatist. Son of a tradesman, he worked as office worker, writer and journalist. He lived in Baku and Tbilisi (Tiflis), from 1905 to 1910 he studied at the Sorbonne; between 1919 and 1926 he lived in the USA, then settled in Yerevan. From 1880 he published stories, novels and plays (in USSR memoirs) and comedies of emigré life. Shirvanzade took up the tradition of G. Sundukian and Raffi (qqv) and worked on the classic of the critical-realist novel and drama in modern

East Armenian literature. The dramatized novels *Namus* (Honour, 1885) and *Xaos* (Chaos, 1898), and the play *Patvi hamar* (For Honour, 1914), are marked by firm composition, an ingenious plot, emotional tension, and are still part of the stock repertoire of the Armenian theatre. Further works: novels, *Artist* (1902, in French trans., *L'Artiste*, Paris 1909); *Vardan Ahrumian* (1899); drama, *Evgine* (1901); *Morgani xnamin* (Morgan's Best Man, 1926, produced in 1963 as a comic opera entitled *Paron Mintoyev Parizum*, Mr. Mintoyev in Paris). LM

Shota Rustaveli, Georgian poet, see **Rustaveli,** Shota.

Shukrī, 'Abd al-Raḥmān (b 1886 Port Said, d 1958 Alexandria), Egyptian poet. Shukrī graduated from the Teachers' Training College in Cairo in 1909, having met there his future colleague, Ibrāhīm 'Abd al-Qādir al-Māzinī. Together they developed an enthusiasm for English literature and poetry, an interest which Shukrī was able to extend when he spent the following three years (1909–12) in England at Sheffield. On his return, he and al-Māzinī joined with 'Abbās Maḥmūd al-'Aqqād to form a new school of poetry known as the 'Dīwān' group. However their co-operation soon broke down as a result of bitter quarrels amongst themselves. Shukrī's own *dīwān* consists of eight parts; it was published in Alexandria in 1960. During the vital formative years of the 'Dīwān' group (1909–1919) he was by far the most prolific member. The prefaces to the individual sections of his poetic works demonstrate how carefully and systematically he had studied English poetry and criticism in the 18th and 19th centuries. RCO

Şinasî, Ibrahim (b 1826, d 1871 Istanbul), Turkish writer. In histories of Turkish literature his translations of French poetry (Racine, Lamartine, La Fontaine, Fénélon and Gilbert) are usually quoted as initiating the European period in modern Turkish literature, but H. W. Duda has shown that Ibrahim Pertev Pasha (1824–1873) had forestalled him with his selection from the work of Victor Hugo, J.-J. Rousseau and Voltaire. The honour of opening this new era, which signified the rejection of the oriental traditions of Turkish literature, must therefore go to the latter. Şinasî translated from the French in order to encourage the renaissance of Turkish literature, proving that the language was quite capable of expressing western ideas, and that it needed to free itself from the bombast of the classical period. He propagated modern, progressive ideas in his contributions to the papers *Tesviri Efkar*, *Tercümani Ehval* and *Ceridei Askeriye*, trying to express himself as simply and comprehensibly as possible. He was the author of one of the first plays, *Bir Şair Evlenmesi* (The Poet's Marriage), on the feminist question. His *Makaleler* (Articles) were published in Latin script in 1960, by Fevziye Abdullah Tansel. The real breakthrough to a modern literature was the work of his disciple Namık Kemal (qv).

HGTM 10–12; H. W. Duda, *Vom Kalifat zur Republik* (Vienna 1948) 86–91; PHTF II, 469 ff.
OS

Sinuhe, courtier under the Egyptian kings Amenemhēt I (qv) and Senwosret I (c1991–1928 BC), hero of and perhaps author of a popular autobiographical tale. Fearing court intrigue after the death of Amenemhēt he fled to Syria and was hospitably received by Semitic nomads; he married a chief's daughter and later himself became chief of a tribe. Many years later he accepted the invitation of Senwosret to return to Egypt, and with his wife and children ended his life in his native land. His style is somewhat studied, but lively; during the New Kingdom (after 1570 BC) it was considered a classic of Egyptian literature, as shown by the numerous fragmentary copies on papyri and ostraka from that period.

A. H. Gardiner, *Notes on the Story of Sinuhe* (Paris 1916); Battiscombe Gunn, *Land of Enchanters* (London 1948) 29–46. JČ

sīra (literally way, path), Arabic term for a biography, especially a biography of the Prophet Muhammad. The earliest, which have not survived, were written as early as the beginning of the eighth century; the earliest which have been preserved are those by Ibn Isḥāq (d 768), edited by Ibn Hishām (d 833), and, on Muhammad's campaigns, by al-Wāqidī (qv, d 823). Innumerable later biographies, right up to the 19th century, were based on these early *sīras*. IH

Sirach, Hebrew collection of proverbs, see **Ecclesiasticus.**

sirat 'Antar, an Ancient Arabian popular epic based on the life of the sixth-century poet 'Antara or 'Antar (qv). Sometimes erroneously described as a romance by European writers, the epic is based on stories told about 'Antar and literary material from Arabic and other sources put together in the 12th century and then developed further. One of the best-elaborated works of this type, it is still popular in the Arab world today. As time passed and cyclization proceeded along genealogical lines, the central themes of 'Antar's love for 'Abla, the obstacle of class distinctions, and the struggle to escape from slavery, were expanded by the introduction of adventures (Iraq, Syria) attributed to 'Antar's descendants. There are also echoes of the Crusades in the epic. The story is told in prose, alternating with poetry and rhymed prose (see *saj'*) and has survived in two versions, the Hijazi and the Syrian; the Iraqi version is based on the latter.

Trans.: Hamilton Terric, *Antar, a Bedueen Romance* (London 1819–20); G. Ronger, *Le roman d'Antar* (Paris 1923).
EI² 518–21; WILA 268. KP

Skantcheli, Armenian historian, see **Koriun.**

Sofronov, Anempodist Ivanovich (b 1886 Zhekhsogonsk, Alexeyev district, d 1935), Yakut writer and dramatist. He was one of the founders of Yakut literature. At the age of sixteen he began to attend the four-year church school in Ytyk-Kel' village, and continued his education with his brother, a teacher on Aldan. He began to write in 1912, in the manner of critical realism. His plays *Jadangy Jaakyp* (Poor Jacob, 1914), *Taptal* (Love, 1915–16, which was influenced by Ostrovskiy), and *Büdürüy-büt kömmöt* (If you Stumble you'll Never Get up, 1917), were intended to rouse public opinion against the inhumanity of the pre-Revolution patriarchal-feudal society of Yakutsk and were sharp in their criticism of village life. In the play *Olokh jeberete* (Game of Life, 1918, 1922) he said: 'all the Russians do is bad for us', and it took him a long time to overcome his narrow-minded nationalism, which he shared with a certain number of bourgeois intellectuals in Yakutsk. It was not until the late 20's or 30's that he became a socialist–realist writer. MM

Solomon (Hebrew: Shelomo), King of Israel, c961–922 BC, described by tradition as a wise man who gathered proverbs together. Later tradition wrongly connects this wise king with Biblical wisdom literature (*Ecclesiastes*, qv, the apocryphal *Wisdom of Solomon*, the *Psalms* and later *Odes of Solomon*, qqv) and also the *Song of Songs* (qv), assigned to this category later. SZS

Song of Songs, book of old Hebrew love poems, songs in the manner of ancient Egyptian love poems, and poems which seem to have been part of the marriage rite. Traditionally linked with the name of King Solomon (qv), the collection was probably not put together until the 5th century BC. Sincere and natural, for all the effective imagery and ardent feeling imbued with chaste restraint, this love poetry nevertheless seemed too far removed from religious thought for the Jews of later times and for the mediaeval Christian church; it was then regarded as allegorical. Herder and Goethe were among those who argued for a straightforward interpretation. Ever since the Middle Ages the *Song of Songs* has inspired both love poems and religious lyrics. See also Five Festal Scrolls and Bible. SZS

Sufism (from *ṣūf*, wool, the material from which the garments of early Muslim ascetics were made), Islamic mysticism, which grew out of constant meditation by the pious on the Qur'anic revelation. The primarily ascetic character became truly mystic after the Iraqi woman saint Rābi'a preached the doctrine of 'pure love'. During the 9th century in Iraq, Khorasan and Egypt stages of mysticism were elaborated. Absolute trust in God, patience, gratitude, etc, can lead the mystic to intuitive knowledge of God and culminate in love which manifests itself in 'joy of suffering'. Repetition of the divine names and religious formulas was practised; *samā'*, mystical concerts and whirling dances helped to induce ecstatic states. The greatest of the early mystics was al-Ḥallāj (executed 922 in Baghdad), the 'martyr of love'. A period of systematization followed; and a large prose literature came into existence.

From the 12th century onwards fraternities or orders of mystics crystallized, transforming Sufism into a mass movement. The veneration offered to the leader often reached dangerous consequences. At the same time, Persian mystical poetry developed both lyrical and epico-didactic forms, and the alternation of the worldly and the divine object of love constitutes the most fascinating aspect of this poetry, which is echoed in Turkey and Muslim India. In these countries popular mystical poetry also grew up, in local languages, conveying the message of love and divine Unity, and the veneration of the Prophet Muhammad, the Perfect Man, to all strata of the population. A highly sophisticated system of theosophy was built up by the Spanish-born Ibn 'Arabī (d 1240); his language coloured the mystical literature of the following centuries. Most Muslim poetry and even the fine arts can scarcely be appreciated without a knowledge of the Sufi influence, which also permeated the daily life of the faithful to a considerable extent. AS

Sulaymōni, Otajon, Tajik poet, see **Payrav.**

Süleyman of Stal (b 1869 Ashaga-Stal, southern Dagestan, d 1937), Lezgian poet. The son of a poor peasant, he was orphaned early and had to earn his own living. He worked for many years far from home, in the Baku oil-fields, on the Samarkand railway, and elsewhere. After the establishment of Soviet Dagestan he put his talent and ability into the service of Dagestan and the Soviet Union. Gorki, who made his acquaintance at the first Congress of Soviet Writers in 1934, called him the Homer of our epoch. He was given the title of National Dagestan poet in 1934; his first poems date from 1900. He was one of the last great folk poets. Most of his poems are in the classical oriental strophic form *koshma*, quatrains rhymed *abab*, *cccb*, etc, the last line of the first quatrain recurring, usually unchanged, as the refrain; his verse abounds in colourful oriental imagery, but in subject reflects the problems and requirements of modern Soviet life. The first major selection of his work appeared in 1934, since when much has been published in the original and in Russian and other Soviet translations. VAČ

Sulkhan Saba, surname Orbeliani (b 1685 Tbilisi, d 1725 Moscow), Georgian writer and diplomat, of an ancient family, closely connected with the Georgian king. He was educated with the royal princes; as a young man he was converted to Catholicism by the mission in Tbilisi (Tiflis). Later he held office at court, and was tutor to the royal princes. He divorced his second wife, probably for political reasons, and entered a monastery. King Vakhtang VI recalled him from his retreat to head a mission to Louis XIV and Pope Clement XI, to persuade them to make a common front with Georgia against Iran. The mission failed, and Sulkhan Saba returned home, to follow his king into exile in Russia, where he later died. The most important of his literary works is the *Sitqviskona* (Dictionary of the Georgian Language), compiled for the king. In his *Mogzauroba Evropashi* (Journey to Europe) he gave a detailed description of his travels and discussed with knowledge and understanding the art and monuments of European civilization. His best known work, however, is *Tsigni sibrdzne-sitsruisa* (Book of Wisdom and Lies), a collection of fables and moral exhortations which includes many fables from oriental and European sources, in the framework of the story of the education of a royal prince.

Trans.: Oliver Wardrop, *The Book of Wisdom and Lies* (London 1894); M. von Tseretheli, *Die Weisheit der Lüge* (Berlin 1933). VAČ

Sultan Veled (b 1226 Larende, Iran, d 1312 Konya), the first West Turkish poet. Son of Jalāloddīn Rūmī (qv), he accompanied his father into Asia Minor and joined the *mevlevi* order of dancing (whirling) dervishes. He continued his father's work, developing and interpreting his ideas. His long mystic *Mesnevî*, also known as *Velednâme* (The Book of Veled), and his great *dīwān* (qv) are written in Persian, with verse interpolations in Greek and Turkish. 156 Turkish lines are the earliest surviving example of the use in West Turkish literature of Persian metres, inaugurating a school of contemplative lyrics distinct from the popular mysticism of the followers of Yunus Emre (qv).

GHOP I 151 ff; PHTF II 407 ff. ZV

Sumerian Disputes, together with proverbs, sayings and riddles, are the main forms of Sumerian wisdom literature. The *Disputes* take the form of a dialogue between two boasters, and are usually introduced by a description of the myth of the creation of the world and of things. Some *Disputes* (eg the calf and the grain, the bird and the fish, the tree and the reed, the spade and the plough) are fables, in which things and animals act as human beings; even abstract concepts may be endowed with human characteristics, as in the dispute between summer and winter. The dispute style has also been used in some myths, eg in the poem *Enmerkar and Suhkeshda'anna* (see Enmerkar) and in *Dumuzi and Enkimdu* (the dispute between the shepherd and the farmer). The majority of the *Disputes,* without the mythological introductions, were composed in the Sumerian schools; they describe the advantages and also the difficulties of the various professions, particularly of that of the scribe (eg the dispute between the teacher and the scribe, or between two pupils). Like the other forms of wisdom literature, the *Disputes* were used in the schools in extensive series, which it has not been possible to reconstruct entirely.

S. N. Kramer, *The Sumerians* (Chicago 1963), 217–23; E. Gordon, *A New Look at the Wisdom of Sumer and Akkad* (Bibliotheca Orientalis 17, Leiden 1960), 122–50. BH

Sumerian Laments are one of the typical literary genres whose prototype goes back to the Ancient Sumerian period (24th century BC), to a text describing the destruction of Lagash. All the later laments over the destruction of Mesopotamian cities (*Lament for the Destruction of Ur, Lament for the Destruction of Nippur, Lament for the Destruction of Sumer*, known as the *Second Lament for Ur*) are based on historical facts, on the invasion of Sumer by alien tribes at the end of the third Ur dynasty (c2000 BC). In these *Laments* the enemy is always described as godless and uncultured, conquering by brute force alone; the fall of Sumer is not blamed on the people, nor on the local gods, but on the supreme gods, especially Enlil, who is said to have vented his wrath on the whole land. The stereotyped style of the *Laments* has found its way into epic poetry too, especially when the subject is war (eg the *Epic of Lugalbanda*, qv). The elegies, the loveliest of which is that on the death of the god Dumuzi, are close to the *Laments* in style and subject. So far there has been only one find (in Nippur) of Sumerian elegies (two) on the death of ordinary citizens; the text is very fragmentary.

S. N. Kramer, *Lamentation over the Destruction of Ur* (Chicago 1940); ANET[3] 455, 611; *Two Elegies on a Pushkin Museum Tablet* (Moscow 1960). BH

Sundukian, Gabriyel (b 1825 Tbilisi, d 1912 Tbilisi), Armenian playwright. He studied history and philology in St Petersburg (with a thesis on Persian prosody) and returned an enthusiastic follower of the Russian revolutionary democrats, an admirer of the Russian realist theatre and of Shakespeare. For over 50 years he was the leading figure in theatrical life of Armenians in Tbilisi (Tiflis). Following the classicist and Romantic period of historical plays,

Sundukian was the first to bring contemporary life to the stage in a realistic form. He wrote (in the dialect of Tbilisi) sketches and comic scenes (some of these under the pseudonyms Hamal and Hadid). The best of his plays, still presented in the theatre, are comedies of everyday Tbilisi life, with the psychology, ethics and ideas of the ordinary people he knew and expressed so well: *Xatabala* (Confusion, 1866); *Ēli mek zoh* (Another Victim, 1870); *Pepo* (1871, trans. Boston 1931); *K'andac ōǰax* (The Overturned Hearth, 1872, German trans. L. Rubenli, ie A. Hovhannissian, in AB II, VII). Later attempts to depict improbably changed bourgeois characters were less successful. Sundukian initiated the realist drama, especially comedy, in modern East Armenian literature, forming a tradition which the next generation (see Shirvanzade) was able to take up and develop.

LLS; HO 7; THLA. LM

Süreya (Seber) Cemal, see **Cemal Süreya.**

Sürurî, Turkish poet, see **Vehbî,** Sünbülzade.

as-Suyūṭī, Jalāladdīn 'Abdarrahmān (1445–1505), Arab encyclopaedist. Born in Cairo, the son of a professor, he studied for many years there and elsewhere in Egypt and in Mecca. After a short period in the justiciary, he became a professor at 22. Thirty-five years later rebellious students had him deprived of his post in consequence of misbehaviour regarding the bursary monies in his charge. He spent the last four years of his life in retirement on the island of ar-Rawḍa in the Nile, devoting himself entirely to scholarly writing. As-Suyūṭī was one of the most prolific writers of Islam, if not indeed the most prolific, in spite of his relatively short life. There is hardly a sphere his writing did not touch on, even to sexology and pornography. Most of his works are compilations, but they often include texts which have not otherwise survived. Their strictly schematic arrangement and clear treatment have ensured a permanent place in the basic literature for his works, on the *Qur'ān* (qv), *Tafsīr al-Jalālayn* (Commentary of the Two Jalals), *al-Itqān fī 'ulūm al-Qur'ān* (The Improvement on the Sciences of the Qur'ān), his biographies, *Tabaqāt al-mufassirīn* (The Classes of Those Who Knew the Qur'ān by Heart), *Lubb al-lubāb fī taḥrīr al-ansāb* (The Quintessence of Writing Nouns of Kinship). *Bughyat al-wu'āh fī tabaqāt al-lughawīyīn wa'n-nuḥāh* (The Classes of the Philologists and Grammarians), *Naẓm al-'iqyān fī a'yān al-a'yān* (The Arrangement of Pure Gold among the Illustrious People, ie Who is Who in the 15th Century), *Tārīkh al-Khulafā* (History of the Caliphs), *Ḥusn al-muḥāḍara* (The Beautiful Lecture on the Chronicle of Egypt and Cairo) and philological studies, *Muzhir fī 'ulūm al-lugha* (The Bright Work on the Sciences of the Language).

Trans.: H. S. Jarrett, *History of the Chalifs* (Calcutta 1881).
BGAL² II 180–204, S II 178–98. RS

Sydykbekov, Tügelbay (b 1912 Ken-Suu), Kirghiz writer. He studied agriculture, worked as a journalist and later as a cattle farming expert. He began as a poet, but later achieved success in epic works inspired by M. Sholokhov. His novel *Ken-Suu* (1937), later rewritten as *Up in the Mountains*, describes the collectivization of Kirghiz agriculture and is considered the first realist prose work in Kirghizia. His work takes its themes primarily from farming. His best known novel is *Bizdin zamandyn kishileri* (People of our Time, 1948). Other works: poetry, *Kürösh* (Struggle, 1933); *Batyrlar* (Heroes, 1936); novels, *Temir* (1940); *Batiyna* (1962); books for young people, *Ala-Toodon altoo* (Six from Ala-Too, 1936); *Too baldary* (Children of the Mountains, 1953); *Dostuk* (Friendship, 1960). LH

T

aṭ-Ṭabarī, Abū Ja'far Muḥammad b. Jarīr (839–923), a great scholar writing in Arabic, primarily a historian from Amul in northern Persia. He is said to have known the

Qur'ān (qv) by heart at the age of seven, and to have gone to Rayy (Teheran) as a student a short time afterwards. His wealthy father allowed him to continue his studies in Baghdad, at that time the centre of Islamic learning. His pursuit of knowledge took him as far afield as the Persian Gulf, Syria, Palestine and Egypt. He returned to Baghdad and settled down to a life of teaching and scholarly writing, interrupted only by two visits to his faraway homeland. Ṭabarī's writings deserve admiration simply on the grounds of their volume: it has been calculated that he must have written about forty pages a day for over sixty years. Much of what he wrote has not survived, eg about the law school he founded, on poetry, grammar, lexicography, mathematics and medicine. Even his greatest work, the first Arabic history of the world, *Tārīkh ar-rusūl wa'l-mulūk* (The History of the Prophets and Kings) has only come down to us in a selection made by the author himself and covering about one tenth of the original; even so it is a work of twelve volumes. His annals started with the patriarchs and earliest rulers, and went on to deal with the Sassanids; then followed the times of the Prophet, the first four Caliphs, the Umayyads and the Abbasids, up to the year 915. Later historians continued his work, but for the periods he dealt with, his successors are almost entirely dependent on his version. His extensive commentary on the *Qur'ān* (*Tafsīr*) is an authentic source of information on early Islamic exegesis, and still awaits a critical appreciation, eg from the aspects of literature, history, dogma, legal theory and language.

Trans. (parts): E. Marin, *The Reign of al-Mu 'tasim* (New Haven 1951); Th. Nöldeke, *Geschichte der Perser und Araber zur Zeit des Sasaniden* (Leiden 1879).
BGAL[2] I 148 f, S I 217 f. RS

Tabidze, Galaktion (b 1892 Tchkviisi, d 1959), Georgian poet. The son of a village schoolmaster, he studied at the theological seminary in Kutaisi and at Tbilisi (Tiflis). In 1915 he made contact with the Russian symbolists in Moscow. After the establishment of Soviet rule in Georgia (1921) he published his own paper, *Galaktion Tabidzis zhurnali* (1922–23). In 1933 he was made a National Artist of the Georgian Republic. Besides writing poetry, he studied Georgian literature. His verse is inspired by Russian and west European symbolism, but in the course of time it became simpler and more monumental, and echoes of classical and folk poetry appeared.

Trans.: UAGP. VAČ

Tabidze, Titsian (b 1895 Shuamta, d 1937), Georgian poet. The son of a priest, he went to the classical secondary school in Kutaisi, where he joined the circle of young poets which included Mayakovskiy. He spent two years in Moscow, in contact with the symbolists, before returning in 1915 to Kutaisi, where with V. Gaprindashvili and P. Iashvili (qv) he founded the symbolist group, Blue Horns. After the group was dissolved he held various functions, and was one of the leaders of the Georgian Writers' Union. In 1937 he was accused of nationalism and of activities against the interests of the people, and executed. In his earliest verse he called himself a dandy, a young Oscar Wilde; his verse was vehement and provocative, in spite of its formal polish and cultivated style. In his later verse he expressed his deep admiration and respect for the Soviet system, especially in his volumes *Rionporti* (The Harbour of Rioni, 1928) and *Samshoblo* (Fatherland, 1936).

John Lehmann, *Prometheus and the Bolsheviks* (London 1937). VAČ

Ṭāhā, 'Alī Maḥmūd (b 1902 Mansura, d 1949), Egyptian poet. He graduated as an architect in Cairo in 1924, and in due course became a member of the Apollo Group. In common with other of the Apollo poets, influences from Western poetry are particularly apparent in his work: he was fond of French Romantic poetry, and translated a number of poems from this period into Arabic. He travelled extensively in Europe, and often wrote down in verse impressions and reminiscences of his journeys. He also experimented with the poetic drama in Arabic, another obvious effect of the European tradition. His most famous *dīwān* (qv), *al-Mallāḥ at-tā'ih* (The Lost Sailor) was published in 1934. RCO

Ṭāhā Ḥusayn (b 1889 Maghāghah), Egyptian polygraph. Born in a provincial, conservative family of modest means, and blind from the age of two, he was nevertheless independent of mind and self-assertive. In 1902 he was sent to Cairo to attend the traditional theological University of al-Azhar, but was soon attracted to secularist modernist circles, especially that of Aḥmad Luṭfī as'-Sayyid. From 1908, he studied under Arab and European professors at the newly created Egyptian University, and was awarded its first doctorate in 1914. Sent to France at the State's expense, he obtained a doctorate from the University of Paris in 1918, and a Doctorat d'État in 1919. He married a Frenchwoman. He became Professor first of Ancient History then of Arabic Literature in the Egyptian University, and was the first Egyptian to be Dean of the Faculty of Arts. He also held high office in the Ministry of Education. His bold stands on social and religious issues and his involvement with various political parties often affected his career in public service; its highest point was during the last Wafdist government (January 1950–January 1952) when, as Minister of Education, he greatly extended the State school system and made it free.

He composed poetry only in his student days, but he has been an extremely prolific and versatile prose-writer. Numerous critical studies, three volumes of autobiography, several novels and many shorter narratives, historical and educational works, social and political essays, epigrams, as well as translations, mostly from the French, have been produced. Volume I of his autobiography, a work of much charm and pathos, was the first modern literary work to attract attention outside the Arab world. In this, as in his novels and short stories, though they show no great power of sustained invention, there are moving intimate pictures of rural and provincial life, which also give voice to deep-felt compassion for the underprivileged.

It is in literary criticism, however, that Ṭāhā Ḥusayn has made his deepest marks. His doctoral dissertation on al-Ma'arrī was the first systematic study in Arabic of a writer's background, life, and work. A later study on pre-Islamic poetry, which proclaimed Cartesian freedom from preconceptions and cast doubt on the authenticity of much early poetry, caused a furore in the late 20's, mainly because of references to Islamic 'myths'. Cumulatively, his many other critical studies, though they stem from no fully thought-out aesthetic theory, combine a deep-rooted interest in the Arab–Islamic heritage with a constant readiness to revalue and revitalize it by new insights. His style, too, stringently classical in syntax yet supple, lively and elegant, has been a model to many. Notable not so much for tight thinking as for his courage and open-mindedness, Ṭāhā Ḥusayn was a leading spokesman in a generation whose educated élite had broad and vital choices to make. Selected bibliography: Autobiography, *al-Ayyām* (The Days, Vol. I, 1927, Vol. II, 1939); *Mudhakkirāt* (Memoirs, 1967); theses, *Dhikrā Abī al-'Alā* (Commemoration of Abu'l-'Alā [al-Ma'arrī], 1915); *Étude Analytique et Critique de la Philosophie Sociale d'Ibn Khaldoun* (1917); criticism, *Fī'sh-shi'r al-jāhilī* (On Pre-Islamic Poetry, 1926); *Hadīth al-arbi'ā'* (Wednesday Discourses; collected articles, 3 vols., 1937–45); narratives, *Adīb* (Cairo), *'Alā hāmish as-sīrah* (On the Margin of the Prophet's Life, 3 vols., 1946–47); *al-Mu'adhahabūna fī'l-arḍ* (Hell upon Earth, 1949); history, *al-Fitnah al-kubrā* (The Great Schism, 2 vols., 1947–53); education, *Mustaqabal ath-thaqāfah fī Miṣr* (The Future of Culture in Egypt, 1938).

Trans.: E. H. Paxton, *An Egyptian Childhood* (London 1932); H. Wayment, *The Stream of Days* (London 1948); Amina and Moenis Taha-Hussein, *Adib ou l'aventure occidentale* (Cairo 1960); G. Wiet, *L'Arbre de la misère* (Cairo 1964); S. Glazer, *The Future of Culture in Egypt* (Washington 1954); R. Francis, *L'appel du Karaouan* (Beirut 1961).

P. Cachia, *Ṭāhā Ḥusayn* (London 1956); R. Francis, *Taha Hussein Romancier* (Cairo 1945). PC

al-Ṭahṭāwī, Rifā'a Rāfi' (b 1801 Tahta, d 1873), Egyptian author. Born in Upper Egypt, al-Ṭahṭāwī studied in Cairo at the

Azhar Mosque from 1817, coming into contact with the renowned scholar, Sheikh Ḥasan al-'Aṭṭār. He was appointed Imām of the first important educational mission which Muḥammad 'Alī sent to France in 1826, where he spent five years. As a result of his careful observations of French society, and his wide and varied reading, al-Ṭahṭāwī's subsequent career and writings exercised a considerable influence on the development of political and social thought in Egypt. In 1836, he became head of the School of Languages created by Muḥammad 'Alī and performed important work as a translator. He also edited the official newspaper *al-Waqā'i' al-miṣriyya* (Egyptian Actualities). His two most important books are *Takhlīṣ al-ibrīz fī talkhīs Bārīz*, published soon after his return from France, and *Manāhij al-albāb al-miṣriyya fī mabāhij al-ādāb al-'aṣriyya*. The former is an entertaining but penetrating account of French manners and customs, while the latter is a much more serious work on his conception of the future social and political development of Egypt.

A. H. Hourani, *Arabic Thought in the Liberal Age* (Oxford, 1962). RCO

takhalluṣ, or **takhalloṣ,** poetic pseudonym, see **qit'a** and **Mo'ezzī,** Amīr 'Abdollāh Moḥammad.

Ṭālebūf, Ḥājjī Mīrzā (b 1855 Tabriz, d 1910 Tamir-Khan-Shura), Iranian man of letters and public figure. By profession a merchant, he gave all his energy to the pioneering work of popularizing scientific, cultural and political ideas among the broad public. This he did in the form of discussions between learned men (a sort of symposium) or of dialogues between a wiser adult and young people. He undoubtedly drew on the *adab* (qv) tradition in Persian literature, but his original contribution lay in the topical nature of his subjects (though he was sometimes very naive in his interpretation) and above all in that he used the simple language of everyday life. The clear simplicity of Ṭālebūf's idiom served as a precedent for later generations of writers. His works: *Ketābe Aḥmad yā safīneye Ṭālebī* (Ahmad's Book, or the Student's Vessel, 1896); *Masālek ol-moḥsenīn* (The Ways of the Charitable, 1905).

RHIL 343; AGEMPL 76–7. KB

Talmud (Hebrew: learning), a collection of discussions by Jewish scholars dealing with the rules laid down in the *Mishnah* (qv) and including both *halakhah* (rules for behaviour and daily life both for the individual and the religious community) and *haggadah* (qv, stories more or less concerned with the rules being discussed); the latter often include interesting material of considerable literary value. The Palestinian of Jerusalem *Talmud* dates from the 3rd–5th century and is written mainly in Palestinian Aramaic; the Babylonian *Talmud*, written in the Aramaic dialect of the Babylonian Jews and completed c500 AD, is fuller and enjoys greater authority. See also *Midrash*.

Trans. and text: I. Epstein, ed., *The Babylonian Talmud*, 64 Vol. (New York 1959).
H. L. Strack, *Introduction to the Talmud and Midrash* (New York 1969). SZS

Taner, Haldun (b 1916 Istanbul), Turkish writer and dramatist. He studied political economy and German at Heidelberg and Istanbul, and is now a university professor. His articles on culture and the arts have been republished in book form, *Devekuşuna Mektuplar* (Letters to an Ostrich, 1960). Taner's realistic short stories, classical in form, give a sensitive portrayal of urban man and his environment. His most important play, *Keşanlı Ali Destan* (Destan about Ali of Keşan, 1964) was the first epic drama to appear on the Turkish stage. Other works: short stories, *Yaşasın Demokrasi* (Long Live Democracy, 1949); plays, *Dışardakiler* (Those Outside, 1957); *Günün Adamı* (The Man of the Day, 1961). LH

at-Tanūkhi, Abū 'Alī al-Muḥassin, (940–994), Arab scholar, son of a Qadi of Basra. He too followed a juristic career and became Qadi in Baghdad and other towns. Soon, however, his career as a government official was affected by political instability. A change of vizier meant that he, too, lost his position and his property; later reappointed, another period of imprisonment followed, and even after release he endured much unpleasantness from political and personal enemies. His three volumes of tales must be seen against the background of his life and experience; in anecdotal form they represent a certain degree of criticism of social conditions, although the theocratic Islamic basis of society is never questioned. The theme of 'Deliverance after Distress' was developed by him into a literary form in his book *Kitāb al-faraj ba'd ash-shidda*. The tales are scenes of public and private life, mainly of his own time and of the world of viziers and qadis, but also of the vast numbers of minor and lowly officials who regarded their positions as gifts and benefices in the Arab sense of munificence, and accepted their life between *dīwān* and prison as ordained by fate.

Trans.: D. S. Margoliouth, *The Table-Talk of a Mesopotamian Judge, being the first part of the Nishwār al-Muḥāḍara or Jāmi' al-Tawārīkh of Abū 'Alī al-Muḥassin al-Tanūkhī* (London 1922, parts 2 and 8, in Islamic Culture, 1929–32). BGAL² I 161 f, S I 252 f. RS

Tarancı, Cahit Sıtkı (b 1910 Diyarbakır, d 1956 Vienna), Turkish poet. He abandoned his studies at the School of Political Science (*Mülkiye Mektebi*), and left for Paris in 1939, where he continued them at the Faculty of Political Sciences. He returned to Turkey in 1940. After serving in the army (from March 1941 to October 1943) he worked as a French translator in *Anadolu Ajansı* and in the civil service until he fell ill in January 1954. His illness lasted two years before he was sent to Vienna for treatment, where he died in hospital in 1956. His works incontestably represent the finest achievement of modern Turkish poetry in syllabic-rhymed verse. C. S. Tarancı writes of death, and its terrors, of old age and solitude, of the search for a lost god and momentary joys, of unfulfilled love. The simplicity and spontaneity of his style recalls Verlaine. His works are the authentic expression of the intellectual and emotional preoccupations of the Kemal generation of intelligentsia. His views on literature are best formulated in letters he wrote to a childhood friend, the writer Ziya Osman Saba, published after his death as *Ziya'ya Mektuplar* (Letters to Ziya, 1957). Collections of poems: *Ömrümde Sükût* (Silence in my Life, 1933), *Otuz Beş Yaş* (35 Years of Age, 1946), *Düşten Güzel* (More Beautiful than a Dream, 1952), *Sonrası* (What Next, 1957). DL

targum (Aramaic and Hebrew: rendering), the Aramaic version of the Hebrew Old Testament (qv) books, of which the earliest known is the *targum* to the *Book of Job* discovered in Qumran, Cave XI, of pre-Christian origin. The earliest *targumim* (Onkelos, Jonathan Neophyti) are on the whole faithful translations, while the later *targumim* are often paraphrastic, and contain some additions. SZS

Ṭarzī, Maḥmūd (b 1868 Ghazni, d 1935 Istanbul), Afghan statesman, propagator of enlightenment, and an important writer in Dari (Persian). His father, the well-known poet Ṭarzīi Afghān, was imprisoned in 1888 and then forced to leave his country. The family lived mainly in Syria, until in 1904 Emir Ḥabibullāh called upon Maḥmūd Ṭarzī to return home. He then became one of the leaders of the Young Afghan movement and founded the fairly advanced paper *Sirāj ul-akhbār* (Lamp of News) which ran from 1911 to 1918. Ṭarzī exerted an influence over the future king Amānullāh (1919–29), during whose reign he was Foreign Minister and then ambassador in Paris. In 1928 he was again forced to leave Afghanistan, and settled in Turkey where he devoted himself to writing. His own works and his critical writings brought about a complete change in the literature and thought of his country. He was the pioneer of poetry on new subjects, his own *ghazals* (qv) dealt with science, coal, roads, telegraphic communication, and so on. He was one of the founders of the new prose, to which he contributed important translations of Jules Verne's novels. He also made an original contribution to philosophy, eg in his *'Ilm wa islāmiyyat* (Science and Islam), *Waṭan* (My Country) etc. As a politician and journalist he condemned the policy of the imperialist powers, and in 1933 wrote a poem attacking the Japanese invasion of Manchuria. He preached the emancipation of women and founded the first women's paper, encouraged the introduction of new teaching methods in schools, fought against tribal separatism and in his paper called for a completely independent Afghanistan. His verse appeared in the volume *Adab wa fann* (Literature and Science, 1915) and *Parākanda* (Scattered); *Rauẓai ḥikam* (Garden of Wisdom) is a collection of essays on political and literary subjects. JB

Ṭarzii Afghān, Afghan poet, see **Ṭarzi,** Maḥmūd.

Tatianos (second half 2nd century AD), Syrian writer on philosophy and religion. Originally from Adiabena, Tatianos was converted to Christianity in Rome, and wrote a Greek apologetic for Christianity. About 172 he was attracted to the ascetic Gnostic sect of the Encratites and went to live in Antioch in Syria. He put together a Syriac harmony of the *Gospels* for which he seems to have used the *Gospel* according to the Hebrews as well as the four gospels of the New Testament (qv). This *Diatessaron* (in Greek: a specific musical interval) has been translated into many European and Oriental languages.

WSHSL. SZS

Tavalloli, Farīdūn (b 1917 Shiraz), Persian poet. His work has an important place in the involved movement for a poetic renaissance represented in its early stages by Nīmā Yūshīj (qv). Tavallolī belongs to those poets who gradually find their own, moderate approach, allowing a measure of compromise with the classical tradition. His favourite form is a simple quatrain rhyming *abcb*, a form popular with poets of this trend, but his rich language, refined imagery and melodic verse place him among the foremost modernists. It is in his lyrics that he achieves an intrinsic personal expression; imbued with a gentle melancholy, their moving descriptions of nature merge with the eternal love-song in one harmonious whole. A note of pessimism seems to be ever growing in his work; fear and horror of death together with the sense of man's helplessness recall analogous motifs in old Persian poetry, particularly that of 'Omar Khayyām (qv). Two volumes of his verse have appeared: *Rahā* (1952) and *Nāfe* (1962).

AGEMPL 226–7. HS

at-Tawḥidi, Abū Ḥayyān (d 1023, perhaps at 90 years of age), one of the leading Arab authors and philosophers, of Persian origin. He spent most of his life in Baghdad, first as a student of law and philology, later as a secretary. Expelled from the city for the propagation of heretical opinions, he wandered through Persia for many years, taking the rough with the smooth. In Baghdad once again, he contented himself with the barest necessities in order to be able to devote his whole time to the society of learned men and to his own writing. He is one of the great figures of classical Arabic prose; brilliant in content, his works have perfect style. His numerous writings cover many subjects, from the art of caligraphy *Risāla fī 'ilm al-kitāba* (Epistle on Penmanship), to a pamphlet of sarcastic satire on two viziers *Kitāb al-wazīrayn* (The Book of the Two Viziers); from friends and friendship *Risāla fi 'ṣ-ṣadāqa* (Epistle on Friendship and Friends) to philosophical treatises and discussion with his contemporaries *al-Muqābasāt* (Doctrines and Opinions), *Kitāb al-imtā' wa'l-mu'ānasa* (The Book of Enlivenment and Good Company). The historical basis of these discussions, in which there are many neo-Platonist ideas, is unusually close to reality. Although not himself a Sufi, his whole way of life was close to that of Sufi circles, as can be seen from another of his works, *al-Ishārāt al-ilāhīya* (The Divine Signs).

Trans.: M. Bergé, *Une anthologie sur l'amitié d'Abū Ḥayyān at-Tawḥīdī* (Bulletin d'Études Orientales, 1958–60).
BGAL[2] I 283, S I 435 f; EI[2] I 126–7. RS

Taymūr, Maḥmūd (b 1894 Cairo), Egyptian story-writer and playwright in Arabic. He was born into a family which included several men-of-letters and his home was the meeting-place of intellectuals. With a natural penchant for literature, and being influenced by this home atmosphere, Taymūr interrupted his studies in economics and turned to writing. Early in his career, he was influenced by French literature (particularly by Maupassant), and from it he borrowed the essentials of several of his plots. Nevertheless, his stories and plays deal mostly with twentieth-century Egyptian life, in a characteristically realistic style. His short stories describe village and town, and their heroes are people from all walks of life.

Taymūr's plays are mainly comedies and social dramas; a few are historical plays. His experience in story writing is reflected in his plays. These usually unfold with a swift pace, well-suited to the stage, and frequently border on social satire. They are generally outspoken, but sometimes elusive. While he occasionally overemphasizes comic situations (bordering on slapstick), he has a natural psychological insight. Taymūr particularly enjoys pricking inflated egos and chastising pompousness, demagogy, and similar foibles. His comedy *Qanābil* (Bombs), for instance, pokes fun at the way people from various social strata behave in an air-raid shelter under threat of danger and starvation. Similarly, *al-Muzaiyyfūn* (The Forgers) flays the lying, hypocrisy, nepotism, and demagogy prevalent among political parties. Of special importance is Taymūr's skill with language. In his stories and plays he uses both literary and colloquial Arabic to suit the situation and the speaker. He frequently employs them alternately, to increase the realistic effect. Indeed, he has written and published some of his plays twice, in literary and colloquial editions. Throughout, he has shown a remarkable mastery of Arabic, which brought him, in 1954, membership in the Academy of the Arabic Language in Cairo. Several of Taymūr's works have been translated into French.

BGAL S III, 217–26, 255–6; LSATC 147–53; GLAIG 205–9; WILA 288–9. JML

taẕkira (Persian *taẕkere*, Turkish *tezkire*, an Arabic word meaning memorandum, record), is the term used in Persian and Turkish literature for a collection of lives, most frequently those of poets, but also of saints, sheikhs or calligraphers. A *taẕkira* of poets, which gives a brief biography and examples of each author's work, may be either general, dealing with all the well-known poets up to the time of the author of the *taẕkira*; or specialized, dealing only with poets of a certain period or type. They may be arranged chronologically, or geographically according to where the poets were born or lived; or the poets may be grouped in generations, according to the type of verse they wrote. There are even some *taẕkiras* arranged lexically. FT

Tcharents, Yeghishe (real name Soghomonian; b 1897 Kars, d 1937 or 1938 Yerevan), Armenian poet. Son of a tradesman, he volunteeered for the front at 15; he was in Moscow during the February and October Revolutions and in the Red Army in the Civil War, and its aftermath of turmoil in political and cultural life. In 1924–25 he served as a Soviet diplomat in Turkey, Italy, France and Germany; in the 30's he was a leading figure in Armenian public life. Tcharents first published lyrics (1912) influenced by symbolism, followed by poetry based on war experiences. After 1917 he sought new aesthetic principles to express the revolutionary nature of the time. His vast and formally varied work laid the foundations of Soviet Armenian literature and aesthetics, comparable in its significance to that of Mayakovskiy in Russian literature. Tcharents wrote both intimate lyrics and propaganda poetry, creating the 'radio-poem' expressive of the emotional heroics of the day, a series of poems on Lenin, satirical poems and a novel about the enemies of the Revolution. He attempted to clothe the new ideas in the conventional forms of European and Oriental poetry (sonnet, rondo, *beyt*, *rubai*, *ghazal*, qqv, etc). The interpretation of Armenian history in his last works provoked persecution and liquidation of the author. Chief works: lyric poetry, *Erek'erg txradaluk ağjkan* (Three Songs to a Pale Girl, 1914); *Tesilažamer* (Time of Delusions, 1915); *Ciacan* (The Rainbow, 1917); *Tağaran* (Hymn-book, 1921); *Epik'akan lusabac* (Epic Dawn, 1930); *Girk' čanaparhi* (Book of the Way, 1934); epic poetry, *Dant'eakan aŕaspel* (Dantesque Legend,

1914); *Soma* (1918); *Amenapoem* (Omnipoem, 1921); *Čarenc-name* (1922); *Lenin* (1924); prose, *Erkir Nairi* (Nairi Land, 1925).

Trans.: J. Gaucheron, in OeO; J. Gaucheron, P. Gamarra and P. Loriol, in ELA.
SLM. LM

Teimuraz I (1589–1663 Astrabad), Georgian king and poet, a member of the Kakhetian branch of the royal Georgian house of the Bagratids; he was twice King of Kakheri (1606–1616 and 1623–32). He was brought up at the Persian court and placed on the throne by the Shah Abbās the Great (1586–1628), but he spent his life fighting against the Persians, who destroyed his whole family. He tried to ally himself with the Turks and later with the Russians, paying a visit to Moscow for the purpose in 1658, but he was unsuccessful; the Persians put him in prison and he died in exile. In his poetry Teimuraz was the most important propagator of Persian models. He used the Georgian *shairi* line, cultivated by Rustaveli (qv), for his poems on Persian subjects like the love of Leylā and Majnūn, Yūsuf and Zuleykhā, The Rose and the Nightingale, The Moth and the Candle, etc. Besides the carefree mirth, the drinking parties, and the romantic love stories, his verse reveals a growing mood of hopeless despair, lamenting the ill-will of the world. Teimuraz expressed his patriotic feelings in the long poem *Tsameba Ketevan dedoplisa* (The Sufferings of Queen Ketevan), on the martyrdom of his mother. In his old age Teimuraz proclaimed himself the greatest poet of Georgia; this has been the cause of much discussion, but there can be no doubt that he opened the way for Persian poetry to penetrate the literature of Georgia, enriching it and extending its opportunities.

Trans.: UGAP.
W. E. D. Allen, *A History of the Georgian People* (London 1932); D. M. Lang, *The Last Years of the Georgian Monarchy* (New York 1957). VAČ

tekerleme, Turkish poetic genre, see **Yunus Emre.**

Telepinu, Hittite god, son of the storm god, the chief god of the Hittite pantheon. From mid-second millennium BC there are several versions of a myth, in Hittite, telling of the disappearance of an angry god, bringing with it stagnation, drought, bad harvest, hunger and desolation for men and the gods. In the best-preserved version this god is Telepinu. Most of the myth is concerned with the vain search for him, in which all the Hittite gods take part; an eagle is sent out by the sun god, and even Telepinu's father himself, the highest Hittite god, goes to seek him. Finally he is found by a bee sent out by the mother-goddess, but the bee angers Telepinu even more by stinging him and waking him from his sleep. Many rites are necessary to pacify him. When finally the god is conciliated, Telepinu's anger is shut away in the underworld, in bronze pots, behind seven locked gates; Telepinu, appeased, returns and with his reappearance the evil effects of his absence disappear; nature revives and the earth is prosperous once more. The number of differing versions of this myth show the broad distribution and the long tradition behind it. It is undoubtedly a folk myth taken over by the Hittites from the previous inhabitants of Asia Minor, the so-called proto-Hittites, for the name of Telepinu is proto-Hittite.

O. R. Gurney, *The Hittites* (London 1969); A. Goetze, in ANET[3] 126 ff; H. G. Güterbock, in *Mythologies of the Ancient World* (New York 1961), 143 ff. VS

Ter-Hovhannisian, Grigor, Armenian novelist, see **Muratsan.**

Terian, Vahan (real name Ter-Grigorian; b 1885 Gandza, d 1920 Orenburg), Armenian poet. Son of a priest, he studied history and philology at Moscow and St Petersburg (as a disciple of N. Marr). As a consultant for the Armenian question he was present at the Brest-Litovsk peace conference. From 1917 he was a communist, active in politics. He published first in 1908 (*Mt'nšaği anurǰner*, Dreams in the Twilight), under the influence of French and Russian symbolism, the depressing social atmosphere in Russia after the defeat of the 1905 revolution, and above all, the tragedy of the Armenian minority in Turkey. Terian's lyrics, especially *Banastegćut'yunner* (Verses, 1912), was written in pure language, free of dialect, remarkably musical thanks to the wealth of rhyme and alliteration; it raised Armenian lyric poetry to a higher level of achievement after the dynamic growth at the turn of the 20th century (see Thumanian, Isahakian). It is the poetic confession of a subtle mind, cast about with melancholy and a dreamy nostalgia, the picture of the situation of the group to which the poet belongs, seen through an involved perspective. After the Revolution Terian sincerely believed in the imminence of a better life for mankind, and optimism, happiness, enthusiasm found their way into his verse. This phase of his work which, with the early work of Tcharents (qv), links pre- and post-revolutionary Armenian poetry, has lost nothing of its sincerity or of its art, for it never sank to the level of mere propaganda. Terian translated Baudelaire, Verlaine, Wilde, Verhaeren, Blok, Bryusov, Rustaveli (qv), Lenin, etc.

Trans.: P. Loriol, in ELA. LM

Theodicy, see **Babylonian Theodicy.**

Thumanian, Hovhannes (b 1869 Dsegh, d 1923 Moscow), Armenian writer and poet. A countryman with no more than secondary education, he was for many years a teacher and public figure. In 1899 he organized the *Vernatun* (Garret) circle of young writers; in 1909 he was imprisoned. From 1912 onwards, as a secretary of the Caucasian Society of Armenian Writers, he was one of the most active organizers of aid for Armenian refugees from Turkey. After the Revolution he worked as chairman of the *Hayartun* (House of Armenian Art). His work includes social, patriotic, love and nature lyrics (*Banastegćut'yunner*, Verses, French excerpts by P. Gamarra in ELA), satirical verses, philosophical quatrains, stories (*Gik'xor*, 1895), tales and fables for children, adaptations of national epics, criticism and other articles, and poetic legends and ballads with historical and folk-lore themes (*Axtamar*, 1891, trans. into 14 languages, Yerevan 1969, French by P. Gamarra in ELA; *P'arvana*, 1902). His best works took the form of lyrical epics revealing the tragic side of life among the patriarchal Armenian mountain villagers on the poet's native district, Lori (*Anuš*, 1890, second version 1902, opera by A. Tigranian; *Maro*, 1887; *Loreci Sak'o*, 1889, etc). His language is rich in local dialect and other vernacular elements, and gives a faithful impression both of the background and of the mentality of the characters. Thumanian translated Pushkin, Lermontov, Nekrasov, Gorki, Byron, Schiller, Heine, Goethe and Russian folk tales and poems. His work marks the peak of Armenian democratic literature in the pre-socialist era.

Further trans.: M. Kudian, *The Bard of Loree* (London 1969, Thumanian's selected works); A. Gaspard, *La geste de David le Sassouniote* (Genève 1945).
HO 7; THLA. LM

Togolok Moldo (Bayimbet Abdrakhmanov; b 1860 Kurtka, d 1942), Kirghiz folk singer, whose work is mostly based on improvisation; his style was influenced by that of his contemporary Toktogul Satylganov. In the moralizing or didactic tradition, after the Revolution his songs turned to political agitation. In 1939 he wrote down all his verse to date, and became a member of the Kirghiz writers' organization. LH

Tosefta (Aramaic: addition), a collection of traditional writings similar to the *Mishnah* (qv) treatises and arranged collaterally to them.

H. L. Strack, *Introduction to the Talmud and Mishna* (New York 1969). SZS

Tsadasa, Hamzat (b 1877 Tsada, Khunsakh Dagestan, d 1951), Avar poet. Orphaned as a child, he had to care for six sisters. At the age of ten he was befriended by an uncle, who sent him to the mosque school where he learned Arabic language and literature. This school qualified him for the religious duties he performed for some years, but the village was too poor to support him, and he went to the Grozny oilfields. Soon a victim of malaria, he returned to the mountains. After the Revolution he worked for the Muslim religious court (*shariat*) and later became editor of various Avar publications. In 1934 he was given the title of National Dagestan poet. His earliest poems were witty Arabic parodies on the *sūras* of the *Qur'ān* (qv); his love poetry was also written in Arabic. As a mature writer he found his medium in satire. He also translated, in particular, the work of Pushkin. He began writing in 1891, while still at school. His poems are all in traditional Dagestan forms, especially the eleven (6+5) and seven syllable unrhymed syllabic lines rich in parallelism and alliteration. As an Avar poet he is scholarly, his verse abounding in Arabisms and allusions to the oriental tradition; this he consciously used for excellent comic effect. His verse presents a commentary on everything that happened in Dagestan. His later verse is classical in its purity and formal balance. VAČ

Tserents (real name Hovseph Shishmanian; b 1822 Istanbul, d 1888 Tbilisi), Armenian novelist. He studied in Venice and Paris, where he experienced the revolution of 1848. He worked as a doctor and journalist in Istanbul and was exiled to Cyprus for three years. As an old man he taught Armenian history and languages in Tbilisi (Tiflis). He wrote three historical novels placed in the period of the Arab invasions and the heyday of Armenian Cilicia, in which he created the type of patriotic hero fighting for Armenian independence in the spirit of the national ideals of the 80's and 90's: *T'oros Levoni* (Toros, Son of Levon, 1877; English trans. Boston 1917), *Erkunk' 9 daru* (Torments at the Birth of the 9th Century, 1879), *T'eodoros Rštuni* (1881).

THLA; HO 7. LM

Tsereteli, Akaki (1840–1915), Georgian poet. Of an ancient princely family, he was brought up in a peasant home, then sent to grammar school in Kutaisi and the Department of Oriental Languages in St Petersburg. In 1862 he returned to Georgian and devoted himself to writing. With I. Chavchavadze (qv) and the educationalist Iakob Gogebashvili, he helped to form modern literary Georgian. His poetry was of the Heine type, but without Heine's irony. All his poems bear a resemblance to folk poetry; he sings of the simple joys of rural life and is one of the first writers to comment on social questions in the countryside. A cultivated vernacular made its first appearance in Georgian literature in his verse, which is melodious and has served as a text for many popular Georgian songs, eg *Suliko*. Much of his love poetry, however, is an allegory portraying his love for his country. His prose is also patriotic in theme. The true idol of his people, his funeral was a national demonstration.

Trans.: UAGP.
D. M. Lang, *A Modern History of Georgia* (London 1962). VAČ

Tsurtateli, Iakob (late 5th century), Georgian hagiographer, confessor to Shushanik, a princess of the princely Armenian family of Mamikonian and wife of the Persian governor (*pitiakhsh*) Varsken. In his *Tsamebay tsmidisa Shushanikisi dedoplisay* (The Martyrdom of St Shushanik) he described her life, and death as a martyr. From allusions in the text, it was probably written between 475 and 484. Although it is one of the earliest original literary works in Georgian, the language is rich and the style polished, which has led to the suggestion that a considerable literary tradition must have been in existence, which has since been lost. In any case, the author must have been an educated scholar with broad political experience. His work is an excellent source of information on fifth century Georgian history.

Trans.: P. Peeters (Latin) *Analecta Bollandiana*, t. LIII, fasc. III and IV, (1935) 5–48; English abridgement: D. M. Lang, *Lives and Legends of the Georgian Saints* (London 1956). VAČ

Tukay, Gabdulla (b 1886 Kushlanych near Kazan, d 1913), Tatar poet. He studied at a traditional *madrasa* and his first poetry was written under the influence of classical Turkish poetry. Coming to Kazan in 1907

he shed his dependence on the old poetic forms. Tukay began to read Russian and European poets (Pushkin, Lermontov, Byron). This was the period of his most important work, bringing the new poetic language of Tatar literature to its full bloom, making possible a modern poetry in the language. Principal works: *Khüriyat khakynda* (On Freedom, 1905); *Shurala* (1907); *Milletchelar* (Nationalists, 1908).

PHTF II 774. NZ

Ṭūqān, Ibrāhīm (b 1905 Nablus, d 1941 Jerusalem), Palestinian Arab poet. It was during his undergraduate years at the American University in Beirut 1926–29 that he published his first poems. He taught first at home, then at the A.U.B. He then entered public service in Palestine, but clashed with the mandatory authorities. After another spell of teaching in Baghdad, his health, which had never been good, deteriorated, and he returned to Palestine for hospitalization. In his lyrical poetry, he favoured short metres and multi-rhyme arrangements showing the influence of the Andalusian *muwashshaḥ* (qv); his love poems are pleasing rather than profound, deftly worded rather than imaginative. He was more famous for his patriotic ardour, and his tireless attempts to mobilize his contemporaries against Zionism. This 'platform poetry' is marked by emphatic assertions, rhetorical appeals, and sometimes sarcasm. Main work: *Dīwān* (Collected Poems, 1955). PC

Turkestani style, in Persian literature, see **Khorasani style.**

Tursunzoda, Mirzo (Russian: Tursunzade; b 1911 Qaratogh), Tajik poet, dramatist and public figure. His father was a small craftsman; he was educated at a boarding-school in Dushanbe, followed by a year at the Teachers' Training College in Tashkent. From 1930 he collaborated with the youth movement paper, and later held many posts in the writers' union; he has now been its President for many years. In 1951 he was made a member of the Tajik Academy of Sciences; he is a member of the Supreme Soviet, holds many positions in international organizations, and has been awarded many prizes and honours. He made his first appearance in periodicals in 1929, and *Bayraqi zafar* (Flag of Victory, 1932) is a collection of his earliest stories, reportages and poems. Although still somewhat schematic, these early efforts show the author's dislike of all that was decadent in the old world, and his interest in ordinary people; he wrote poems to the Five Year Plans, to the ardent young Komsomols, etc. The play *Hukm* (Judgment, 1934), attacking the enemies of the Sovietization of Tajikistan, reveals his literary talent.

Tursunzoda is primarily a lyric poet, however. Simple language and style are characteristic of his verse, as is his warm affection for folk poetry, which he also collected and edited. In his own verse he used images current in folk poetry; whole themes, as in the libretto for the opera *Tohir va Zūhra* (1944) are also based on folk poetry. He is well read in classical Persian poetry, and an admirer of Neẓāmī (qv). Tursunzoda wrote many individual poems and great *dostons* (see *dāstān*); the cycles *Qissai Hinduston* (Indian Ballad, 1947) and *Sadoi Osiyo* (The Voice of Asia, 1956) both deal with subjects surpassing the author's own land. The story of Tajikistan and of the young Tursunzoda is given in the versified history *Hasani arobakash* (Hasan the Carter, 1954). The ballad *Arūs az Moskva* (The Bride from Moscow, 1945) is dedicated to the last days of war and the beginning of peace; the musical comedy *Arūs* (The Bride, 1947) was based on it. He also wrote the libretto for the first Tajik opera, *Shurishi Vose'* (The Vose Rebellion). One of his most recent *dostons* is *Roja dar Kreml'* (Rajah in the Kremlin, 1971), which carries on from *Qissai Hindustan* with the story of the Indian revolutionary Pratap, who went on foot to see Lenin. In his poetry Tursunzoda is a philosopher, a lyricist, and a political and public figure. He has also written much in the field of literary criticism, literary history, and journalism. Tursunzoda is one of the outstanding figures in Tajik cultural life. Further works: doston, *Pisari Vatan* (Son of his Country,

1942); the well-known volume *Man az Sharqi ozod* (I, a Man from the Free East, 1951); *Charoghi abadī* (Eternal Light, 1958), etc.

RHIL 577–9. JB

tuyugh, a poetic form used in Turkish literature of the classical period and employing the metre – ∪ – – / – ∪ – – / – ∪ –. The independent quatrains rhyme *aaba*, the rhyming words being homonyms. LH

Twelve Lesser Prophets, the last twelve books of the Old Testament (qv), the shorter prophesies of Hosea, Joel, Amos, Obadiah, Jonah, Micah, Nahum, Habakkuk, Zephaniah, Haggai, Zechariah and Malachi (qqv). See *Prophets* and Bible. SZS

U

Ulfat, Gulpāchā (b 1909 Laghman), Afghan writer and philosopher writing in Pashto. Of a *sayyid's* family, he studied theology in Nangrahar and then in Kabul. He edited a number of journals and then held various posts, being president of the Pashto Academy (*Pashto ṭolena*) and from 1963 head of the Department for Tribal Affairs. He was the first chairman of the Society for Afghan-Soviet Friendship, was elected to parliament several times, and in 1956 was delegate to the *Loya jirga* (Great Assembly). He became acquainted with European literature and at the turn of the 40's and 50's was a member of the *Wēsh zalmiyān* movement (Youth Awakes). He has written many poems, stories, essays and philosophical and philological articles. In his verse and prose, which is often philosophical, he praises the courage of the Afghans, lauds education, introduces occasional social motifs as in the allegorical poem *De rama khāwanda* (Owner of Herds). His *Psarlay naghma* (Spring Melody), a collection of quatrains and *maṣnavī* (qv) sings the praises of nature. Other volumes of verse are *De zre waynā* (The Voice of the Heart, 1962), *Ghwara ash'ār* (Selected Poems, 1955), while his prose writings have been published in *Ghwara naṣrūna* (Selected Prose, 1963), *'Ālī afkār* (Noble Thoughts) etc. He has made a valuable contribution to Pashto literary history and criticism in Afghanistan, in such works as *Neway sabk aw neway adab* (A New Style and a New Writing, 1949), *Līkwālī* (The Writer's Craft, 1960), *Adabī baḥsūna* (Literary Conversations, 1953), *Millī qahramān Khushḥāl Khān Khaṭak* (The National Hero Khushhal Khan Khatak, 1965) ao. JB

Ulughzoda, Sotim (b 1911 Varzik), Tajik writer and dramatist. His father was a poor peasant, and he lost both parents when he was ten. He went to boarding school and then to the Tashkent Teachers' Training College. He became a teacher, then a journalist, and translated from Russian. During the war he wrote for the Tajik and Uzbek military papers. After the war he was elected to the Supreme Soviet of the Tajik Republic and became a member of the Academy of Sciences. From 1930 onwards he wrote for the papers and at the end of the 30's published two plays: *Shodmon* (1939) about setting up kolkhoz farms, and *Kaltakdoroni surkh* (Reds Armed with Sticks, 1940), about the fighting against the Basmachis, a play which is still performed. The play *Dar otash* (Under Fire, 1944) was popular during the war, with its fighting theme; *Rūdakī* was the first Tajik historical drama. After the war Ulughzoda wrote many novels and short novellas, most of which have been translated into other languages of the Soviet Union and other countries. His first novella *Yoroni bohimmat* (Noble Friends, 1947) is about women and their loyalty to their men at the front. The novel *Navobod* (Land Reborn, 1948) deals with kolkhoz farms, the feminist question, and others. He wrote *Subhi javonii mo* (The Morning of Our Lives, 1954), on the model of S. Aynī's (qv) autobiographical novels, and at the same time described how Soviet power established itself in the Tajik countryside in the 20's. He continued to write plays (*Juyandagon*, The Seekers, 1951), and also wrote film scenarios. From 1961 to 1967 he was engaged on an historical novel,

Vose', about the Tajik leader of the 19th century peasant rising. One of his most recent works is the play *Temurmalik* (1969), about the heroic defender of the town of Khujand against Chinggiskhān. In the 30's his literary criticism was widely read and in 1940 he took an active part in the publication of *Namunahoi adabiyoti tojik* (Anthology of Tajik Literature); he is also the author of the first literary history of *Ahmad Donish* (qv). His witty comedy *Gavhari shabchirogh* (Precious Jewel, 1961) won the Rūdakī prize.

RHIL 572–4. JB

'Umar b. Abī Rabī'a (b c643 Mecca, d 719 or 712), Arab poet of the Umayyad period, the finest writer of urban love poetry. Of a wealthy merchant family, he spent most of his life living in Mecca, enjoying wild adventures. Later Arabic literary tradition made him the symbol of amorous frivolity, and spurious legends have since accrued to his name. He did not write political poetry, the other type of verse coming to the fore at that time. His *dīwān* (qv) comprises for the most part the somewhat frivolous love verse which he made typical of his time and environment. He deserted the traditional Arabic *qaṣīda* (qv) form, in which love was only one of the customary themes, and wrote in a malleable, unconventional style, even introducing dialogue into his verse. He was also a pioneer in metre, and his usage had a strong influence on later love poetry. Many of his poems have been set to music and are known throughout the Arab world, where he is still considered the greatest love poet in Arabic, along with Imru' al-Qays (qv). His *dīwān* has often been published in Europe and in the East (eg P. Schwarz, *Der Diwan des 'Umar ibn Abi Rebi'a*, Leipzig 1901, 1909).

BGAL I 45–7, SI 76–7; EI[1] III 979; NLHA 237; AAP 40; WILA 48. KP

'Umar Khayyām, Persian poet, see **'Omar Khayyām.**

Ünaydın, Ruşen Eşref (b 1892 Istanbul, d 1959) Turkish writer. He worked as a teacher of French and Turkish after studying at the Arts Faculty in Istanbul. He entered politics during the national liberation war, and later became a diplomat. His main work is in essay form, especially interviews with artists, entitled *Diyorlar ki* (They Say . . ., 1918). Famous as the author of the first newspaper article about Kemal Atatürk. Other works: *Damla Damla* (Drop by Drop, 1929, poems in prose); memoirs. LH

'Unṣuri, Abulqāsim, Persian poet, see **'Onṣori,** Abolqāsem Ḥasan b. Aḥmad.

'Urfi, Persian poet, see **'Orfi.**

Usāma b. Munqidh (b 1105 Shayzar, d 1188 Damascus), Arab writer. Of a noble Syrian family, he led an adventurous life devoted to war, hunting and literature, at various Syrian and Egyptian courts. He took part in many military campaigns especially against the Crusaders, but often met them under peaceful conditions as well. The whole of his family perished in the earthquake of 1157. Usāma b. Munqidh is the author of one of the earliest genuine Arabic autobiographies, *Kitāb al-i'tibār* (Book of Instructions), which gives a lively and sometimes humorous account of his war experiences and hunting adventures, as well as the court intrigues; it also describes the cultural and literary atmosphere of his times. Although his account is episodic and a patchwork, it nevertheless gives a vivid picture of the social and cultural conditions in Syria and Egypt during the Crusades, and is of considerable value as a source for the history of the period.

Trans.: P. K. Hitti, *An Arab-Syrian Gentleman and Warrior in the Period of the Crusades; Memoirs of Usamah ibn Munqidh* (New York 1929). IH

Uyghun, Rahmatullo (b 1905 South Kazakhstan), Uzbek lyric poet, dramatist and critic. He went from the secondary school to the Teachers' Training College at Tashkent, and on to the Pedagogical Academy of Samarkand. He began to teach. He is the representative of a new generation, endeavouring to find new poetic images to express new ideas in poetry. His literary criticism deals with the development of Soviet Uzbek literature. During the war he turned to drama; *Alisher Navoiy* was written in collaboration with I. Sultonov. He has also produced sensitive translations of Pushkin, Lermontov, Chekhov and Tolstoy. Main works: volumes of verse, *Bahor sevinchlari* (Joys of Spring, 1929), *Ikkinchi kitob* (Second Book, 1933), *Uyghun she'rlari* (Poems of Uyghun, 1936), *Muhabbat* (Love, 1939), *Zafar taronalari* (Songs of Victory, 1942), *Ghazab va muhabbat* (Anger and Love, 1943), *Hayot ishqi* (Love of Life, 1948), *Tinchlik kuylari, She'rlar* (Songs of Peace, Poems, 1960), *Ukrayina yelleri* (Wind in the Ukraine, 1942); plays, *Alisher Navoiy* (1943), *Navbahar* (Spring, 1949), *Hayot koshighi* (Song of Life, 1947).

PHTF II 712. NZ

V

Vagif, Molla Pänah (b 1717 Salahly, d 1797 Karabagh), great Azerbaijan lyric poet. Of a peasant family, he studied at the Muslim university and became a teacher. His learning and his poetry were highly appreciated, and he was admitted to the Shah's court in Karabagh, where he ultimately became Chief Vizier. He used his poetry, like his political work, to raise the cultural level of his people. His extensive *dīwān* (qv) includes poems in folk forms as well as classical *ghazals* (qv). His poetry is musical, evoking feelings of happiness and the enjoyment of life. He was also one of the first among his people to sing openly of women and their beauty in a way which could no longer be construed as mystical allegory. His verse is still popular today, for its vivid themes and musical appeal.

PHTF II 658. ZV

Vahyan, Vahe (real name Sargis Aptalian; b 1907 Kürin, Turkey), an Armenian poet in the diaspora. Originally an architect, he has been a teacher, editor and journalist. Since the thirties he has published in the Armenian press in Beirut, Paris and Yerevan. He made his debut with lyrical poetry in the manner of Metsarents and Varuzhan (qqv), in order to encourage the Armenian minorities in their difficult life (*Arevanjrev*, Sun-rain, 1933). After the revolution he began building a 'bridge' between the Armenians in the diaspora and their native country (this is shown in the symbolic title of his most popular volume of verse, *Oski kamurǰ*, Golden Bridge, 1946). In his latest volume, *Matyan siroy ew mormok'i* (The Book of Love and Defiance, 1968) he also included translations from Whitman, Mayakovskiy, Eluard and others.

LM

Vardan Aygektsi (d about 1250), Armenian historian and writer of fables, born in Marata near Aleppo. A monk and wandering preacher, he settled in old age in the Aygek monastery where he wrote a history (up to 1267), and an interpretation of the Bible (qv), epistles for the edification of young monks, sermons, etc. To make his works livelier and easily memorable he included short anecdotes and fables from folk sources and of his own invention. After undergoing many changes in form and interpretation, these became the material for popular collections, such as *Arakk' Vardanay* (The Fables of Vardan, 1668, French trans. Paris 1825).

HO 7. LM

Varuzhan, Daniel (real name Tchpukkiarian; b 1884 Brgnik near Sivas, d 1915), Armenian poet. Brought up in the countryside, he studied at Istanbul, Venice and Gand (social sciences), and worked as a teacher of Armenian and French in Istanbul. Like many other Armenian intellectuals (eg poets Siamantho, and R. Sevak, qv) he was a victim of the 1915 massacres. His first volume of lyrics *Sarsurner* (Frights, 1906), was inspired by the involved tragedy of minority life in Turkey, characterized by religious and national oppression, constant insecurity by being cut off from the eastern branch of the nation on the one hand, while on the other there was direct contact with the political and cultural trends of Europe. Varuzhan's poetry (*J̌ard*, Massacre, 1907; *Ceġin sirt*, Heart of the Race, 1909; *Het'anos erger*, Pagan Songs, 1912; *Hacin erg*, Songs about Bread, 1921) combines faith in the ultimate unifying of his people with bitter despair at their present life, a hatred of violence, and commitment for the ideas of the working class movement and socialist revolution, and at the same time a longing for escape which led to his idealization of the past and the exaggerated picture of idyllic patriarchal life. Varuzhan is a master of West Armenian, his poetry is dynamic and lively, with a wealth of imagery. It is a remarkable synthesis of the literary and philosophical influences of Europe (especially Kant and Maeterlinck, whom he translated), Armenian folk poetry and the work of his predecessors (Durian, Metsarents, qqv); it stands at the peak of the development of the lyrics in modern West Armenian.

Trans.: V. Godel, *La concubine* (Paris 1956); *Le chant du pain* (Paris 1959); ELA.
HO 7; THLA. LM

Vāṣefī, Zeynoddīn Maḥmūd b. 'Abdoljalīl (b 1485 Herat, d between 1551/66), Central Asian Persian poet and literary scholar. Until 1505 he was teacher at the court of the Timurid, Ḥoseyn Bāyaqrā; he later went to Samarkand, Bukhara and Tashkent. He was a master of the poetic techniques in the tradition of his times. He wrote panegyric *qaṣīdas* (qv), *chīstān* (puzzles) and *mo'ammā* (verbal rebus) but his chief work is the long prose *Badā'ī' al-vaqā'ī'* (Remarkable Tales), with passages in verse, which is very popular in Central Asia. It includes anecdotes, fairy-tales, biographical passages about Jāmī, Navā'ī, Benā'ī (qqv), his own and others' poems, and autobiographical remarks. The book is a valuable source of information on life in Central Asia at the turn of the 16th century. Its principal value lies not so much in exact historical fact as in the way it reflects life and events in the minds of the characters.

RHIL 501–2. JB

Vazha-Pshavela (real name Luka Razikashvili; b 1861 Chargali, d 1915 Tbilisi), Georgian poet. Born in an inaccessible mountain village as the son of a priest, he learned to read and write at home before going to the seminary in Gori. This was followed by several years attending lectures at the university in St Petersburg. Returning to his native village, he lived a simple mountain peasant's life. He took no part in the movement for national awakening headed by Chavchavadze and Tsereteli (qqv), and the dialect in which he wrote, that of the Pshavs, also distinguishes him from the movement. The most important of his poems are the epics, based for the most part on Pshav and Khevsur folklore. He took a tragic view of the world, seeing life as a constant battle between man and nature, the individual and the community, one tribe and another. Yet it is often two truths that come into conflict: Mindia, in his poem *Gvelismchameli* (The Snake-Eater), had to choose between his love for his family, left to starve by the community, and his responsibilities towards the community which depended on his wisdom to lead them. He decides to feed his family, but this robs him of his supernatural powers, and in the ensuing battle his community is defeated. Joqola in *Stumarmaspindzeli* (The Guest and His Host) true to the laws of hospitality protects his guest against the community, but calls for his head according to the law of blood feuds. Aluda Ketelauri, the hero of the poem of the same name, freeing his village from an evil enemy yet gives the dead man an honourable burial as a brave fighter; this goes against customs and he is driven from the village in shame. The laws governing one and the same society may conflict, and then the individual must decide for himself; whatever his decision, he suffers in one way or the other. He can therefore only decide in accordance with the dictates of his own conscience, and neither society nor any supernatural powers can take the responsibility from him.

Vazha-Pshavela's lyric poetry expresses the same attitude towards the world. Nature is alive with demons, lying in wait for man in every hidden nook, and only the truly manly can stand up to them. The plants and animals, subservient to the same inexorable laws, are his allies. Above them all rise the chill, noble mountains, like warriors turned to stone. Vazha-Pshavela also wrote stories and tales in a lyrical mood, and ethnographical studies of life in his native Pshav region. He esteemed the function of folklore in the national culture, regarding it as the raw material from which the poet can make a thing of beauty and value. All Vazha-Pshavela's poems are written in the style and form of the mountain folk songs, using the sixteen syllable *shairi* line (see Rustaveli), with a caesura after the eighth syllable. The rhyme scheme in the strophes of varying length is irregular. The outward form is less important, however, than the central idea to which it is subordinated. The works of Vazha-Pshavela have inspired many artists and musicians, and their involved original thought has made them a constant source of interest to literary scholars and philosophers. As the esteem in which the ideas of his contemporaries are held has gradually lessened, Vazha-Pshavela has come to be accepted as the greatest Georgian poet of the new age. Some of his poems have become folk songs.

Trans.: UAGP. VAČ

Vazurgmihr, son of Bukhtak (late 6th century), vizier of the Sassanian kings, Husrav Anōsharvān and Ōhrmazd IV. Vazurgmihr was an outstanding writer, adviser to kings, interpreter of dreams, inventor of parlour games, and clever diplomat. In the Iranian version of his life story he is the model minister of state. *Pandnāmak i Vazurgmihr* (Book of Vazurgmihr's Advice) is attributed to him, and he himself is the hero of the Book of Chess (*Vicharishn i chatrang u nihishn i nēvartakhshēr*).

RHIL 55 ff; TMSLZ. OK

Vehbî, Sünbülzade (b c1717 Maraş, d 1809 Istanbul), Turkish poet. Of an official's family, he was official and judge in Stara Zagora in Bulgaria, on Rhodes and elsewhere, and travelled on official business throughout Iran. His *dīwān* (qv) includes two famous *kasides* (see *qaṣīda*), on his travels in Persia and on the Sultan's revenge on the last Khan of the Crimea. His *Lütfiye*, verse advice to his son Lutfullah, is an imitation of Nabî's (qv) *Hayriye*. His frivolous verse, especially of the time in Stara Zagora when his friend was the poet Sürurî, forms a contrast. Vehbî also wrote introductions to the poetics of Persian and Arabic which were in use in schools up to recent times.

GHOP IV 242 ff; PHTF II 451 ff. ZV

Veli, Kanık Orhan, Turkish poet, see **Kanık, Orhan Veli.**

Veṣāl, Mīrzā Moḥammad Shafī' (1779–1846), Persian poet. He was thoroughly educated, and talented in both calligraphy and music. Highly thought of by his contemporaries, his place in the history of Persian poetry is due only to his excellence as an epigone of the old masters, eg Sa'dī and Ḥāfeẓ (qqv). Innumerable parallels to their poems can be found in his *dīwān* (qv). Apart from his lyrics, his large output of verse also include two *masnavīs* (qv). His six sons, all of whom inherited his poetic talent to a greater or lesser degree, were also men of letters.

RHIL 331–2. HS

Visramiani, a Georgian prose version of the romantic epic by the Persian poet, Fakhroddīn Gorgānī (qv), *Vīs o Rāmīn*, the oriental parallel to the story of *Tristan and Isolde*. The author was Sargis Tmogveli (early 12th century). It differs from the Persian model in that it has a framework in which the original author appears as Parpur Gorganeli. The polished style and pure classical language made it a model for prose writing up to the 19th century. Several poetic versions were produced in the seventeenth century.

Trans.: Oliver Wardrop, *Visramiani* (London 1914). VAČ

Vosifī, Zaynuddīn, Persian-Tajik poet, see **Vāsefī,** Zeynoddīn Maḥmūd.

Vurghun, Sämäd (b 1906 Salahly, d 1956 Baku), Azerbaijan poet and dramatist. A village teacher, he studied the history of literature in Moscow, and returned home to devote himself entirely to literature and active politics. From short love poems he turned to longer poems on social themes towards the end of the 20's; he replaced the traditional figure of woman, the beauty, by the new working woman, the fighter. In his verse plays Vurghun gives a broad picture of 18th century Azerbaijan. Satire plays a less important part in his work. He made an important formal contribution to Azerbaijan literature, combining classical and folk elements in a new metre, thus creating a new form for modern Azerbaijan poetry. Like other poets he translated works from Russian. Poems: *Baki* (Baku, 1946), *Lirika* (Lyrics, 1957); dramas: *Vagif* (1938), *Hanlar* (1939), *Färhad vä shirin* (1941).

PHTF II 689. ZV

W

al-Wāqidi, Muḥammad b. ʻUmar (b 747 Medina, d 823 Baghdad), Arab historian. Little is known of his life; he held office in Baghdad, where he died a judge. His *Kitāb al-maghāzī* (Book of Campaigns) is one of the earliest surviving biographies of the Prophet Muhammad; it deals only with the period after 622, and particularly with military campaigns and political events, as well as the Arab expansion under the first Caliphs. The work is important for its full collection of the historical material and exact indications on the sources used. Although some of the facts given are mere legend, the work belongs to the most important monuments of the early Arab historical writing.

Trans.: J. Wellhausen, *Muhammad in Medina. Das ist Vākidī's Kitāb al-Maghāzī* (Berlin 1882, London 1966). IH

Y

Yaghmā, Mīrzā Abu 'l-Ḥasan (b 1782 Khūre Biyābānak, near Yazd, d 1859 ibid.), Persian poet. His life and work correspond to the troubled atmosphere of Iran in those years. Coming from the country, as a young man he became secretary to an important noble. Against his will he entered the army, becoming a general's scribe. Six years later he was framed by envious enemies and sent to prison, his property being confiscated. On his release he wandered about the country as a dervish and visited many of the holy places. Settling in Teheran, for a time he was the spiritual adviser of the all-powerful vizier, Ḥājjī Mīrzā Āqāsī, and later governor of the province of Kashan. Here his sense of justice again roused the enmity of powerful men, and he was removed from office, probably taking up his dervish staff again. Nothing is known of the last years of his life. As a poet he was one of the most striking personalities of the first half of the 19th century, famous mainly for two contradictory sides of his work. His satires on his personal enemies and on social injustice were often unrestrained in their choice of expression. Written in *ghazal* form (qv), they are arranged in cycles on individual subjects, eg the well-known *Sardāriyye* against the general under whom the poet served. At the other end of the scale are his lovely philosophical and religious elegies, in simple and sometimes even colloquial language. Yaghmā appears to have invented the form of elegy known as *Nouḥeye sīnezanī* (Lamentation Accompanied by Beating of the Breast); it was often imitated by the revolutionary poets of the late 19th century. Yaghmā also wrote traditional *ghazals*, *robāʻis* and *qitʻa* (qqv) and a collection of letters where the attempt to avoid Arabic expressions is even more marked than in his poetry.

BLHP IV 337–44; RHIL 333–4. VK

Yahvalaha III (real name Markos; b 1244 Kuosheng, China, d 1317), subject of a Syrian biography by an unknown author. A monk in the Nestorian monastery near

Peking, he was made Metropolitan of China while travelling to Palestine. The biography gives a vivid description of Mongolian political life, and of the diplomatic mission undertaken in 1287 by Yahvalaha's teacher, Sauma, to persuade the ruling heads of Europe to help the Mongols in their struggle against the Arabs.

Trans.: J. A. Montgomery, *The History of Yaballaha III* (New York 1927); E. A. Wallis Budge, *The Monks of Kublai Khan* (London 1928). SZS

Yahya Bey (d c1572), Turkish poet. Of an Albanian family, he conscripted as a janissary, was educated in Istanbul, and as an officer took part in many campaigns. Forced to retire to his Bosnian estate in old age, he died at over 80. His *dīwān* (qv) comprises mainly lyrics, meditations on the meaning of life, and mystic-erotic verse which was given an important place in the poetry of the time. He wrote five epics, declaring that he would not eat the khalva left by dead Persians; these *mesnevi* (see *maṣnavī*) follow an old tradition in their subjects, but are written in fresh, simple language, with descriptions of Constantinople and of nature; there is originality in the imagery and the characterization. The best is *Şâh ü Gedâ* (The Shah and the Beggar), the love of a beggar for a beautiful youth, mystically passing into pure love for Allah. The celebration of manly beauty was common, love for a woman being considered too intimate a matter for literature; love epics (eg that of Yusuf and Züleyha) were presented as mystical allegories, although it seems unlikely that readers and listeners accepted them only as such. Firmly opposed to the intrigues of the court, Yahya Bey expressed views which were unique in his days. Other works: *Usulnâme* (Book of Principles); *Gencînei Râz* (Treasury of Secrets); *Süleymannâme* (Book of Süleyman).

GHOP III 116 ff; PHTF II 427 ff. ZV

Yāqūt, al-Ḥamawī, Shihābaddīn Abū (or b.) ʻAbdallāh (1179–1229), Arab geographer and biographer. He was a slave and prisoner-of-war when he came to Baghdad as a child, perhaps of Byzantine parents. He was exceptionally talented and hardworking, however, which soon brought him to the position of secretary and plenipotentiary of the Baghdad merchant who had bought him. After his master's death he earned a living as a merchant, bookseller, and copyist, on long journeys which took him to northern Iraq, Syria, Palestine, Egypt and northern Persia. Outstanding among Yāqūt's writings are the Geographical Dictionary (*Muʻjam al-buldān*) of about 14,000 entries and almost 4,000 pages; and the Dictionary of Learned Men (*Irshād al-arīb ilā maʻrifat al-adīb*) containing over 1,000 biographies and more than 2,700 pages. The *Geographical Dictionary* summarizes the contemporary state of knowledge of the world, and is remarkable for the methodical and systematic arrangement, the penetrating observation and clear formulations of the unusually well-read author. Besides such obvious facts as the extent, position and distance of a place, he also noted the history and legends surrounding it, and the great men it had produced. The *Dictionary of Learned Men*, with its detailed account of the life and work of its individual subjects, gives a deep insight into the daily life of Islam.

R. M. N. E. Elahie, *The Life and Works of Yāqūt ibn ʻAbd Allāh al-Ḥamawī* (Lahore 1968); W. Jwaideh, *The Introductory Chapters of Yāqūt's Muʻjam al-buldān* (Leiden 1959); R. Sellheim, *Neue Materialien zur Biographie des Yāqūt* (Wiesbaden 1967); BGAL² I 630 ff, S I 880. RS

Yāsami, Rashīd (whole name, Rashīd Gholām Rezā Yāsamī; b 1896 Kirmanshah, d 1951 Teheran), Iranian historian and writer. Of Kurdish origin, he received a traditional education, and a French education at the Lycée St Louis in Teheran. He held high posts in several government departments and at the court; later he was made Professor of Pre-Islamic Iranian History at Teheran University. He wrote many works on cultural subjects and literary history (eg *La poésie contemporaine iranienne*, in *L'âme de l'Iran*, 1951); he translated Vol. IV of E. G. Browne's *Literary History of Persia*, and several important European literary works (eg Bourget, *Le Disciple*). He made verse translations of many French poems, and was a poet himself, writing mainly intimate and nature lyrics in the classical forms of the *ghazal*, *robāʿī* (qqv) and *mosammat*. This makes him one of the many Iranian intellectuals (including men of exacting professions like that of university teacher, doctor, as well as men of renown in political and cultural life) who write and publish verse, and on private social occasions also recite it.

MLIC 168–70. VK

Yasavî, Ahmet (d 1166), the first Turkestan mystic, saint, and founder of a school of mysticism. He was the author of the religious poems in the collection *Hikmat* (Wisdom), which enjoyed a great popularity among the simple people and exercised a long-lived influence on the work of poets of the Turkish peoples.

PHTF II 272 ff; GHOP I 71 ff. LH

Yashin, Komil (b 1909 Andizhan), Uzbek poet, dramatist and theatre critic. He returned to Andizhan as a secondary shool teacher after studying in Leningrad. As a student he published poems and sketches, but later turned to the drama. His first play, *Ikki kommunist* (Two Communists), appeared in 1929. He drew his themes from the Civil War and the early years of postwar socialism and, in his later plays, from kolkhoz life, woman's fight for equality, etc. Verse: *Kuyosh* (Sun, 1930); *Kurash* (Struggle, 1931); plays and opera librettos: *Nomus va muhabbat* (Honour and Love, 1934); *Hamza* (1940); *Gulsara* (1934); *Bŭron* (Storm, 1937); *Ulugh Kanal* (The Great Canal, 1940); *Ŭlim boskinchlarga* (Death to the Invaders, 1942). NZ

Yasna, part of the **Avesta.**

al-Yāziji, Nāṣīf (b 1800 Kfar Shima, d 1871 near Beirut), last and most famous of the court poets of the Amīr Bashīr al-Shihābī, who ruled in Bayt ad-Dīn in Lebanon until his deposition in 1840. Through his numerous poems, *maqāmāt* (narrative in rhymed prose), and works on grammar and rhetoric, all written in a polished though highly conventional style, al-Yāzijī contributed greatly to the re-establishment of Arabic as a literary medium of dignity and beauty. After leaving the Amīr's court, he lived near Beirut and worked as a teacher of Arabic, having a considerable influence on many of the younger writers from Syria and Lebanon who were to lay the foundations of modern Arabic literature. RCO

Yeghishe (Eliseus; c420–470), Armenian historian. A contemporary of the Armenian war against Persia in 451, he was perhaps a soldier and secretary to the commander; he ended his life in a monastery, writing up the events he had witnessed: the causes, course and consequences of that struggle, with a good knowledge of the facts, historical erudition and patriotic fervour. He praised the heroes and condemned the traitors, giving a fascinating account of the fighting, especially the decisive battle of Avarayr, and of outstanding personalities like that of Vardan Mamikonian, whose family led the revolt. His work has also expressive lyrical intermezzi, for instance the hymn to the moral courage and nobility of the Armenian women. Being one of the earliest texts in classical Armenian, his *Vasn Vardanay ev Hayoc Paterazmin* (About Vardan and the War of the Armenians) belongs to the treasury of Armenian literature, and is a source of inspiration for Armenian writers (Alishan, Demirtchian, qqv) and artists.

Trans.: C. F. Neuman (London 1880); *The History of Vartanank* (New York 1952); *The Epic of St Vardan the Brave* (New York 1951); K. Sarafian, *Battle of Vardanantz and Vardan Mamikonian* (Fresno 1951); V. Langlois, in LCHAMA II.
J. Nirschl, *Patrologia III* (Mainz 1883); NACL; AAPJ. LM

Yunus Emre, Turkish poet, born near the Sakarya river in central Anatolia. He was, according to his own poems, the disciple of a certain Tapduq Emre, and lived till c1321. Many places in Anatolia claim to be his burial place. Yunus Emre played the decisive role in shaping a mystical Turkish folk-poetry in Anatolia. Although Ahmet Yasavî (qv) in Central Asia had already written in Turkish, Yunus was the first poet to express the doctrines of traditional Sufism (qv) in Turkish verse, partly in Persian metres, partly in the popular Turkish syllabic metres. Besides the traditional vocabulary inherited from mystical writing in Persian, Yunus' vocabulary contains many images taken from daily life in Anatolia, and he excels in expressing his mystical experiences, his fear and hope, his 'strife with God', his love and longing and, finally his union with Him in unforgettable verse. The simplicity of his style, the colourful images and the freshness of feeling have made him the most imitated master of mystical folk poetry. He deeply influenced Bektashi poetry. Yunus also wrote a famous *tekerleme* which was commented upon, eg by Niyazî Misrî (d 1697), ie a 'nonsense-poem' that could be mystically interpreted; this was a genre which was apparently popular in dervish circles (a fine example is the poetry of the 15th century Bektashi Kaygusuz Abdal). Those of Yunus' poems which speak of the *dhikr*, 'recollection of God', may partly be meant to be sung as hymns at meetings of the dervishes. Many of his poems have been put to music; a modern Turkish composer, Adnan Saygun, has written an oratorio on words by Yunus. Turkish children today still learn some of his sweet and touching songs.

Trans.: Yves Regnier, *Le divan de Yunus Emre* (Paris 1963).

J. K. Birge, *Yunus Emre* (The Macdonald Presentation Vol., Princeton 1933); PHTF II 408 ff. AS

Yurdakul, Mehmed Emin (1869–1944 Istanbul), patriotic Turkish poet. From 1890 to 1909 he was a Customs officer, and became a governor in Hijaz (1909), Sivas (1910) and Erzerum (1911). In 1913 he was a member of Parliament for Mosul and, after the liberation, member for other Turkish provinces. His poetry was not written in the classical *arūḍ* (qv) but in purely Turkish syllabic verse. His language was simple and natural, close to popular speech; it was easily understood by the uneducated, for whom he most wished to write. His poems call upon the peasants and small craftsmen, beginning with *Cenke Giderken* (On the Way to Battle, 1897) and *Türkçe Şiirler* (Turkish Verses, 1899); during World War I his volumes of poetry *Türk Sazı* (Turkish Lyre) and *Ey Türk Uyan* (Arise, Turk!) were greeted with enthusiasm. He wrote as a patriot, worked for the Turkish national movement and published the paper *Türk Yurdu* (Turkish Homeland).

Trans.: A. Fischer, *Dichtungen Mehmed Emins* (Leipzig 1921).

O. Hachtmann, *Die türkische Literatur des 20. Jahrhunderts* (Leipzig 1916); PHTF II 548–51. OS

Yūshij, Nīmā (b 1897 Yūsh, d 1960 Teheran), Iranian poet. He came from the countryside, bringing with him a deep love of nature. He attended the French *lycée* in Teheran where he acquired a thorough knowledge of French literature, particularly of the French Romantics, who exercised a strong influence over his early work. From the outset he broke away from the principles of classical Persian rhetoric. He pioneered a new trend in Persian poetry, that of free verse (*she're āzād*), considering the classical *'arūḍ* (qv) inadequate to express the emotional experience of modern man. He aimed at the most effective poetic expression of his ideas, choosing rhythm and rhetorical media according to the general mood of the poem. In his later poems he found a very effective and original manner; he had many followers among the younger poets. He put forward his view of poetry in a number of theoretical studies which still play an important role today in establishing a new poetics. His works: *Afsāne* (Romance, 1922), *Khānevādeye sarbāz* (Soldier's Family, 1925).

RHIL 405; AGEMPL 192–4; MLIC II 159–80. KB

Yūsuf al-Khāl (b 1918), Lebanese poet and journalist. He studied Arabic literature at the American University in Beirut, where he taught the same subject later. He published a number of significant periodicals concerned with poetry, at one time for UNO in New York; the most important was *ash-Shi'r* (Poetry), which appeared from 1957 to 1964, and was the main tribunal of modern Arabic writing at that time. His verse is decidedly philosophical in character, and is often on religious and ethical subjects. He was the first Arabic poet to make considerable use of Christian themes, drawing principally on the Bible, and taking as his main subject the regeneration of man. He draws on the poetic example of such English poets as Ezra Pound, T. S. Eliot, Edith Sitwell and others, translating their work into Arabic. Like other modernists (expecially those of Iraq, such as Badr Shākir as-Sayyāb, Nāzik al-Malā'ika, and 'Abdalwahhāb al-Bayātī, qqv) he discarded the metrical unity of the poem and used varying feet. He also wrote verse which does not employ rhyme. By his editions, his own verse, and his theoretical studies, he has done much to fix modern forms firmly in Arabic poetry. Besides translations (*Anthology of American Poetry*, 1958; *Carl Sandburg*, 1959; *Robert Frost*, 1962) he has published several plays: *Herod* (1954), *Three Plays* (1959), and volumes of verse: *al-Ḥurrīya* (Freedom, 1944), *al-Bi'r al-majhūra* (The Deserted Well, 1958), *Qaṣā'id fi'l-arba'īn* (Poems at Forty, 1960), *Qaṣā'id mukhtāra* (Selected Poems, 1958).

CAR 240; ALAC-P 168. KP

Z

Zahrat (real name Zareh Yaldizchian; b 1924 Istanbul), one of the most interesting Armenian poets in Turkey, a rebel against tradition, expressing the feelings of city dwellers and their ironic turn of mind: *Mec k'aġak'* (The City, 1960), *Gunavor sahmanner* (Coloured Boundaries, 1968), *Bari erkink'* (Good Sky, 1968). His stand against myth, and his experiments with poetic form, exert certain influence especially upon the youngest generation of Soviet Armenian poets.

Trans.: S. Boghossian, in ELA. LM

zajal, a strophic poetic form frequent in Arabic literature in Muslim Andalusia. Written in colloquial Arabic it employed other metres than classical poetry and the *muwashshaḥ* (qv) strophe-form. KP

Zākānī, Persian poet, see **'Obeyd Zākānī.**

az-Zamakhsharī, Abu 'l-Qāsim Maḥmūd b. 'Umar (1075–1144), Arab philologist and scholar from Zamakhshar in Khwārazm. He travelled widely during his studies, although he had a wooden leg, passing through Bukhara and Nishapur to Baghdad, and spending some time in Mecca. Back home, he achieved great success as a professor for several decades, until his sudden death in Gurganj castle (Ürgenj). His works reveal him as master of several languages, and as a sensitive, witty, critical and original philologist, particularly in his famous grammar, *al-Mufaṣṣal* (The Detailed Book), his much-used dictionary *Asās al-balāgha* (The Basis of Eloquence) and his well-known commentary on the *Qur'ān* (qv) *al-Kashshāf* (The Unveiler), which also reflects his Mu'tazilite views on religion. In his *Muqaddimat al-adab* he attempted to introduce Arabic words to non-Arabs by a mnemotechnic method, but the unique value of the work for linguistics lies in the fact that it was early provided with Persian, Khwarazmi and Turkish glosses.

J. Benzing, *Das chwaresmische Sprachmaterial einer Handschrift der 'Muqaddimat al-Adab' von Zamaxšarī*, 1: *Text* (with German trans., Wiesbaden 1968); BGAL² I 344–50, S I 507–13. RS

Zarathushtra (probably before 560 BC), religious leader of ancient Persia; his birthplace is unknown. A wandering preacher, he finally after much tribulation won Prince Vishtāsp to his teaching (probably in Merv, in eastern Iran). He is thought to have perished, possibly in Balkh, during a nomad invasion. He stressed the importance of farming and animal husbandry, worshipping light and fire, the truth and order; he fought against magicians, those who sacrificed cattle, and the nomads. Seventeen chapters of religious hymns, known as *gathas* (sung verses) and preserved in the *Avesta* (qv) are attributed to him. As literature they come in the category of reflective lyrics. The *gathas* are versified sermons; the lines are arranged in strophes, and the metre is based on stress. They comprise invocations of the spirits of the heavenly world, and brief items of religious intent, a kind of concentrated essence of his teachings. It is impossible to say whether Zarathushtra himself gave their present form to all these verses, or whether his disciples and followers did so. The reader is held at once by the subjective, personal, impassioned tone, the attitude of a strong personality who believes, worships, hopes, fights, threatens his enemies, laments over difficulties, sometimes even to the point of despair, only to find hope again and be restored to belief in the significance and validity of his mission. As poetry the *gathas* have considerable merit.

J. Duchesne-Guillemin, *Zoroastre* (Paris 1948); W. B. Henning, *Zoroaster, Politician or Witch-Doctor?* (Oxford 1951); RHIL 5 ff. OK

Zariadres and Odatis (before 300 BC), an ancient Persian tale which is the sole example of light prose in ancient Persian literature. It tells of the power and magic of fate, and the wakening love in King Zariadres for the Scythian princess Odatis in a dream; she is the daughter of Omart, and the most beautiful woman in Asia. Zariadres asked for her hand but was rejected. Later her father arranged a banquet at which Odatis was to give a goblet to the man she would like to marry; Zariadres arrived and the couple recognized each other from their dreams; he took the cup and carried off the maiden. This story was recorded by Chares, a courtier of Alexander the Great, and was incorporated by Athenaios in the Sophist Banquet. OK

Zarian, Nairi (real name Hayastan Yeghiazarian; b 1900 Kharakonis, Turkey, d 1969 Yerevan), Soviet Armenian writer. He came to eastern Armenia with thousands of fugitives and was brought up in an orphanage. He studied history in Yerevan and literary history in Leningrad. His work began in 1926 with lyrics on the building of socialism, followed by an epic on the collectivization of Armenian village life *Rušani k'arap'* (The Rock of Rushan, 1930), and a novel *Hacavan* (1947). During the war he took his inspiration from Armenian history for his greatest work, the verse tragedy *Ara Geġecik* (Ara the Fair, 1945). After writing travel books and journalism he returned to lyric poetry in *Du inj kp'ntres* (You will Search for Me, 1965).

Trans.: L. Zelikoff, in SLM; P. Gamarra, in ELA; L. Mardirossian, in OeO. LM

Zaydān, Jurjī (b 1861 Beirut, d 1914 Cairo), Lebanese journalist, novelist, critic, philologist and historian. He was one of the first to enrol in the medical school of the American University in Beirut. He was interpreter to Gordon's forces in the Sudan, and visited London. He had experience of teaching and of journalism in Egypt and Lebanon before finally settling in Cairo, where he founded, in 1891, the very successful *al-Hilāl* journal and publishing house. Zaydān was among the most prolific and enterprising propagators of European ideas. In him, confessional loyalties give way to pride in the Arab-Islamic cultural heritage. His history of Arabic literature (*Tārīkh ādāb al-lughah al-'Arabiyyah*, 4 vols., 1911–14) is the first of the kind, and his 21 historical novels and one romance, though long on didacticism and short on characterization, are a landmark in the development of the genre. *Tārīkh at-tamaddun al-islāmī* (History of Islamic Civilization, 5 vols., 1902–6). PC

Zechariah, Hebrew book of prophecies. The first part, dating from 6th century BC, is a nocturnal vision of hope for the restoration of Judah's independence; the other two parts, of later date, prophesy the salvation and purifying of Jerusalem. See also Bible, and Twelve Lesser Prophets. SZS

Zephaniah (7th century BC), book of Hebrew prophecies, prophesying the 'Day of the

Lord', the day of judgment and retribution. It inspired the mediaeval sequence *Dies irae, dies illa* (probably by Thomas de Celano). See also Bible, and Twelve Lesser Prophets. SZS

Zeyn'olābedin, Ḥājjī Marāghe'ī (d 1910 Istanbul), Iranian writer. Of a wealthy merchant's family, he was given a European-type education and spent most of his life outside his home country. His three-volume novel was the first attempt at this genre in Persian. Presenting the sharply critical reactions of a young Persian educated abroad to the social life of Iran in the second half of the 19th century, it is written in the form of a travel diary. The standard of writing is not constant throughout the book, the last volume being decidedly weaker. Nevertheless, the bold and on the whole objective criticism of society the work offers, and the simple language which was an innovation in its day, make the trilogy *Siyāḥatnāmeye Ebrāhīm Beg* (The Travel Diary of Ibrahim Beg, 3 vols., undated, 1909) a valuable contribution to Persian literature.

Trans.: Ibrahim Beg, *Zustände im heutigen Persien* (Leipzig 1903).
RHIL 369; KMPPL 17–20; AGEMPL 77–8. KB

Zhghenti, B., Georgian critic, see **Chikovani,** Svimon.

Zhabayev, Zhambyl (1846–1945), Kazakh folk singer; he was an *akyn*, ie in the rich stream of individual creative literature that together with the anonymous folk poetry was the only form of literature known to the nomads. He spent most of his life wandering, especially in the Zhetisuv (Semirechiye) region, and was soon very popular. His lyrics and epic ballads were on topical social and political themes, and before the Revolution he was persecuted; the Soviet regime has paid him great honour. His poetry has been published several times, eg 1955 in three volumes. LH

Ziya Paşa (b 1825 Istanbul, d 1880 Adana), Turkish poet and writer of the *Tanzimat* period. In 1853 he became a secretary in Saraye; he was governor of the *sanjak* of Cyprus (1861) and Amasya (1863). He went to Paris and London with Namık Kemal (qv), and returned to Istanbul after Abdülaziz was dethroned, to be State Secretary in the Ministry of Education; later he held the office of *vali* (governor) in Konye and Adana. With Sinasî and Namık Kemal (qqv), Ziya Pasha founded the West European trend in Turkish literature. His ideal was a bourgeois liberalism on the French model, with rights, freedom, the homeland and democracy as its slogans. The Young Turk movement found in him an energetic supporter. He criticized the old style and tried to write simple, flowing prose; in his *Şiir ve Inşa* he attempted to come closer to the language of the common people. Yet he wrote his Arabic-Persian-Turkish anthology *Harabat* (The Inn, 3 vols., 1874), and his *Terkibi Bend*, in the classical style. His translations of Rousseau, Molière and La Fontaine were very influential.

H. W. Duda, *Vom Kalifat zur Republik* (Vienna 1948) 88–92; PHTF II 475–8. OS

Ziya, Uşaklıgil Halit (b 1866, d 1945 Istanbul), Turkish writer. An attractive and fertile writer, he edited the paper *Serveti Fünun* in 1894, the time when the influence of West European literature began to gain ground, and the 'art for art's sake' movement began; it was the period of Tevfik Fikret (qv) and Halit Ziya. Ziya was the first writer to use the modern colloquial language, in his novels *Mavi ve Siyah* (Blue and Black, 1897), *Aşki Memnu* (Forbidden Love, 1900), *Nemide* (Frivolous Hopes, 1889), and *Bir Ölünün Defteri* (Diary of a Dead Man, 1889). Two of his best novellas are *Solgun Demet* (Fading Bouquet, 1901) and *Bir Yazın Tarihi* (History of a Summer, 1900). He published his memoirs (*Kırk Yıl*) in five volumes in 1936.

KLL I 1014, 1648; IV, 518, 1865; V, 374; PHTF II 529 ff. OS

Ziyāda, Mārī (Maryam), known as Mayy (1895–1941), Arab writer and critic. Mayy was born to a Catholic family in Nazareth, but moved with her parents to Lebanon, where she was educated. She then migrated to Egypt, where she studied at Cairo University and produced most of her literary work. She began writing at an early age and by 1911 had already published her first article in the weekly *al-Maḥrūsa* (Cairo), which was followed through the years with numerous others. A collection of these was published in Cairo, under the title *al-Jazr wa'l-madd* (Ebb and Flow, 1924). She wrote mainly of society and literature, including sensitive reviews of new Arabic poetry. She also translated several works from French, English and German into Arabic. Her Arabic writing is a combination of ardent patriotism, compassion for the poor and unfortunate, and appreciation of western humanism. Her prose is both erudite and rhythmic. Perhaps the very first woman writer of stature in the modern Arab East, Mayy has served as a model and inspiration to other women active in the field of Arab letters.

BGAL S III 259–62; GLAIG 205. JML

Zohrap, Grigor (b 1861 Istanbul, d 1915?), Armenian writer. An architect and lawyer, well-known as an orator and defence counsel (he defended Dreyfus in public). One of the opposition to Sultan Abdul Hamid, he became a member of Parliament for the Young Turk party, where he defended the interests of the Armenian minority. Arrested in 1915 with other intellectuals as hostages, he refused plans to escape and perished on the way to exile. Zohrap wrote short stories and novellas, novels, travel sketches, verse, philosophical, critical and political articles and essays. He exposed the inhumanity of bourgeois society, notably the rich people of the cities and the rural poor as their victims. The environment and the types he describes were well known to him and he drew them with brief and effective strokes; his love stories are very sentimental and the women characters too superficial and one-sided. Being the founder of the realistic story in modern West Armenian literature, Zohrap is one of the most widely read authors in Armenia. Almanacs: *Xǵčmtank'i jayner* (Voices of Conscience, 1909), *Kyank' inčpes or ē* (Life as it is, 1911); *Luŕ caver* (Dumb Pains, 1911).

HO 7; THLA. LM

Zorian, Stephen (b 1890 Gharakilisa, now Kirovakan, d 1967 Yerevan), Armenian novelist. Of a peasant family, he worked in Tbilisi (Tiflis) as an editor, translator and writer, moving to Yerevan after the establishment of Soviet power; he became active in public life. He made his debut with a collection of short stories influenced by Chekhov and Turgenev, *Txur mardik* (Sad People, 1918), drawn from provincial life. Under Soviet rule he turned to contemporary themes (*Heǵkomi naxagah*, Chairman of the Revkom, 1923; *Gradarani aǵjik*, Girl from the Library, 1925; *Vardajori komun*, The Vardadzor Commune, 1925) and wrote one of the first autobiographical novels in Soviet Armenian literature, *Mi kyank'i patmut'yun* (The Story of a Life, 2 vols., 1935–39). Most popular are his historical novels *Pap t'agavor* (King Pap, 1944), *Amiryanneri entanik'* (The Family of Amiryans, 1959) and *Hayoc berd* (The Armenian Fortress, 1959), in which, with Demirtchian (qv), he began the tradition of this form in Soviet Armenian literature. He translated works by Tolstoy, Turgenev, Mark Twain, Zweig, Sienkiewicz, and others.

Trans.: A. Ingman, in SLM; A. Yéghiaian, in ELA; L. Butkiewicz, in OeO. LM

Zoroaster, see **Zarathushtra.**

Zuhayr (6th–early 7th century), Arabian poet ranked with the six finest poets of the pre-Islamic era. He is reputed to have lived to a hundred and is therefore also placed among the 'long-lived' poets (*mu'ammarūn*). There seems to have been widespread poetic talent in his family, for both his sister al-Khansā' and his son Ka'b were considerable poets. A serious didactic tone pervades his work and in his famous long poem (*mu'allaqa*, qv) he comes forward as the arbiter between contending tribes. His *dīwān* (qv) has survived and was printed in 1889 (Landberg).

BGAL I 23, S I 47; EI[1] IV 1338; BGLA 269; NLHA 116; WILA 30. KP

LIST OF NATIONAL LITERATURES

ABKHAZIAN LITERATURE

Gulia, Drmit
Narts

ADYGHEI LITERATURE, see CIRCASSIAN LITERATURE

AFGHAN LITERATURE, see PASHTO AND DARI LITERATURES

AKKADIAN LITERATURE

Atra-hasīs
Babylonian Theodicy
Enuma elish
Etana
Gilgamesh
Ishtar's Descent to the Nether World
Ludlul Bel Nemeqi

ALGERIAN LITERATURE

Dib, Mohammed
Haddad, Malek
al-Ibrāhīmī, Bashīr
al-'Īd, Muḥammad 'Alī Khalīfa
Kateb, Yacine
Mammeri, Mouloud
See also Arabic (classical) literature

ARABIC (CLASSICAL) LITERATURE

Abu 'l-Atāhiya
Abu 'l-Fidā', Ismā'īl
Abū Nuwās
Abū Tammām
adab
al-Aḥwaṣ al-Anṣārī
al-Akhṭal
Alf layla wa layla
'Antara b. Shaddad
'arūḍ
al-A 'shā Maymūn
al-Ash 'arī, Abu 'l-Ḥasan
ayyām al-'Arab
Bashshār b. Burd
al-Bīrūnī, Abu 'r-Rayḥān
al-Buḥturī
dīwān
Durayd b. aṣ-Ṣimma
al-Fārābī, Abū Naṣr
al-Ghazzālī, Abū Ḥāmid
ḥadīth
al-Ḥarīrī, Abū Muḥammad
Hassān b. Thābit
İbn 'Abd Rabbihi, Aḥmad
Ibn Bājja, Abū Bakr
Ibn Baṭṭūṭa, Muḥammad
Ibn al-Fāriḍ, Sharaf ad-Dīn
Ibn Fāris, Abu 'l-Ḥusayn
Ibn Ḥazm, 'Alī Aḥmad
Ibn Khaldūn, Abū Zayd
Ibn Khallikān
Ibn al-Muqaffa'
Ibn al-Mu'tazz, Abu 'l-'Abbās
Ibn Qutayba
Ibn Rushd, Abu 'l-Walīd
Ibn Sīnā, Abū 'Alī
Ibn Taymiyya, Taqī ad-Dīn
Ibn Ṭufayl, Abū Bakr

al-Idrīsī
Imru' al-Qays
al-Iṣfahānī, Abu 'l-Faraj
al-Jāḥiẓ, Abū 'Uthmān
Jamāluddīn Afghānī
kalām
al-Khansā'
al-Khwārizmī, Abū 'Abdallāh
al-Kindī, Abū Yūsuf
Labīd b. Rabī'a
al-Ma'arrī, Abu 'l-'Alā
maghāzī
Majnūn
maqāma
al-Maqrīzī, Abu 'l-'Abbās
al-Mas'ūdī, Abu 'l-Ḥasan
mu'allaqāt
al-Mutanabbī
muwashshaḥ
an-Nābigha adh-Dhubyānī
nasīb
qaṣīda
al-Qazwīnī, Abū Yahyā
al-Qur'ān
ar-Rāzī, Abū Bakr
saj'
ash-Shanfarā'
sīra
sīrat 'Antar
as-Suyūtī, Jalāladdīn
aṭ-Ṭabarī, Abū Ja'far
at-Tanūkhī, Abū 'Alī
at-Tawḥīdī, Abū Ḥayyān
taẕkira
'Umar b. 'Alī Rabī'a
Usāma b. Munqidh
al-Wāqidī, Muḥammad
Yāqūt, al-Ḥamawī
zajal
az-Zamakhsharī, Abu 'l-Qāsim
Zuhayr

ARAMAIC LITERATURE

Achikar
Apocalypse
Apocrypha
Ginza
Marqeh
Midrash
Talmud
targum
See also Hebrew

ARMENIAN LITERATURE

Abovian, Khatchatur
Addarian, Garnik
Agathangeghos
Alishan, Ghevond
Bakunts, Aksel
Davoyan, Razmik
Demirtchian, Derenik
Durian, Petros
Emin, Gevorg
Frik
Ghazaros Pharpetsi
Grigor Narekatsi
Grigoris Aghthamartsi
Haik, Vahe
Hakobian, Hakob
Isahakian, Avetikh
Kaputikian, Silva
Khrakhuni, Zareh
Koriun
Kostandin Yerznkatsi
Mahari, Gurgen
Mathevosian, Hrant
Metsarents, Misak
Mkhithar Gosh
Mkhithar Sebastatsi
Mkrtitch Naghash
Movses Khorenatsi
Muratsan
Nahapet Khutchak
Nalbandian, Mikhayel
Nar-Dos
Nerses Shnorhali

Paronian, Hakob
Patkanian, Raphayel
Phaphazian, Vrthanes
Phavstos Buzand
Proshian Pertch
Raffi
Sasuntsi Davith
Sayath-Nova
Sevak, Paruir
Shiraz, Hovhannes
Shirvanzade
Sundukian, Gabriyel
Tcharents, Yegishe
Terian, Vahan
Thumanian, Hovhannes
Tserents
Vahyan, Vahe
Vardan Aygektsi
Varuzhan, Daniel
Yegishe
Zahrat
Zarian, Nairi
Zohrap, Grigor
Zorian, Stephan

ASSYRIAN LITERATURE see AKKADIAN

AVAR LITERATURE

Hamzatov, Rasul
Mahmud of Kahabroso
Tsadasa, Hamzat

AZERBAIJAN LITERATURE

Akhundzadä (Akhundov)
Füzuli, Mehmed
Hüsäin, Mähti
Ibrahimov, Mirzä
Jabbarly, Jäfär
Mammadguluzadä, Jälil
Mir Jälal Pashaoghly
Nesimi, Sayyid Imadeddin
Rza, Räsül
Sabir, Mirzä Aläkbär
Vagif, Molla Pänah
Vurghun, Sämäd
See also Turkic and Turkish

BABYLONIAN LITERATURE see AKKADIAN

BALOCHI LITERATURE see VOL. II

CARTHAGINIAN LITERATURE see PHOENICIAN

CHAGHATAY LITERATURE see UZBEK

CHUKOT LITERATURE

Rytgev, Yuriy

CIRCASSIAN (ADYGHEI) LITERATURE

Charasha, Tembot
Narts

COPTIC LITERATURE

Shenute

DARG LITERATURE

Batiray

DARI (AFGHAN PERSIAN) LITERATURE

Bētāb, Sūfī 'Abdulḥaqq
Ḥabībī, 'Abdulḥay
Ḥamīd Kashmīrī
Jamāluddīn Afghānī
Khalīlī, Khalīlullāh
Mustaghnī, 'Abdul 'alī
Pazhwāk, 'Abdurraḥmān
Qārīzāda, Ziyā
Tarẓī, Maḥmūd
See also Persian (Ancient and Middle), Persian (New Persian), Tajik and Indo-Persian (Vol. II)

EGYPTIAN (ANCIENT) LITERATURE

Akhnaton
Akhtoy
Amenemhēt I
Amenemope
Book of the Dead
Ipuwēr
Ptahhotep
Sinuhe

EGYPTIAN (ARABIC) LITERATURE

'Abdalmu'tī Ḥijāzī
'Abdaṣṣabūr, Ṣalāḥ
'Abduh (Abdo), Muḥammad
Abū Mādī, Īlīyā
Abū Shādī, Aḥmad Zakī
al-'Aqqād, 'Abbās Maḥmūd
al-Bārūdī, Maḥmūd Sāmī
al-Ḥakīm, Tawfīq
Ḥaqqī, Yahyā
Haykal, Muḥammad Ḥusayn
Ibrāhīm, Ḥāfiẓ
Idrīs, Yūsuf
al-Khamīsī, 'Abdarraḥmān
Luṭfī as-Sayyid, Aḥmad
Maḥfūz, Najīb
Mandūr, Muḥammad
al-Manfalūṭī, Muṣṭafā
al-Māzinī, Ibrāhīm
Mūsā, Salāma
Muṭrān, Khalīl
al-Muwayliḥī, Ibrāhīm
an-Nadīm, 'Abdalla
Ṣabrī, Ismā'īl
ash-Sharqāwī, 'Abdarraḥmān
Shawqī, Aḥmad
Shukrī, 'Abdarraḥmān
al-Shidyāq, Aḥmad Fāris
Tāhā, 'Alī Maḥmūd
Tāhā Ḥusayn
al-Ṭahṭāwī, Rifā'a Rāfi'
Taymūr, Maḥmūd
al-Yāzijī, Nāṣīf
Ziyāda, Mārī
See also Arabic (classical)

GEORGIAN LITERATURE

Abasheli, Aleksandre
Abashidze, Grigol
Amirani
Aragvispireli, Shio
Archil
Baratashvili, Nikoloz

Besiki
Chakhrukhadze
Chavchavadze, Aleksander
Chavchavadze, Ilia
Chikovani, Svimon
Chiladze, Tamaz
Gamsakhurdia, Konstantine
Grishashvili, Joseb
Guramishvili, Davit
Iashvili, Paolo
Ioane-Zosime
Javakhishvili, Mikheil
Kartlis Tskhovreba
Kldiashvili, Sergo
Leonidze, Giorgi
Lortkipanidze, Niko

Margiani, Revaz
Merchuli, Georgi
Mirtskhulava, Alio
Orberliani, Grigol
Qazbegi, Aleksandre
Rustaveli, Shota
Shatberashvili, Giorgi
Shavteli, Ioane
Sulkhan Saba
Tabidze, Galaktion
Tabidze, Titsian
Teimuraz I
Tsereteli, Akaki
Tsurtateli, Iakob
Vazha-Pshavela
Visramiani

HEBREW LITERATURE

Amos
Apocalypse
Apocrypha
Bible
Chronicles
Daniel
David
Deborah
Ecclesiastes
Ecclesiasticus
Esther
Ezekiel
Ezra
Five Festal Scrolls
Habakkuk
haggadah
Haggai
Hodayot
Hosea
Isaiah
Jeremiah
Job
Jonah
Joel
Joshua
Judges

Kings I and II
Malachi
Micah
Midrash
Mishnah
Moses
Nahum
Nehemiah
New Testament
Obadiah
Old Testament
Pentateuch
Prophets
Proverbs
psalm
pseudepigrapha
Ruth
Samuel
Solomon
Song of Songs
Tosefta
Twelve Lesser Prophets
Zechariah
Zephaniah
See also Aramaic

HITTITE LITERATURE

Illuyanka
Kumarbi

Telepinu

IRAQI LITERATURE

Ayyūb, Dhū'n-Nūn
al-Bayātī, 'Abdalwahhāb
al-Jawāhirī, Muḥammad Mahdī
Nāzik al-Malā'ika
as-Sayyāb, Badr Shākir
See also Arabic (classical)

JEWISH-PERSIAN LITERATURE

Ketābe Anūsī
'Omrānī ('Imrānī)
Shāhīn of Shiraz

JORDANIAN LITERATURE

Ziyāda, Mārī

KAZAKH LITERATURE

Auezov, Mukhtar
Kunanbayev, Abay
Mukanov, Säbit
Müsrepov, Ghabit
Mustafin, Ghabiden
Seyfullin, Säken
Zhabayev, Zhambyl
See also Turkic

KIRGHIZ LITERATURE

Aytmatov, Chingiz
Bayalinov, Kasymaly
Manas
Sydykbekov, Tügelbay
Togolok Moldo
See also Turkic

KURDISH LITERATURE

Gōrān, 'Abdullāh
Hazhār
Khānī, Aḥmade
Kōyī, Hājī Kādyr
Malāyē Jizrī
Pīramērd, Tawfīq
Shamo, Arabe

LEBANESE LITERATURE

'Aql, Sa'īd
Ba'albakkī, Laylā
al-Bustānī, Buṭrus
Fākhūrī, 'Umar
Hāwī, Khalīl
Jibrān, Jibrān Khalīl
Nu'ayma, Mikhail
ar-Rayḥānī, Amīn
Yūsuf al-Khāl
Zaydān, Jurjī
See also Arabic (classical)

LEZGIAN LITERATURE

Süleyman of Stal

MANDAIC LITERATURE see ARAMAIC

MOROCCAN LITERATURE

Chraïbi, Driss
al-Fāsī 'Allāl
See also Arabic (classical)

OSSETIAN LITERATURE

Gaediaty, Taomaq
Khetaegkaty, Kosta
Narts
Niger
Qamberdiaty, Mysost

PALESTINIAN LITERATURE

Ṭūqān, Ibrāhīm
See also Arabic (classical)

PASHTO LITERATURE

'Abdulḥamīd Mōhmand
'Abdulqādir Khān Khaṭak
'Abdurraḥmān Mōhmand
Afẓal Khān Khaṭak
Ākhund Darwēza
Ashraf Khān Khaṭak
Bāyazīd Anṣārī
Bēnawā, 'Abdurraūf
Ḥabībī, 'Abdulḥay
Kāẓim Khān Khatak 'Shaydā'
Khādim, Qiyāmuddīn
Khushḥāl Khān Khaṭak
landey
Mustaghnī, 'Abdul'alī
Pazhwāk, 'Abdurraḥmān
Rishtīn, Ṣadīqullāh
Ulfat, Gulpāchā
See also Pashto in Vol. II.

PERSIAN LITERATURE (ANCIENT AND MIDDLE)

Artāk Virāz Nāmak
Avesta
Avyātkār I Zarērān
Bahrām Gūr
Bārbud
Bundahishn
Dēnkart
Drakht(i) Āsūrīk
handarz
Husrav I Kavātān U Rētakē
Kārnāmak I Artakhshēr I Pāpakān
Khvatāy Nāmak
Mānī
Matagdān I Yavisht I Friyān
Mazdak
Vazurgmihr
Zarathushtra
Zariadres and Odatis

PERSIAN (JEWISH) LITERATURE, see JEWISH-PERSIAN

PERSIAN LITERATURE (NEW PERSIAN)

Abū Sa'īd
Afghānī, 'Alī Moḥammad
'Alavī, Bozorg
Āle Aḥmad, Jalāl
Amīr Khosrou
Amīrī, Mīrzā Ṣādeq Khān
Anṣārī, Haravī
Anvarī, Ouḥadoddīn

'Āref, Mīrzā Abolqāsem
'arūḍ, ('arūẓ)
Asadī, Abū Manṣūr 'Alī
'Aṭṭār, Farīdoddīn
Bābā Ṭāher 'Oryān
Bahār, Moḥammad Taqī
Bal'amī, Abū 'Alī Moḥammad
Bēdil, 'Abdolqāder
Behāzīn, Moḥammad E'temādzāde
Benā'ī, Kamāloddīn
Beyhaqī, Abu'l-Fazl Moḥammad
al-Bīrūnī, Abu'r-Rayḥān
Bosḥāq, Aḥmad Abū Eṣḥāq
Chūbak, Ṣādeq
Daqīqī, Abū Manṣūr
Dashtī 'Alī
dāstān
Dehkhodā, 'Alī Akbar
dīwān
Doulatshāh, b. Alā'oddoule
Ebne Yamīn
'Erāqī, Fakhroddīn Ebrāhīm
'Eshqī, Moḥammad Reẓā
Farrokhī, Abu'l-Ḥasan 'Alī
Farrokhzād, Forūgh
Farzād, Mas'ūd
Ferdousī, Abolqāsem
al-Ghazzālī, Abū Ḥāmid
Gorgānī, Fakhroddīn
Ḥāfeẓ, Moḥammad Shamsoddīn
Ḥātefī, 'Abdollāh
Hedāyat, Ṣādeq
Ḥejāzī, Moḥammad
Helālī, Badroddīn
Ḥoseyn Vā'eẓe Kāshefī
Ibn al-Muqaffa'
Ibn Sīnā, Abū 'Alī
Indian Style
Īraj Mīrzā
Iraqi Style
Jamāluddīn Afghānī
Jamālzāde, Moḥammad 'Alī
Jāmī, Moulānā 'Abdorraḥmān
Kalīle o Demne
Kamāloddīn Esmā'īl
Kātebī Torshīzī
Kesā'ī, Abu 'l-Ḥasan(?)
Khāqānī, Afẓaloddīn Badīl
Khorasani Style
Khvājūye Kermānī
Lāhūtī, Abolqāsem
Malkom Khān Mīrzā
Manūchehrī, Abū'n-Najm
maṣnavī
Mas'ūd Dehātī, Moḥammad
Mas'ūd b. Sa'de Salmān
Mo'ezzī, Amīr 'Abdollāh Moḥammad
Mushfiqī, 'Abdurraḥmān
Nāderpūr, Nāder
Nafīsī, Sa'īd
Nāsere Khosrou, Abū Mo'īn
Nāṣeroddīn Shāh
Naṣīroddīn, Abū Ja'far Moḥammad
Neẓāmī 'Arūẓī
Neẓāmī, Neẓāmoddīn Ilyās
Neẓāmolmolk, Ḥasan b. 'Alī
'Obeyd Zākānī
'Omar Khayyām
'Onṣorī, Abolqāsem Ḥasan
'Orfī
'Oufī, Moḥammad
Parvīn E'teṣāmī
Qā'ānī, Ḥabībollāh Fārsī
qasīda
Qatrān, Abū Manṣūr
qit'a
Rashīd Vaṭvāṭ
Rashīdoddīn, Faẓlollāh
ar-Rāzī, Abū Bakr
Reẓāqolī Khān Hedāyat
robā'ī
Rūdakī, Abū 'Abdollāh
Rūmī, Jalāloddīn (Moulavī)
Sa'dī, Sheykh Moṣleḥoddīn
Sā'eb, Mīrzā Moḥammad
saj'
Salmāne Sāvajī, Jamāloddīn
Sanā'ī, Abu 'l-Majd
Ṣan'atīzāde Kermānī
Shabestarī, Maḥmūd
Shādmān, Fakhroddīn
Shahriyār, Moḥammad Ḥoseyn
Sheybānī, Abū Naṣr Fatḥollāh
Sufism
Ṭālebūf, Ḥājjī Mīrzā
Tavallolī, Farīdūn

tazkira	Yūshīj, Nīmā
Vāṣefī, Zeynoddīn	Zeyn'olābedīn, Hājjī, Marāghe'ī
Veṣāl, Mīrzā Moḥammed	See also Persian (Ancient and Middle),
Yaghmā, Mīrzā Abu 'l-Ḥasan	Dari, Tajik and Indo-Persian literature
Yāsamī, Rashīd	(Vol. II)

PHOENICIAN LITERATURE

Hanno	Sanchuniaton

SUDANESE LITERATURE

al-Faytūrī, Muḥammad Miftāh	Ṣāliḥ, aṭ-Ṭayyib

SUMERIAN LITERATURE

Enmerkar	Lugalbanda
Gudea	Sumerian Disputes
Ishtar's Descent	Sumerian Laments

SYRIAC LITERATURE

Acts of Thomas	Odes of Solomon
Afrem	Tatianos
Barhebraeus, Grigor Abu 'l-Faraj	Yahvalaha III

SYRIAN LITERATURE

Baghdādī, Shawqī	Qabbābī, Nizār
al-Kawākibī, 'Abdarraḥmān	See also Arabic (Classical)

TAJIK LITERATURE

Aynī, Sadriddin	Niyozī, Foteh
Donish, Ahmad	Payrav Sulaymonī
Fitrat, Abdurrauf	Sayyidā (Sayyido), Mīrābīd
Ikromī, Jalol	Tursunzoda, Mirzo
Jalil, Rahim	Ulughzoda, Sotim
Lohutī (Lāhūtī), Abulqosim	See also Persian (Ancient and Middle),
Mirshakar, Mirsaid	Persian (New), Dari and
Mushfiqī, 'Abdurraḥmān	Indo-Persian (Vol. II)

TATAR LITERATURE

Bashirov, Gomar	Nasirī, Kayum
Jalil, Musa	Tukay, Gabdulla
Merjani, Shihabuddin	See also Turkic

TUNISIAN LITERATURE

ad-Du'ājī, 'Alī	Memmi, Albert
al-Fersī, Muṣṭafā	ash-Shābbī, Abu'l-Qāsim
Khraïef, al-Bashīr	See also Arabic (Classical)
al-Mas'adī, Maḥmūd	

TURKIC LITERATURE

Codex Cumanicus
Dede Korkut
Yasavî, Aḥmet

TURKISH LITERATURE

Abasıyanık, Sait Faik
Adıvar, Halide Edip
Ahmedî, Taceddin Ibrahim
Ali, Sabahattin
Anday, Melih Cevdet
âsik
Bakî, Mahmud Abdülbakî
Behramoğlu, Ataol
Berk, İlhan
Beyatlı, Yahya Kemal
Bilbaşar, Kemal
Burhaneddin, Kadi Ahmed
Cansever, Edip
Çelebi Dede, Süleyman
Cem Sultan
Cemal Süreya (Seber)
Dağlarca, Fazıl Hüsnü
Ekrem Recaizade
Evliyá, Çelebi
Fikret, Tevfik
Fuat, Memet
Füzuli, Mehmet
Galib Dede, Şeyh
Gökalp, Mehmed Ziya
Güntekin, Reşat Nuri
Gürpınar, Hüseyin Rahmi
Haci Halifa, Katib Çelebi
Hamdî, Mehmed Hamdüllah
Hamit, Abdülhak Tarhan
Haşim, Ahmet
Ḥikmet Ran, Nazım
İzzet Molla Keçecizade
Kanî, Abu Bekir
Kanık, Orhan Veli
Karacaoğlan
Karagöz
Karaosmanoğlu, Yakup Kadri
Kemal, Orhan
Kemal, Yaşar
Kırk Vezir Hikayasi
Lâmiî, Şeyh Mahmud
Makal, Mahmut
Midhat, Ahmet
Nabî, Yusuf
Naci Muallim
Namık, Kemal
Nasreddin Hoca
Nayır, Yaşar Nabi
Necatî Bey
Nedim, Ahmed
Nef'î
Nesin, Aziz
Özel, Ismet
Rifat, Oktay
Sabit, Alauddin
Seyfettin, Ömer
Şinasî, İbrahim
Sultan Veled
Ganer, Haldun
Tarancı, Cahit Sıtkı
tuyugh
Ünaydın, Ruşen Eşref
Vehbî, Sünbülzade
Yahya Bey
Yunus Emre
Yurdakul, Mehmed Emin
Ziya Paşa
Ziya Uşaklıgil Halit
See also Turkic

TURKMEN LITERATURE

Kerbabayev, Berdi
Mahtumkuli
See also Turkic

UGARITIC LITERATURE

Baal
Ilumilku
Keret

UZBEK LITERATURE

Aynī Sadriddin
Bābur, Muḥammad Ẕahīruddīn
al-Bīrūnī, Abu 'r-Rayhān
Fitrat, 'Abdurraūf
Furqat, Zokirjon
Ghulom, Ghafur
Muqimī, Amin Mirzo
Navā'ī, Alīshīr
Niyoziy (Niyazī), Hamza Hakimzoda
Olimjon, Hamid
Oybek, Muso Toshmuhammadov
Qahhor, Abdullo
Qodiriy, Abdullo
Uyghun, Rahmatullo
Yashin, Komil
See also Turkic

YAKUT LITERATURE

Achchygyya, Amma
Elley
Kulakovskiy, Aleksey
Oyuunskay B.
Sofronov, A. I.